LEARNING services

01209 722146

Duchy College Rosewarne
Learning Centre

This resource is to be returned on or before the last date stamped below. To renew items please contact the Centre

Three Week Loan

Water Use Efficiency in Plant Biology

Biological Sciences Series

A series which provides an accessible source of information at research and professional level in chosen sectors of the biological sciences.

Series Editor:

Professor Jeremy A. Roberts, Plant Science Division, School of Biosciences, University of Nottingham, U.K.

Titles in the series:

Biology of Farmed Fish
Edited by K.D. Black and A.D. Pickering

Stress Physiology in Animals
Edited by P.H.M. Balm

Seed Technology and its Biological Basis
Edited by M. Black and J.D. Bewley

Leaf Development and Canopy Growth
Edited by B. Marshall and J.A. Roberts

Environmental Impacts of Aquaculture
Edited by K.D. Black

Herbicides and their Mechanisms of Action
Edited by A.H. Cobb and R.C. Kirkwood

The Plant Cell Cycle and its Interfaces
Edited by D. Francis

Meristematic Tissues in Plant Growth and Development
Edited by M.T. McManus and B.E. Veit

Fruit Quality and its Biological Basis
Edited by M. Knee

Pectins and their Manipulation
Edited by Graham B. Seymour and J. Paul Knox

Wood Quality and its Biological Basis
Edited by J.R. Barnett and G. Jeronimidis

Plant Molecular Breeding
Edited by H.J. Newbury

Biogeochemistry of Marine Systems
Edited by K.D. Black and G. Shimmield

Programmed Cell Death in Plants
Edited by J. Gray

Water Use Efficiency in Plant Biology
Edited by Mark A. Bacon

Water Use Efficiency in Plant Biology

Edited by

MARK A. BACON
Department of Biological Sciences
University of Lancaster, UK

Blackwell
Publishing

CRC Press

© 2004 by Blackwell Publishing Ltd

Editorial Offices:
Blackwell Publishing Ltd, 9600
Garsington Road, Oxford OX4 2DQ, UK
  Tel: +44 (0)1865 776868
Blackwell Publishing Asia Pty Ltd,
550 Swanston Street, Carlton, Victoria
3053, Australia
  Tel: +61 (0)3 8359 1011

ISBN 1-4051-1434-7

Published in the USA and Canada (only)
by CRC Press LLC, 2000 Corporate
Blvd., N.W., Boca Raton, FL 33431, USA
Orders from the USA and Canada (only)
to CRC Press LLC

USA and Canada only:
ISBN: 0-8493-2354-1

This book contains information obtained
from authentic and highly regarded
sources. Reprinted material is quoted with
permission, and sources are indicated.
Reasonable efforts have been made to
publish reliable data and information, but
the author and the publisher cannot
assume responsibility for the validity of
all materials or for the consequences of
their use.

Trademark notice: Product or corporate
names may be trademarks or registered
trademarks, and are used only for
identification and explanation, without
intent to infringe.

First published 2004

Library of Congress Cataloging-in-
Publication Data:
A catalog record for this title is available
from the Library of Congress

British Library Cataloguing-in-Publication
Data:
A catalogue record for this title is
available from the British Library

Set in 10/12 Times
by MHL Production Services Ltd,
Coventry, UK
Printed and bound in Great Britain by
MPG Ltd, Bodmin, Cornwall

The publisher's policy is to use permanent
paper from mills that operate a sustainable
forestry policy, and which has been
manufactured from pulp processed using
acid-free and elementary chlorine-free
practices. Furthermore, the publisher
ensures that the text paper and cover
board used have met acceptable
environmental accreditation standards.

For further information on
Blackwell Publishing, visit our website:
www.blackwellpublishing.com

Contents

Contributors

Dr Mark Bacon, Department of Biological Sciences, University of Lancaster, Lancaster LA1 4YQ, UK

Professor Manuela Chaves, Instituto Superior de Agronomia, Tapada da Ajuda, 1349-017 Lisboa, Portugal

Professor Bill Davies, Department of Biological Sciences, University of Lancaster, Lancaster LA1 4YQ, UK

Professor Peter Gregory, School of Human and Environmental Sciences, Department of Soil Science, The University of Reading, PO Box 233, Whiteknights, Reading RG6 6AH, UK

Dr Linda Handley, Scottish Crop Research Institute, Invergowrie, Dundee DD2 5DA, UK

Professor Hamlyn Jones, Department of Biological Sciences, University of Dundee, Dundee DD1 4HN, UK

Professor Brian Loveys, Horticulture Unit, CSIRO Plant Industry, PO Box 350, Glen Osmond, South Australia 5064, Australia

Dr Júlio Osório, Universidade do Algarve, FERN, Campus de Gambelas, 8000-117 Faro, Portugal

Professor John Passioura, CSIRO Plant Industry, GPO Box 1600, Canberra ACT 2601, Australia

Dr J.S. Pereira, Instituto Superior de Agronomia, Tapada da Ajuda, 1349-017 Lisboa, Portugal

Professor John Raven, Division of Environmental and Applied Biology, School of Life Sciences, University of Dundee, Dundee DD1 4HN, UK

Dr M. Stoll, Horticulture Unit, CSIRO Plant Industry, PO Box 350, Glen Osmond, South Australia 5064, Australia

Professor Roberto Tuberosa, Department of Agroenvironmental Sciences and Technology, Via Fanin, 44, 40127-Bologna, Italy

Dr Sally Wilkinson, Department of Biological Sciences, University of Lancaster, Lancaster LA1 4YQ, UK

Dr Bernd Wollenweber, Department of Plant Biology, The Danish Institute of Agricultural Sciences, Research Centre Flakkebjerg, DK 4200 Slagelse, Denmark

Dr Jianchang Yang, Department of Agronomy, College of Agriculture, Yangzhou University, Yangzhou, Jiangsu, China

Dr Jianhua Zhang, Department of Biology, Hong Kong Baptist University, 224 Waterloo Road, Hong Kong

Preface

This is the first volume to provide comprehensive coverage of the biology of water use efficiency at molecular, cellular, whole plant and community levels. While several works have included the phenomenon of water use efficiency, and others have concentrated on an agronomic framework, this book represents the first detailed treatment with a clearly biological focus.

The volume sets out the definitions applicable to water use efficiency, the fundamental physiology and biochemistry governing the efficiency of carbon gain *vs* water loss, the environmental regulation of this process and the detailed physiological basis by which the plant exerts control over such efficiency. Chapter 1 offers a general overview of the volume. Chapter 2 provides sound definitions of water use efficiency on which subsequent chapters build. Chapter 3 considers the 'carbon compromise' – the inevitable loss of water incurred at leaf level to gain carbon, and the way in which biochemistry and physiology combine to preserve water use efficiency in an ever-changing environment. Chapter 4 provides in-depth coverage of the environmental control of water use efficiency at leaf level and the role of hydraulic and chemical signalling, which regulate gas exchange, growth and development within the plant. The volume also discusses the role of nutrition in governing water use efficiency, with detailed coverage in chapter 5 of the effects of plant nutritional status on water use efficiency at the single plant and ecosystem levels. Chapters 6 and 7 consider the exploitation of this understanding in agriculture, using agronomic and physiologically-based approaches. This section of the book concludes with a case study demonstrating the application of an understanding of plant water use efficiency in the growing of rice crops in China (chapter 8). The molecular basis of water use efficiency is detailed in chapter 9, which examines our increasing ability to identify water use efficiency traits and phenotypes and to introduce such traits into crop species, using traditional and emerging methodologies. The book concludes with chapter 10, which sets our understanding of the subject, from molecular to community level, in the context of delivering increased water use efficiency in agriculture.

The contributors to this volume represent some of the most prominent researchers in this subject area, who have worked within Europe, Asia, the Americas, Africa and Australia. It is hoped, therefore, that readers throughout the world will be able to relate to the coverage in the context of their native agricultural systems.

The book is aimed at researchers and professionals in plant molecular biology, agriculture, plant developmental biology, plant biotechnology, plant biochemistry and ecology. It will also inform those involved in formulating research and development policy in this topic, in all parts of the world. The book will prove useful for agricultural engineers wishing to gain a better understanding of how potential agronomic and biotechnological advances can be married effectively with agricultural engineering for the efficient use of scarce resources of water.

This book has been made possible by the willingness of contributors to participate in the project and it is a pleasure to acknowledge their professionalism and continued dedication to this area of plant biology. I also wish to thank colleagues and friends who have provided help and advice in the editing of this book and the publication team who have provided support over the last year.

Mark A. Bacon

1 Water use efficiency in plant biology

Mark A. Bacon

1.1 Introduction

This chapter sets the scene for the volume, by reviewing our current understanding of the term 'water use efficiency' in the context of plant biology and the opportunities for exploiting such understanding. A key theme will be the need to integrate our understanding of water use efficiency at the molecular, physiological, biochemical, whole plant and ecosystem levels, because in doing so, we further our empirical understanding and therefore the opportunities for exploitation in modern agriculture.

1.1.1 The global perspective

Global agriculture now accounts for 70 per cent of the amount of water used by humans, with many parts of the world using even more (Figure 1.1). In October 1999, the six billionth person was added to the planet. Rapid global population growth, diminishing agricultural lands due to unsustainable practices, and global climate change mean that, now more than ever before, there is a need to provide technological solutions to achieve sustainable and efficient use of water. Increasing the water use efficiency of crops is one way in which to achieve this (Anderson et al., 1999).

In developed agriculture, losses due to poor nutrition and plant health are greatly reduced (Passioura, 2002) to the extent that crop losses relating to water availability (and failed water use efficiency, in terms of productive yields) continue to exceed those from all other causes (Kramer, 1980). Boyer (1982) successfully illustrated this point by surveying the causes and costs of crop losses in developed agriculture in the United States from 1939 to 1978 (Tables 1.1 and 1.2). By comparison with native populations, Boyer (1982) also established that a large genetic potential for yield can be better realised if breeding programmes develop varieties better adapted for the environments in which they grow – a philosophy that is now firmly embedded in the breeding programmes of the twenty-first century (see below).

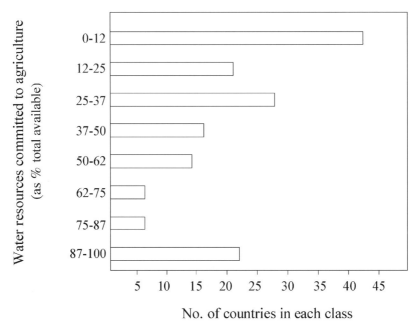

Figure 1.1 Analysis of the extent of water use committed to agriculture (as a percentage of total available water) across 156 countries. Source: United Nations Environment Programme, Global Environmental Outlook Data Portal (http://geodata.grid.unep.ch).

1.1.2   Definition of water use efficiency

Water use efficiency (WUE) does not have a single precise definition. Its definition depends upon the particular context in which it is being discussed, including where the water is in relation to the plant (i.e. inside the plant or in its environment), the time scale over which efficiency is measured (e.g. instantaneous exchange of water vapour for carbon dioxide gas versus biomass accumulation or yield) and the precise measure of efficiency in relation to carbon gain (i.e. carbon dioxide influx, biomass accumulation or economic yield). In most of the current literature, WUE is discussed either in terms of an instantaneous measurement of the efficiency of carbon gain for water loss; or as an integral of such an efficiency over time, (commonly expressed as ratio of water use to biomass accumulation, or harvestable yield). Subsequent chapters will define WUE for the particular context of their discussion.

1.1.3   Historical perspective

According to Stanhill (1986) the first scientific investigation of plant growth and performance in relation to water use was that of Woodward in the

Table 1.1 Area of United States with land subject to environmental limitation (from Boyer, 1982).

Environmental limitation	Area affected (%)
Drought	25.3
Shallowness	19.6
Cold	16.5
Wet	15.7
Alkaline soils	2.9
Saline or no soil	4.5
Other	3.4
None	12.1

Table 1.2 Distribution of insurance payments for crop losses in the USA from 1939–1978. Redrawn from Boyer (1982).

Cause of crop loss	Proportion of payments (per cent)
Drought	40.8
Excess water	16.4
Cold	13.8
Hail	11.3
Wind	7.0
Insects	4.5
Disease	2.7
Flood	2.1
Other	1.5

Philosophical Transactions of the Royal Society, which described the growth increase in spearmint in relation to the expense of water. Lawes (1850) (see Stanhill, 1986) conducted the first set of true experiments to assess arable crop water use, via gravimetric water loss measurements from large containers sown with cereals, pulses and clover and was one of the first to recognise the clear relationship between transpiration and biomass production (see Figure 1.3, p. 8). A landmark in the study of plant water use efficiency came in the early 1900's with the work of Briggs and Shantz (1913) in Akron, Colorado. They determined the transpiration ratio (the ratio of water transpired to dry weight produced – the reciprocal of water use efficiency) for 62 different plant species. Indeed, it would take another 50 years for researchers to realise that the low transpiration ratios recorded for maize, sorghum and millet, could be explained by the existence of C4 photosynthesis (see below). While the validity of the Briggs and Shantz approach was questioned, particularly against emerging meteorological techniques (e.g. Penmon, 1948), reanalysis of the 'Akron series' demonstrated that the analysis was valid and that it was possible to extrapolate the results from container experiments to the field (de Wit, 1958). These studies demonstrated the strict positive relationship between the total amount of water transpired by a crop and its yield (see Figure 1.3, p. 8) by assessing crop yield and water use in response to varying amounts of water (see Kramer and Boyer, (1995) (e.g. Day *et al.*, 1978; Innes and Blackwell, 1981, see also Jones, 1992).

In the latter half of the twentieth century, the discovery that the carbon-isotope fractionation capability of the photosynthetic process, and resultant ratio of stable carbon isotopes within plant tissues could be used to assess both an instantaneous or integrated measure of plant water use efficiency (see Farquhar and Richards, 1984), moved the discussion of water use efficiency firmly within the realms of modern plant science (see below).

1.2   Carbon metabolism and WUE

1.2.1   WUE and the regulation of assimilation

Water use efficiency can be defined as the ratio of CO_2 assimilation into the photosynthetic biochemistry (A) to water lost, via transpiration, through the stomata (T). A and T are regulated by stomatal conductance (g_s) to water and CO_2 and the respective concentration gradients in water vapour (w_i-w_a) and CO_2 (c_i-c_a) between the inside (w_i and c_i, respectively) and outside of the leaf, (w_a and c_a, respectively). Assuming w_i/w_a is independent of c_i/c_a, the so-called 'intrinsic' water use efficiency (W_T) is a negative function of c_i/c_a. Under any particular set of conditions, the driving force for CO_2 uptake will be enhanced by lowering c_i, while the driving force for water loss will remain relatively unchanged, leading to an increase in water use efficiency.

A plant can achieve a lower c_i/c_a ratio (with a concomitant increase in WUE) by decreasing stomatal aperture (lowering c_i by limiting CO_2 diffusion into the leaf interior); increasing photosynthetic capacity for CO_2 (lowering c_i by increasing carboxylation) or more likely, a combination of the two. Indeed, it is commonly observed that stomatal movements can conserve proportionality between c_i and c_a (Wong *et al.*, 1978) with continued debate over if and how photosynthetic capacity, carbon dioxide concentrations and transpiration are sensed and integrated to produce an optimal stomatal aperture (e.g. Cowan, 1982; Farquhar and Sharkey, 1982; Farquhar and Wong, 1984; Jarvis and Davies, 1998).

1.2.2   Photosynthetic biochemistries and WUE

A low mesophyll resistance to CO_2, created by high carboxylation efficiency in the bundle sheath cells of C4 species, ensures a low c_i/c_a and a significant driving force for CO_2 uptake and assimilation. Consequently C4 species can achieve comparable assimilation rates to C3 species at lower stomatal conductances, increasing their water use efficiency. It would also appear that the stomata of C4 species have evolved a reduced direct response to high light, such that a lower internal c_i (and higher WUE) can be maintained under high light conditions, which would usually favour stomatal opening in C3 species (Huxman and Monson, 2003).

With high productivity (under high light and warm temperature environments) there is clear interest in introducing C4 characteristics into C3 crop species, to enhance productivity and WUE. Traditional breeding approaches, involving the hybridisation of C3 and C4 species of *Atriplex*, were less than successful, with independent inheritance of C4 characteristics such as PEP carboxylase and Krantz anatomy (e.g. Björkman *et al.*, 1971). Recombinant DNA technology has enabled the introduction of C4 characteristics into C3 species, including tobacco, potato and rice with some success (Leegood, 2002). This has been achieved primarily via the introduction of enzymatic components of C4 biochemistry rather than attempting to introduce the far more complex leaf structural characteristics of C4 species. Several attempts to introduce C4 characteristics have resulted in clear increases in WUE (Jeanneau *et al.*, 2002a). Transgenic maize lines over-expressing the C4 PEP carboxylase gene resulted in a two-fold increase in PEP carboxylase activity and a 30 per cent increase in the intrinsic water use efficiency (Jeanneau *et al.*, 2002b). For a detailed discussion of molecular engineering of C4 photosynthesis, see Matsuoka *et al.* (2001).

In a similar manner, species exhibiting crassulacean acid metabolism (CAM) have dramatically increased water use efficiencies (10–20 times that of C3 species), by fixing CO_2 during the night when the driving force for water loss is significantly lower. Several species denoted as facultative or inducible CAM exhibit a degree of plasticity in carbon fixing biochemistry, via the use of the C3 pathway when water is sufficient to maximise growth, while switching to CAM metabolism when water supply becomes limiting and evaporative demands are high. Plant families which exhibit such facultative behaviour include the *Crassulaceae, Portulaceae* and *Vitaceae*. In such species, a complex series of gas exchange patterns and stomatal movements can be observed, leading to sustained carbon assimilation (albeit at a much reduced rate) and enhanced water use efficiency as the environment becomes increasingly arid (Cushmann and Borland, 2002). While CAM traits include enhanced WUE, the CAM phenotype is not desirable for introduction into C3 or C4 crop species, due to the low rates of overall biomass accumulation and productivity. However, with increasing atmospheric CO_2 concentrations, and concomitant global warming, the proportion of arid areas of the world is likely to increase, raising the possibility that CAM species may be increasingly cultivated in the future, in order to maintain agricultural productivity.

1.2.3   Isotope discrimination and WUE

The driving force for water loss to the atmosphere from the sub-stomatal cavity is typically 100 times greater than the driving force for CO_2 uptake into the sub-stomatal cavity and photosynthetic biochemistry. In a typical C3 plant, outward fluxes of water vapour may reach 2000–3000 $\mu mol\,m^{-2}\,s^{-1}$, compared to maximal inward fluxes of CO_2 between 20–30 $\mu mol\,m^{-2}\,s^{-1}$. There is therefore, an inherent dominance of water loss to carbon gain in all plants. Plants that can generate increased diffusion gradients for CO_2 will therefore enhance their water use efficiency (at constant stomatal conductance).

Two stable isotopes of CO_2 exist and are naturally abundant: $^{13}CO_2$ and $^{12}CO_2$. The $^{13}CO_2$ isotope has a heavier molecular mass, and although its chemical properties are unchanged, its mass dictates a slower molecular speed and diffusion rate into the leaf and a concomitant increased likelihood of collision with other molecules (such as water molecules travelling in the opposite direction under a significant driving force). While the ratio of $^{13}CO_2$ to $^{12}CO_2$ is approximately 1:99 in atmospheric air (O'Leary, 1993), the ratio of $^{13}CO_2$ to $^{12}CO_2$ in plant tissues is reduced even further. Due to its molecular properties, $^{13}CO_2$ is discriminated against along its pathway of diffusion from the atmosphere up to (and including) the point at which it becomes incorporated into organic carbon molecules. As a result, CO_2 arriving at the site of fixation and the carbon in subsequently formed photosynthate is significantly depleted in ^{13}C. For a detailed review of the biochemical basis of carbon isotope fractionation see O'Leary (1993).

The ratio of ^{12}C to ^{13}C can be measured using mass spectroscopy to determine a so-called carbon-isotope signature or level of discrimination against ^{13}C ($\Delta^{13}C$). Graham Farquhar and co-workers have, from the earlier 1980s, pioneered this technique and established that the extent of $\Delta^{13}C$ varies according to the partial pressure of CO_2 both inside (c_i) and outside (c_a) leaves and concomitant driving force for CO_2 uptake. A higher c_i/c_a results in a greater $\Delta^{13}C$, while a lower c_i/c_a reduces $\Delta^{13}C$. Consequently, carbon isotope signatures demonstrating a greater $\Delta^{13}C$ are diagnostic of a CO_2 fixation environment in which c_i/c_a is relatively high, with signatures exhibiting a lower $\Delta^{13}C$, diagnosing a CO_2 fixation environment with a relatively low c_i/c_a. A low c_i/c_a is diagnostic of a higher WUE, due to the relatively greater driving force for CO_2 uptake generated by a low c_i/c_a, (Figure 1.2). The relationship between $\Delta^{13}C$ (and c_i/c_a), is a function of photosynthetic capacity of the mesophyll and stomatal conductance, with $\Delta^{13}C$ reflecting the indicative set point (or optimum) of these two parameters in any particular instance. A healthy C3 plant will have a c_i/c_a ratio of about 0.7 (Farquhar *et al.*, 1989).

In C4 plants, the primary carboxylating enzyme, phosphenol-pyruvate (PEP) carboxylase has a significantly reduced ability to discriminate against

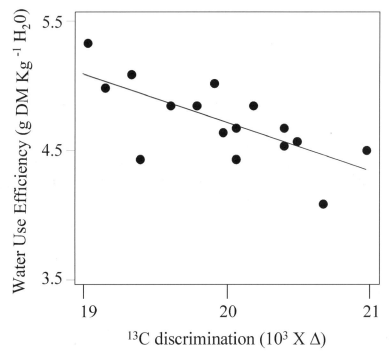

Figure 1.2 Illustrative relationship between $\Delta^{13}C$ and WUE in differing genotypes grown under well-watered conditions (adapted from Condon *et al.*, 1990).

$^{13}CO_2$. Coupled with the ability of C4 photosynthesis to establish a significantly greater driving force for CO_2 uptake (when compared to C3 species), $\Delta^{13}C$ in C4 plant tissue is significantly lower.

Farquhar and co-workers have demonstrated the value of this approach in several crops, by determining the carbon isotope signature and relating a low level of discrimination with enhanced WUE (e.g. Condon *et al.*, 1987; Hubrick *et al.*, 1986; Hubrick and Farquhar, 1989). Importantly, this relationship holds whether measured instantaneously via gas exchange analysis (e.g. Evans *et al.*, 1986) or from plant material integrating carbon fixation over the lifetime of the tissue (e.g. Farquhar and Richards, 1984). But do genotypes expressing enhanced WUE deliver enhanced productivity (i.e. WUE expressed on a yield basis)? This question forms the basis of one of the case studies highlighting how an understanding of WUE in plant biology can deliver significant agricultural benefits (see below).

While the fundamental understanding of $^{13}CO_2$ discrimination in plants has delivered tangible benefits via new cereal varieties with higher water use efficiencies (see below) using $\Delta^{13}C$ alone does not give any information on

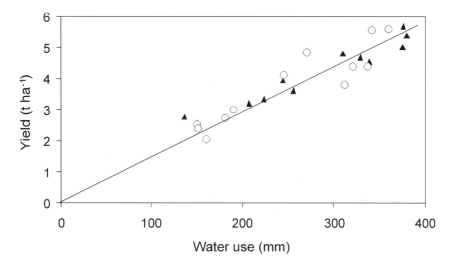

Figure 1.3   Grain yield as a function of water use under a range of irrigation treatments for barley in 1976 (O) and wheat in 1970 (▲) in south-eastern England (data from Day *et al.*, 1978; Innes and Blackwell, 1981). The slope of the line through the origin indicates the WUE for any treatment.

the relative contribution of photosynthetic or stomatal control of c_i and thus the contribution of stomatal regulation in determining water use efficiency. If high WUE is conferred by low stomatal conductances, productivity under well-watered environments may not be as great as that from a genotype with a lower WUE, a consequence of the strict relationship between transpiration and productivity (Figure 1.3). The use of stable isotopes of oxygen may however, be beginning to provide an ability to assess whether highly WUE genotypes (as identified by $\Delta^{13}C$) are also those with high levels of transpiration (e.g. Barbour *et al.*, 2000).

     When water evaporates from the sub-stomatal cavity, the leaf becomes enriched in the O^{18} isotope of oxygen, due to the enhanced evaporation of H_2O^{16} molecules, relative to heavier H_2O^{18} molecules (a consequence of differences in their molecular mass). At a stable vapour pressure deficit, stomatal opening will cause a drop in internal partial pressure of H_2O, and thus increase the driving force for water loss. Under such conditions, enrichment at the site of evaporation is reduced (in an analogous manner to $\Delta^{13}C$, in which an enhanced driving force for CO_2 uptake increases ^{13}C enrichment of tissues). However, total leaf tissue H_2O appears less enriched than that at the site of evaporation, and this disparity increases with transpiration (Flanagan *et al.*, 1994). The reason for this increasing discrepancy is thought to result from diffusion of enriched water (specifically H_2O^{18}) close to the site of evaporation into the leaf away from the sites of evaporation, being opposed by the flux of water from the xylem into the leaf via transpiration flow (the so-called Péclet effect). Increased stomatal conductances will increase the transpirational flux, increasingly opposing the diffusion of H_2O^{18} into the leaf and thus reducing the overall enrichment of bulk leaf tissue. The Péclet effect therefore has the same effect on bulk leaf enrichment as that caused by the decreased fractionation of H_2O^{16} and H_2O^{18} under enhanced evaporative demand. The overall effect is that increased stomatal conductance results in a decrease in H_2O^{18} enrichment. The dominance of both the Péclet effect and fractionation at the sites of evaporation changes depending on climatic conditions such as relative humidity and/or temperature.

The oxygen isotope signature is ultimately stored in leaf carbohydrates, via the molecular exchange of oxygen between water and carbonyl groups within leaf carbohydrates, although the signature is dampened by the fact that the exchange of oxygen molecules between water and carbonyl groups exhibits reduced discrimination against O^{18}. It is possible to model and test the relationship between O^{18} discrimination in bulk leaf tissue and stomatal conductance and the relationships between C^{13} and O^{18} discrimination (Barbour et al., 2000). It can therefore be predicted that as stomata close, ^{13}C discrimination would decrease and ^{18}O discrimination increase and several authors do report this negative relationship (e.g. Saurer et al., 1997). As with measurement of carbon ratios in bulk tissue, oxygen ratios can therefore provide an integrated measure of transpiration, with low levels of enrichment indicative of high rates of transpiration throughout the lifetime of the tissue sampled. The clear relationship between seasonal transpiration and productivity (see above), therefore makes screening of genotypes with low levels of ^{18}O enrichment in tandem with low ^{13}C enrichment, a powerful tool to deliver high yielding, highly water use efficient crop varieties (see below).

1.3 Stomata and WUE

From the discussion above, it is clear that the regulation of stomatal aperture is central to the water use efficiency of plants. Light, temperature, humidity and carbon dioxide concentrations will all act in some way either directly or indirectly on the stomatal aperture, together with internal circadian rhythms, leaf water status and xylem borne signals (e.g. cytokinins, abscisic acid, etc.). At any point in time all of these signals must in some way be integrated to deliver a particular aperture under a particular set of environmental conditions. While work at the cellular and molecular level is beginning to illustrate how such integration may occur (e.g. Webb and Hetherington, 1997) uncertainty on how (or if) stomatal guard cells actually sense some of these environmental signals (particularly CO_2), has remained a source of active debate (e.g. Ball and Berry, 1982; Raschke, 1986; Mansfield *et al.*, 1990; Kearns and Assmann, 1993; Jarvis and Davies, 1998) particularly in relation to the indirect role of mesophyll photosynthesis in controlling stomatal conductance (see Farquhar and Wong, 1984).

Irrespective of these uncertainties it seems intuitive (and clearly advantageous) that stomatal guard cells continually integrate environmental signals from both the aerial environment and the soil, to generate a stomatal aperture, which optimises water loss and carbon dioxide gain under a given set of environmental conditions. Optimisation theory of stomatal aperture (see Cowan, 1982; Farquhar and Sharkey, 1982) predicts that stomatal aperture varies during the day to ensure minimum water loss for maximum carbon gain. As such, while any increase in stomatal conductance generates a proportionally greater increase in transpiration than assimilation, optimisation theory predicts that such variation will keep the ratio of such changes in transpiration and assimilation rate constant, thus preserving intrinsic WUE (see Jones, 1992). Cowan (1982) provides a detailed discussion of optimisation theory, illustrated by considerations of optimisation in relation to diurnal fluctuations in leaf microenvironment and soil water supply, over time periods of relevance to the overall WUE of a plant throughout its development.

While the role of stomata in governing the driving force for CO_2 influx is well appreciated, stomata do have a limited ability to change the driving force for water loss via transpirational leaf cooling. An enhanced ability to achieve leaf cooling (in the absence of any changes in stomatal function), will reduce the internal partial pressure of water vapour and the concomitant driving force for water loss via evaporation. While only a minor trait conferring a theoretical degree of water use efficiency, novel thermal imaging technology can now detect such subtle differences in leaf temperature in both laboratory and field environments (Jones *et al.*, 2003)

As will be discussed in Chapter 2, it is important to recognise that changes in stomatal conductance (and resultant changes in the efficiency of water use) may not necessarily scale to the crop and ecosystem level, due to a series of crop level factors (i.e. canopy boundary layer conductance and temperature due to latent heat of evaporation from the crop surface) which reduce and 'decouple' the stomatal influence on transpiration (see Jones, 1993 and Chapter 2). These factors are also central to understanding the discrepancies often observed between predicated and realised increases in water use efficiency which may be achieved as global atmospheric CO_2 concentrations rise (see e.g. Polley, 2002 and Chapter 3).

1.4   Leaf growth and WUE

While changes in leaf size may also change the CO_2 and H_2O fluxes into and out of the leaf, due to modification to the leaf boundary layer, differences in leaf thickness will also have a significant effect on water use efficiency, with thinner leaves (with a lower ratio of internal volume to leaf surface area) predicted to exhibit lower water use efficiencies than comparable thicker leaves (Stanhill, 1986).

It has also been suggested that rapid leaf development in annuals contributes to the efficiency of water use in the soil. By establishing a high specific leaf area quickly, evaporation of water from the soil is minimised and 'stored' close to the plant, such that it can be drawn upon later in development when water may become limiting (e.g. López-Castañeda et al., 1996). This trait has recently been exploited with some success to deliver more water use efficient and high yielding wheat lines, (see Asseng et al., 2003). As well as the potential direct effects on soil water evaporation, Blum (1996) also suggests that such an adaptation will also minimise the potential for surface roots to come into contact with drying soil, and reduce the likelihood of initiating root-borne signals inhibitory to stomatal conductance. Conversely however, rapid leaf area development may actually prevent soil interception of precipitation, enhancing the rate at which soil water is depleted (Blum, 1996). Clearly this will depend upon the rainfall patterns in a particular environment. The development of a WUE phenotype, well adapted to the environment, is discussed in Chapter 10.

Increasing the photosynthetic capacity of the mesophyll will enhance water use efficiency. However, increased photosynthetic capacity is often associated with a decrease in leaf size (Bhagsari and Brown, 1986), reducing whole plant transpiration and light interception, such that WUE on a plant biomass basis, over time, may actually decrease. The decline in leaf area is likely to occur if any increase in photosynthetic capacity results from an increase in the concentration of enzymes associated with the photosynthetic biochemistry. Under such circumstances, enhanced photosynthetic capacity and limited nitrogen resources are optimised, such that specific leaf area declines (i.e. there is an increase in the dry matter content of leaves on a leaf area basis).

When soil water availability is limited, leaf expansion rates are commonly observed to decline (see Bacon, 1999) in line with transpiration. The innate relationship between transpiration and yield (Figure 1.2) would suggest that this leads to an overall decline in biomass production (and yield productivity of commercial crops). In a majority of cases this is certainly the case. However, under some circumstances, yield and WUE can be sustained or even enhanced even though there is a decline in biomass accumulation at the whole plant level. This would certainly appear to be the case in some cultivated species exhibiting excessive leaf area development when irrigated (e.g. *Vitis vinifera*), which can withstand declines in biomass production (particularly vegetative leaf biomass) due to decreased water availability, without any negative effects on yield (see Chapter 5).

1.5 Roots, hydraulic conductivity and WUE

Deep and expansive root systems are an apparent strategy to ensure maximal water use efficiency in terms of water extraction from the soil, particularly when water availability in soils may decline. An ability to continue to develop deep and advantageous roots when soil conditions become increasingly limiting to root growth will enhance the ability of plants to extract available water efficiently. It is a commonly observed phenomenon that as soil water availability declines, the ratio of roots to shoots typically increases. There are very few data that suggest that root growth can actually be increased by soil drying. Those that do (e.g. Sharp and Davies, 1979), attribute such effects to a stress of particular magnitude which results in increased availability of assimilates to roots, as shoot growth is limited by water deficit in the absence of any effect on carbon gain. More recently, however, Mingo (2003) has reported that under particular circumstances, root growth can be stimulated when roots are rehydrated after a drying episode, relative to roots in moist soil.

Water use efficiency in terms of ability to sustain water extraction from the soil, becomes increasingly difficult as the soil dries. Soil drying places a number of different constraints on the growth and functioning of roots and most of these are poorly understood due to the highly heterogeneous nature of the rooting environment, the delicate nature of the relationship between roots and soil structure and the difficulty of investigating root growth and functioning without disrupting this relationship. One of the common responses to soil drying is that roots show enhanced geo-tropism (e.g. Sharp and Davies, 1985). An increased rooting depth can significantly increase water uptake by root systems even when relatively few roots are involved. The adaptive significance of sustaining root growth (even if at a reduced rate) is only clear, however, if plants are competing in natural communities for different soil water resources. There would appear to be nothing to be gained by plants in a monoculture investing increased carbohydrate into deeper rooting when all plants in the stand are competing for the same reserves of soil water (Bacon *et al.*, 2003). As soil water potentials fall, in substrates with a low mechanical impedance (i.e. roots can penetrate the substrate easily), roots have been observed to thin, an adaptation presumably to commit limited carbohydrate supply to extension growth and allow plants to explore deeper water reserves (Sharp *et al.*, 1988). However, in most soils, decreasing water potentials are commonly associated with increased mechanical impedance, such that roots

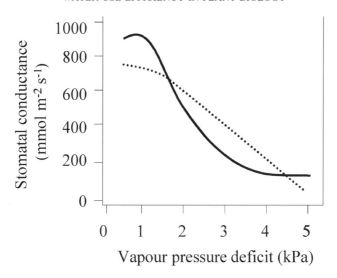

Figure 1.4 Two illustrative responses of stomatal conductance to increasing vapour pressure deficit (adapted from a figure in Atwell *et al.* (1999) which used unpublished data of D. Eamus).

have been shown to swell as soil dries, particularly behind the root apex (Spollen *et al.*, 2000). The prevalence of this phenomenon may allow roots to continue to penetrate the soil as its mechanical impedance increases on drying which may be related to a capacity to generate high turgors in root tips (see Richards and Greacen, 1986; Atwell and Newsome, 1990). Roots of many plants in compacted soils are restricted to cracks in the soil structure. As a result, roots will often grow down these fissures causing substantial localised drying, even when the water content of the bulk soil is still substantial.

Overall transpirational flux through a plant is determined by the characteristics of a plant's hydraulic architecture. Any change in these characteristics could lead to an alteration in the response of open stomata to transpiration rate (see Figure 1.4) and an effect on water use efficiency. This may be particularly important when considering water use efficiency over a prolonged period of time during development. Narrower xylem vessels in roots will result in an overall increase in the hydraulic resistance to water flow throughout the plant (Richards *et al.*, 2002). Consequently, efficiencies in water use may be gained by restricting water uptake early in the development of the plant, to ensure sufficient soil water is available during the reproductive stage. This trait has been confirmed as advantageous in breeding programmes for xylem diameter in wheat (Richards and Passioura, 1989) in which introduction of decreased xylem diameter as a selectable trait increased yields by about 7 per cent.

Sperry and co-workers have recently developed a hydraulic model of water transport through the soil-plant-atmosphere continuum in relation to the component driving forces (and the encountered resistances) to predict transpirational rates, on the basis that plants have evolved hydraulic stomatal optimisation mechanisms to ensure water loss does not exceed uptake by the roots (see Sperry *et al.*, 2002). However, the growing evidence of the root-sourced signals emanating from the root eliciting control of transpiration, does question such a purely hydraulic model. It is becoming increasingly clear that root-sourced signals appear to play a key role in regulating stomatal aperture in response to soil water availability, such that these signals may provide the means by which water supply would appear to regulate water loss.

When soil water availability falls below a certain level, root water potentials and turgors can reach very low values and stimulate the synthesis of several plant growth regulators including abscisic acid (ABA) (Wright, 1977). It is now well established that the production and export from roots of ABA can be related to soil water status and may act as a suitable candidate messenger to ensure the demand for water from the plant is closely controlled by water supply from the roots. The exact roles of hydraulic and chemical regulation of transpiration water loss will continue to attract serious debate and be the subject of future research, with potentially significant impact on understanding (and exploitation of) the physiological basis to water use efficiency. Wilkinson (Chapter 4) extends the discussion of chemical messengers and environmental perturbations which generate them, in relation to the control of plant gas exchange and growth, particularly when water availability declines.

Very interestingly, Holbrook and co-workers have shown that the concentration of potassium ions moving through the xylem can influence the hydraulic conductivity of the transport pathway, perhaps by affecting the nature of the pit membranes within xylem vessels (Zwieniecki *et al.*, 2001), such that a root-sourced chemical signal can influence the properties of the water transport pathways through the root and therefore influence the hydraulic signalling between the roots and shoots. This interesting observation marries exclusively hydraulic- and chemical-based signalling hypotheses together, in a way which brings new knowledge to the understanding of how transpirational water flux may be regulated – with profound effects on the overall water use efficiency of the plant.

Of some increasing interest is the functional significance of aquaporins, hydrophobic proteins which facilitate the movement of water across plasma membranes (Tyerman *et al.*, 2002). It is estimated that plant aquaporins transport the highest total mass of any substance through a plant, when the volume of transpirational water loss is considered, illustrated by the significant reduction in transpirational flux induced by aquaporin inhibitors (see Tyerman *et al.*, 1999). While not contributing directly to the apoplastic flow of water, aquaporins will regulate the cellular flow of water, which becomes increasingly important as water availability declines. It is currently difficult to predict however, what would be the exact involvement of aquaporins in regulating plant water use efficiency. One may envisage that when stomata are closed (as soil water availability declines), decreased aquaporin activity may prevent loss of water to the soil or increased water flow into cells with an enhanced ability to store water (Tyerman *et al.*, 1999). The ability of aquaporins to potentially regulate the turgor of cells in roots and leaves would offer the potential to place the hydraulic conductivity of a plant under metabolic control. In terms of water use efficiency, this may have profound consequences, such that significant xylem tensions, which develop as soil water availability declines, could be partially relieved. Under such circumstances, if stomatal closure in response to soil drying can, in part, be attributable to prevention of xylem cavitation (see Jones, 1993), it may be suggested that such a relief of xylem tension would permit an enhanced level of stomatal conductance and continued fixation of carbon, as water availability declines.

1.6   Uncovering the genetic basis to WUE

An early observation by Martin and Thorstenson (1988) demonstrated that crossing a variety of tomato (*Lycopersicon esculentum* L.) exhibiting a low water use efficiency, with a wild relative (*Lycopersicon pennellii* L.) with high water use efficiency, produced hybrids exhibiting intermediate WUE (Figure 1.5). Differences in WUE between the parental lines and the F1 hybrids were correlated with restriction fragment maps of the tomato DNA and shown to associate closely with three loci within the tomato genome (Martin *et al.*, 1989).

While these initial observations suggest a relatively small number of genes implicated in the genotypic variation, it is readily conceivable that many of the traits conferring WUE will be determined by multiple genes, making breeding programmes for WUE potentially complex. Recently Rebetzke *et al.*, (2003) have explored the inheritability of stomatal conductance traits in wheat cultivars, revealing complex additive and non-additive effects important in the expressed conductance phenotype.

The availability of linkage maps based on molecular markers facilitates the genetic analysis to complex physiological traits such as water use efficiency. Methodologies such as quantitative trait loci (QTL) analysis permit the assignment of variation evident between either environmental or genetic factors to an estimate of the number and location of genetic loci controlling a trait of interest, together with the relative contribution of component traits in an overall phenotype and the development of linked molecular markers. Such markers can be used as selectable markers in conventional breeding programmes and offer the possibility of identifying (and modifying) particular

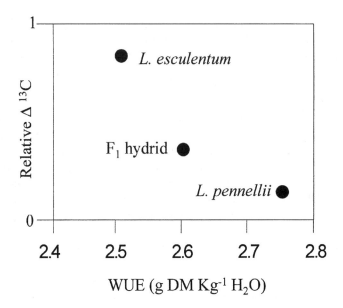

Figure 1.5 The heritability of carbon isotope discrimination and WUE. The F1 progeny of a cross between a domesticated tomato (*Lycopersicon esculentum*) with low WUE (and high $\Delta^{13}C$) and wild tomato relative (*Lycopersicon pennelli*) with high water use efficiency (and low $\Delta^{13}C$), show intermediate behaviour (adapted from a figure in Kramer and Boyer (1995) which used data from Martin and Thorstenson (1988)).

genes generating the trait of interest, by assessing DNA sequence homology of candidate genes with that of sequence close to identified genetic loci or sequencing and identification of new genes at the locus of interest.

As discussed earlier, the development of an efficient root system will increase soil water use efficiency. Via linkage to a molecular marker Champoux *et al.* (1995) have shown that the QTL for root characteristics in rice mapped close to those appearing to regulate drought resistance via enhanced soil water use efficiency. Similar traits in maize have also been reported (Lebreton *et al.*, 1995). With the clear interest in the value of carbon and oxygen isotope discrimination in predicting WUE, several attempts have also been made to establish the loci determining such discrimination. Accordingly, using wheat/barley addition lines (in which individual chromosomes of barley are isolated in a wheat background), Handley *et al.* (1994) investigated the effects of each of the seven barley chromosomes in conferring reduced ability to discriminate against $^{13}CO_2$. Only chromosome 4 was found to have any effect on isotope discrimination. In a similar manner, QTLs for reduced carbon isotope discrimination (and inferred water use efficiency properties) have been identified on chromosomes 1BS and 6BS in the genome of hexaploid wheat (Quarrie *et al.*, 1999). Physiological traits associated with carbon isotope discrimination and WUE are beginning to emerge, with Teulat *et al.* (2002) recently reporting eight QTLs for carbon-isotope discrimination with a physiological basis, including osmotic adjustment.

Identification of putative loci for WUE using QTL analyses is an attractive proposition. This is often coupled with a strategy which involves the matching of so-called 'candidate genes' to such loci in the hope of confirming some envisaged significance of such genes. Such an approach makes an *a priori* assumption as to the significance of candidate genes, which may limit discovery of important sets of genes crucial to the water use efficient phenotype, particularly those which may not intuitively be associated with conferring enhanced WUE. Chapter 9 deals with the genetic basis to WUE further and offers some useful insights into strategies, which can be adopted in pursuit of its elucidation.

Other recent molecular approaches have included differential display and cDNA library screening techniques to establish differences in the expressed genome between genotypes with differing WUE. One such attempt (Zhu *et al.*, 1998) resulted in the isolation of 4 cDNA clones isolated from cDNA of a genotype expressing high WUE. Sequence comparison of one of these clones has revealed it to have close homology to the rubisco activase gene of tobacco. As rubisco activase proteins catalyse an increase in the efficiency of rubisco carboxylation and reduced mesophyll resistance, its enhanced activity will deliver an increase in WUE, offering a potential target for exploitation.

1.7   Adaptations to drought and water use efficiency

Water use efficiency has often been discussed in relation to plant performance when water becomes limiting (e.g. Jones, 1993). While this chapter has already dealt with a discussion on the role of stomata in mediating increases in water use efficiency in response to soil drying and a further chapter in this volume considers this regulation in more detail, it is worth noting other adaptations to reduced water availability other than those already discussed and questioning their relevance in the context of attempts to manipulate water use efficiency in breeding programmes.

Osmotic adjustment within roots, as soil water content declines, has received some significant attention (see Turner and Jones, 1980), in that it may provide an adaptive response to sustain root water potentials to such an extent that the hydraulic driving force for water uptake and transport through the plant can be maintained. The value of osmotic adjustment in achieving an efficiency in limited water use and transport has however been challenged (Munns, 1988), particularly in relation to its suitability as a desirable trait in breeding programmes. Munns (1988) argues that genotypes expressing significant osmotic adjustment are likely to divert carbohydrates away from growth related processes, resulting in drought tolerant genotypes with low growth rates and poor biomass realisation. However, in those cases where reduced water availability results in reduced rates of transpiration and sustained biomass accumulation, water use efficiency will be significantly increased.

Whether these and other adaptations to drought actually offer the potential to inform strategies for the increase in water use efficiency in crop species remains debatable (Blum, 1996). Several examples would suggest that traits conferring higher water use efficiencies and sustained yield in drought are fundamentally opposed. A good example to illustrate this is the interest in manipulating ABA signalling and accumulation in plants. As discussed in an earlier section, the role for ABA in the control of gas exchange and growth, as water becomes limiting, is a key adaptation to drought resistance. But should breeding strategies to deliver high yielding highly water use efficient varieties under water limiting conditions, be selecting for enhanced ABA accumulation? While ABA accumulation may actually increase WUE (in terms of its effects on stomatal limitation), WUE based on yield biomass is likely to be reduced due to the reduced growth rates which accompany ABA accumulation in leaves. Such a striking disparity between drought adaptation and WUE in terms of economic yield, therefore questions the relevance of programmes selecting for enhanced ABA accumulation as a trait to deliver high yielding drought resistant genotypes. Indeed, some would suggest that selection for high water use efficiency may even increase susceptibility to drought (e.g. Read *et al.*, 1993). As such it is now becoming increasingly well accepted that traits that confer drought resistance, may have little impact on water use efficiency in terms of yield (Passioura, 2002). Breeding programmes for high yielding varieties under drought stress are therefore beginning to focus more on empirical traits for water use efficiency, rather than physiological traits identified via observation of plant adaptation to drought (see Passioura, 1977 and Chapter 10). To this end Passioura (2002) has banished the vocabulary of 'drought tolerance' and 'drought resistance' to discussions of survival in natural environments rather than production under cultivation.

1.8 Phenology, the environment and agronomy

Different interpretations can be applied to changes in WUE depending upon the expression of carbon gain (instantaneous assimilation verses biomass accumulation and yielding) relative to water use and the time scale over which efficiency is assessed (instantaneous versus seasonal efficiency). Any discussion of WUE as applied to reproductive success is particularly reliant on an understanding of seasonal water use efficiency. If the phenology of a plant is well matched to the environmental availability of water, to the point that successful reproduction (or yielding) is achieved, high water use efficiency is evident, when considering the efficiency of water use both within the plant and the environment. Integrating an appreciation of water use efficiency over these scales, pioneered by John Passioua (see Chapter 10) unites an understanding of plant water use efficiency at the physiological and environmental levels and underpins much of the exploitation of our understanding evident today (see below). The value of this unity is a key theme of this volume. Evolution would appear to have achieved this unity much earlier, such that seemingly unrelated physiological and phenological traits such as carbon isotope discrimination and early flowing may be intrinsically related at the genetic level, with McKay *et al.* (2003) providing evidence to suggest that the genetic basis to reduced carbon isotope discrimination (a physiological trait for WUE) and propensity of arabidopsis to flower at an optimal time in the season (a phenological trait) are indeed closely related and controlled via a pleotropic gene effect.

1.9 Delivering enhanced WUE into agriculture and horticulture

The release of the 'Drysdale' wheat (Fifield, 2002) for use in the northern Australian wheat belt and development of PRD (partial rootzone drying) as an agronomic technique in irrigated agriculture in large parts of the world, represent two of the most significant advances realised by a fundamental understanding of the physiological basis to water use efficiency.

1.9.1 Drysdale Wheat

Using carbon isotope discrimination methodologies, Tony Condon and co-workers, working at CSIRO Plant Industry in Australia, have developed new higher yielding lines of wheat for growth in Australia, characterised by high water use efficiencies. Several years of study have now identified a strong relationship between $\Delta^{13}C$ and WUE (see Condon *et al.*, 2002). However, as discussed earlier in this chapter, while $\Delta^{13}C$ is a good measure of water use efficiency, the methodology does not provide insight into whether such efficiency (inferred by low $\Delta^{13}C$) is due to stomatal or mesophyll (photosynthetic capacity) control. This important limitation has been reviewed by Condon *et al.* (2002) by comparing the yield performance of two wheat genotypes, differing in the WUE as defined by $\Delta^{13}C$, at two sites in Eastern Australia, differing in rainfall frequency, such that while one site was well supplied with rainwater the other site experienced prolonged drought for much of the season, relying on stored soil water for crop production. Interestingly, at the well-watered site, the genotype experiencing a higher WUE realised a relatively poor yield, compared to the genotype with a lower WUE. Conversely, under drier conditions, the high WUE genotype realised relatively higher yields than the corresponding genotype with low WUE. The reason for these differences lay in the nature of the high WUE exhibited. Stomatal conductances of the high WUE genotype were consistently lower, when compared to the low WUE genotype (about 33 per cent). Due to the strict relationship between transpiration and growth (see above), under well-watered conditions, a high WUE via reduced stomatal conductance, does not confer any yield advantage over a genotype exhibiting a higher stomatal conductance, providing water supply is not limiting. However, under water-limiting conditions, a lower stomatal conductance, will conserve soil water availability (via reduced transpiration), conferring an advantage over genotypes with higher stomatal conductances, which may exhaust soil water before or during yield realisation.

In environments that are not necessarily water limited, increases in WUE due to changed mesophyll photosynthetic capacity may offer a greater advantage. Such a characteristic would be equally advantageous under water-sufficient or limiting environments, although may not be fully realised due to the potential decrease in leaf area which can occur with increased mesophyll photosynthetic activity (see above). Indeed, in water-sufficient environments, such as irrigated agriculture, it may be better to actually select against high WUE (low $\Delta^{13}C$), particularly if lower water use efficiency is associated with higher growth rates.

These observations have been used to develop an extensive breeding programme for wheat in the northern wheat cropping zone in Eastern Australia, which although rainfed, suffers frequent incidence of drought and water limitation. This was achieved by comparing the yielding of 30 back-cross lines of Hartog (a conventional variety used in the northern wheat belt of Australia) exhibiting high WUE, relative to 30 lines exhibiting low WUE (determined via ΔC) grown in several sites in Eastern Australia (differing in the degree of reliance on stored soil water) together with several other sites in Westetrn Australia, which were frequently rainfed. Condon *et al.* (2002) demonstrated that, even in environments were water was sufficient (Eastern Australia), the low ΔC lines realised a greater yield advantage over those expressing higher ΔC (Figure 1.6). Selected lines have now undergone extensive field trials as potential new varieties for the northern wheat belt of Australia. The 'Drysdale' wheat variety is the first of several lines released in 2002 (Fifield, 2002).

1.9.2   Partial rootzone drying

Plants in contact with drying soil are known to produce chemical signals that travel in the xylem and affect a restriction in stomatal aperture and leaf

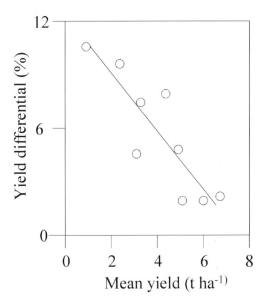

Figure 1.6 Illustrative relationship of variation in mean yield differences between backcross breeding lines of wheat selected for either high WUE (low $\Delta^{13}C$ discrimination) or low WUE (high $\Delta^{13}C$) (adapted from Condon *et al.*, 2002).

expansion rate (see above). For the reasons already discussed, such an adaptation may not be conducive to high yielding (or water use efficiency expressed on a economic yield basis), unless the crop of interest possessed excessive rates of transpiration and vegetative vigour. The fact that this would appear to be the case in several irrigated crops has led to the successful implementation of an irrigation system which exploits this chemical signalling capability to reduce transpirational water loss and crop water requirement.

Early experimental work on root signals in the early 1990s (e.g. Gowing *et al.*, 1990) established the use of a 'split-pot' system, in which the root systems of experimental plants were split between two pots. This artificial system enabled partial watering of the root system to deliver sufficient water to the shoots (to sustain photosynthesis, water relations etc.), while allowing a proportion of the roots to dry the soil in which they were rooted. In this way, scientists could isolate the response of plants to drying soil due to changes in shoot water supply, from those due to the contact of roots with drying soil and resultant root-to-shoot signalling. The principle of the split-pot technique to minimise the negative effects of deficit irrigation regimes, by sustaining shoot water relations, was cleverly exploited in the early 1990s by Brian Loveys and co-workers, at CSIRO Plant Industry, who were interested in controlling the transpirational water loss of grapevines. The 'split-pot' technique was developed into a field-based deficit irrigation system called 'partial rootzone drying' (PRD), to reflect the alternate drying and re-wetting of two halves of the root system in established vines, to sustain shoot water relations, yet generate root-sourced signals to limit transpirational water loss by allowing proportions of the root to periodically dry the soil in which they were rooted. Experimental trials were successful in restricting transpirational water loss and while also restricting leaf area development, no negative effects on yield were recorded (e.g. Dry et al., 1996). Indeed, reduction in the vegetative vigour appeared to enhance the quality of the yield, by opening the canopy and increasing sunlight penetration to the developing grapes. The ability to enhance the WUE of the crop, reduce the amount of scarce and expensive water applied and the unexpected increases in economic yield quality has stimulated increased consideration of such technology in global viticulture as well as in several cereal cropping regions in China and Eastern Mediterranean, cotton crops in Turkey and Australia and *Citrus* and other tree crops in the Mediterranean region.

The precise physiological basis explaining the enhanced water use efficiency of such crops, their yield maintenance and quality enhancement continues as a subject of scientific investigation and is discussed in more detail elsewhere in this volume (see Chapters 5 and 8).

1.10 Summary

This chapter has introduced the topic of water use efficiency in plant biology and established the context for subsequent chapters that explore some of the topics in further detail. In particular, it has referred to the extent to which physiological processes may govern water use efficiency within plants and the recognition that diverse physiological mechanisms, together with evolutionary plant strategies for development and reproduction will integrate, to deliver water use efficiency at scales ranging from the molecule to the ecosystem. While worthy of purely academic discussion, this chapter has also established how a molecular, biochemical, physiological and ecological understanding of water use efficiency in plant biology can be used to deliver significant advances in global agriculture and horticulture, and, in doing so, it has introduced a key theme to this volume.

References

Andersen, P.P., Lorch, M.W. and Rosegrant, M.W. (1999) World food prospects: critical issues for the early twenty-first century, in *Food policy report*. IFPR, Washington D.C.
Asseng, S., Turner, N.C., Botwright, Y. and Condon, A.G. (2003) Evaluating the impact of a trait for increased specific leaf area on wheat yields using a crop simulation model. *Agronomy Journal*, **95**, 10–19.

Atwell, B.J. and Newsome, J.C. (1990) Turgor pressure in mechanically impeded lupin roots. *Australian Journal of Plant Physiology*, **17**, 49–56.

Atwell, B., Kriedemann, P. and Turnbull, C. (1999) *Plants in Action: adaptation in nature performance in cultivation.* Macmillan Education Australia Pty Ltd.

Bacon, M.A. (1999) Biochemical control of leaf expansion. *Plant Growth Regulation*, **29**, 101–112.

Bacon, M.A., Davies, W.J., Mingo, D. and Wilkinson, S. (2003) Root signals. In: *Roots: the hidden half* (eds Y. Waisel, A. Eshel and U. Kafkafi) 3rd edn, Marcel Dekker, Inc., USA, pp. 460–471.

Ball, J.T. and Berry, J.A. (1982) The Ci-Ca ratio: A basis for predicting stomatal control of photosynthesis. *Carnegie Institute Washington Yearbook*, **81**, 88–92.

Barbour, M.M., Fischer, R.A., Sayre, K.D. and Farquhar, G.D. (2000) Oxygen isotope ration of leaf and grain material correlates with stomatal conductance and grain yield in irrigated wheat. *Australian Journal of Plant Physiology*, **27**, 625–637.

Bhagsari, A.S. and Brown, R.H. (1986) Leaf photosynthesis and its correlations with leaf area. *Crop Science*, **26**, 127–132.

Björkman, O., Nobs, M., Pearcy, R., Boynton, J. and Berry, J. (1971) Characteristics of hybrids between C_3 and C_4 species of *Atriplex*. In: *Photosynthesis and photorespiration* (eds M.D. Hatch, C.B. Osmond and R.O. Slayter), Wiley-Interscience, New York, pp. 105–119.

Blum, A. (1996) Crop responses to drought and the interpretation of adaptation. *Plant Growth Regulation*, **20**, 135–148.

Boyer, J.S. (1982) Plant productivity and the environment. *Science*, **218**, 443–448.

Briggs, L.J. and Shantz, H.L. (1913) *The water requirements of plants. II. A review of literature.* Bulletin, USDA, Bureau of Plant Industry No. 285.

Champoux, M.C., Wang, G., Sarkarung, S., Mackill, D.J., O'Toole, J.C., Huang, N. and McCouch, S.R. (1995) Locating genes associated with root morphology and drought avoidance in rice via linkage to molecular markers. *Theoretical and Applied Genetics*, **90**, 969–981.

Condon, A.G., Farquhar, G.D. and Richards, R.A. (1990) Genotypic variation in carbon isotope discrimination and transpiration efficiency in wheat. Leaf gas exchange and whole plant studies. *Australian Journal of Plant Physiology*, **17**, 9–22.

Condon, A.G., Richards, R.A. and Farquhar, G.D. (1987) Carbon isotope discrimination is positively correlated with grain yield and dry matter production in field growth wheat. *Crop Science*, **27**, 996–1001.

Condon, A.G., Richards, R.A., Rebetzke, G.J. and Farquhar, G.D. (2002) Improving intrinsic water use efficiency and crop yield. *Crop Science*, **42**, 122–131.

Cowan, I.R. (1982) Regulation of water use in relation to carbon gain in higher plants. In: *Encyclopedia of Plant Physiology* (eds O.L. Lange and J.D. Bewley). Volume 12B, pp. 535–562.

Cushman, J.C. and Borland, A.M. (2002) Induction of Crassulacean acid metabolism by water limitation. *Plant, Cell and Environment*, **25**, 295–310.

Day, W., Legg, B.J., French, B.K., Johnson, A.E., Lawlor, D.W. and Jeffers, W. de C. (1978) A drought experiment using mobile shelters: the effect of drought on barley yield, water use and nutrient uptake. *Journal of Agricultural Science*, **91**, 599–623.

deWit, C.T. (1958) Transpiration and crop yields. *Verslagan Landbouwkundie Onderzoekingen*, **64**, 1–88.

Dry, P., Loveys, B., Botting, D. and Düring, H. (1996) Effects of partial root-zone drying on grapevine vigour, yield, composition of fruit and use of water. *Proceedings of Ninth Australian Wine Industry Technical Conference*, pp. 129–131.

Evans, J.R., Sharkey, T.D., Berry, J.A. and Farquhar, G.D. (1986) Carbon isotope discrimination measured concurrently with gas exchange to investigate CO_2 diffusion in the leaves of higher plants. *Australian Journal of Plant Physiology*, **13**, 281–292.

Farquhar, G.D. and Sharkey, T.D. (1982) Stomatal conductance and photosynthesis. *Annual Review of Plant Physiology*, **33**, 317–345.

Farquhar, G.D., and Richards, R.A. (1984) Isotopic composition of plant carbon correlates with water-use efficiency of wheat genotypes. *Australian Journal of Plant Physiology*, **11**, 539–553.

Farquhar, G.D. and Wong, S.C. (1984) An empirical model of stomatal conductance. *Australian Journal of Plant Physiology*, **11**, 191–210.

Farquhar, G.D., Ehleringer, J.R. and Hubrick, K.T. (1989) Carbon isotope discrimination and photosynthesis. *Annual Review of Plant Physiology and Plant Molecular Biology*, **40**, 503–537.

Fifield, (2002) CSIRO media release, October 2002.

Flanagan, L.B., Phillips, S.L., Ehleringer, J.R., Lloyd, J. and Farquhar, G.D. (1994) Effects of changes in leaf water oxygen isotopic composition in discrimination against $C^{18}O^{16}O$ during photosynthetic gas exchange. *Australian Journal of Plant Physiology*, **21**, 221–234.

Gowing, D.J., Davies, W.J. and Jones, H.G. (1990) A positive root-sourced signal as an indicator of soil drying in apple, Malus domestica Borkh. *Journal of Experimental Botany*, **41**, 1535–1540.

Handley, L.L., Nevo, E., Raven, J.A., Martinez-Carrasco, R., Scrimgeour, C.M., Pakniyat, H. and Forster, B.P. (1994) Chromosome 4 controls potential water use efficiency ($\delta^{13}C$) in barley. *Journal of Experimental Botany*, **45**, 1661–1663.

Hubrick, K.T. and Farqhuar, G.D. (1989) Carbon isotope discrimination and the ratio of carbon gained to water lost in cultivars of barley. *Plant, Cell and Environment*, **12**, 795–804.

Hubrick, K.T., Farquhar, G.D. and Shorter, R. (1986) Correlation between water use efficiency and carbon isotope discrimination in diverse peanut (*Arachis*) germplasm. *Australian Journal of Plant Physiology*, **13**, 803–816.

Huxman, T.E. and Monson, R.K. (2003) Stomatal responses of C3, C3–C4 and C4 *Flaveria* species to light and intercellular CO_2 concentration: implication for the evolution for stomatal behaviour. *Plant, Cell and Environment*, **26**, 313–322.

Innes, P. and Blackwell, R.D. (1981) The effect of drought on the water use and yield of two spring wheat genotypes. *Journal of Agricultural Science*, **96**, 603–610.

Jarvis, A.J. and Davies, W.J. (1998) The coupled response of stomatal conductance to photosynthesis and transpiration. *Journal of Experimental Botany*, **49**, 399–406.

Jeanneau, M., Gerentes, D., Foueillassar, X., Zivy, M., Vidal, J., Toppan, A. and Preez, P. (2002a) Improvement of drought tolerance in maize: towards the functional validation of the *Zm-Asr 1* gene and increase of water use efficiency by overexpressing C4–PEPC. *Biochimie*, **84**, 1127–1135.

Jeanneau, M., Vidal, J., Gousset-Dupont, A., Lebouteiller, B., Hodges, M., Gerentes, D. and Perez, P. (2002b) Manipulating PEPC levels in plants. *Journal of Experimental Botany*, **53**, 1837–1845.

Jones, H.G. (1992) *Plants and Microclimate: a quantitative approach to environmental plant physiology*. 2nd edn, Cambridge University Press, UK.

Jones, H.G. (1993) Drought tolerance and water-use efficiency. In: *Water Deficits: plant responses from cell to community* (eds. J.A.C. Smith and H. Griffiths), BIOS Scientific Publishers Ltd, UK.

Jones, H.G., Archer, N., Rotenberg, E. and Casa, R. (2003) Radiation measurement for plant ecophysiology. *Journal of Experimental Botany*, **54**, 879–889.

Kearns, E.V. and Assman, S.M. (1993) The guard cell-environment connection. *Plant Physiology*, **102**, 711–715.

Kramer, P.J. (1980) Drought, stress and the origin of adaptations. In: *Adaptation of Plants to Water and High Temperature Stress*. (eds. N.C. Turner and P.J. Kramer), John Wiley & Sons, New York.

Kramer, P.J. and Boyer, J.S. (1995) *Water Relations of Plants and Soils*. 2nd edn, Academic Press Ltd., UK.

Lawes, J.B. (1850) Experimental investigation into the amount of water given off by plants during their growth, especially in relation to the fixation and source of their various constituents. *Journal of the Horticultural Society of London*, **5**, 38–63.

Lebreton, C., Laziæ-Janèiæ, V., Steed, A., Pekiæ, S. and Quarrie, S.A. (1995) Identification of QTL for drought response in maize and their use in testing casual relationships between traits. *Journal of Experimental Botany*, **46**, 853–865.

Leegood, R.C. (2002) C_4 photosynthesis: principles of CO_2 concentration and prospects for its introduction into C_3 plants. *Journal of Experimental Botany,* **53**, 581–590.

López-Castañeda, C., Richards, R.A., Farquhar, G.D. and Williamson, R.E. (1996) Seed and seedling characteristics contributing to variation in early vigour between wheat and barley. *Crop Science*, **36**, 1257–1266.

McKay, J.K., Richards, J.H. and Mitchell-Olds, T. (2003) Genetics of drought adaptation in *Arabidopsis thaliana*: I. Pleiotropy contributes to genetic correlations among ecological traits. *Molecular Ecology*, **12**, 1137–1151.

Mansfield, T.A., Hetherington, A.M. and Atkinson, C.J. (1990) Some current aspects of stomatal physiology. *Annual Review of Plant Physiology and Plant Molecular Biology*, **41**, 55–75.

Martin, B. and Thorstenson, Y.R. (1988) Stable isotope composition (δC^{13}), water use efficiency, and biomass productivity of *Lycopersicon esculentum, Lycopersicon pennellii* and the F_1 hydrid. *Plant Physiology*, **88**, 213–217.

Martin, B., Nienhuis, J., King, G. and Schaefer, A. (1989) Restriction length polymorphisms associated with water use efficiency in tomato. *Science*, **2343**, 1725–1728.

Matsuoka, M., Furbank, R.T., Fukayma, H., and Miyao, M. (2001) Molecular engineering of C4 photosynthesis. *Annual Review of Plant Physiology and Plant Molecular Biology*, **52**, 297–314.

Mingo, D. (2003) Regulation of tomato plant growth in relation to soil water availability. PhD Thesis, Lancaster University, UK.

Munns, R. (1988) Why measure osmotic adjustment? *Australian Journal of Plant Physiology*, **15**, 717–726

O'Leary, M.H. (1993) Biochemical basis of carbon isotope discrimination. In: *Stable Isotopes and Plant Carbon-water Relations* (eds. J.R. Ehleringer, A.E. Hall and G.D. Farquhar), Academic Press Ltd., UK, pp. 19–28.

Passioura, J.B. (1977) Grain yield harvest index and water use of wheat. *Journal of Australian Institute for Agricultural Research*, **43**, 117–120.

Passioura, J.B. (2002) Environmental biology and crop improvement. *Functional Plant Biology*, **29**, 537–546.

Penmon, H.L. (1948) *Proceedings of the Royal Society Series A*, **193**, 120–145.

Polley, H. W. (2002) Implications of atmospheric and climate change for crop yield and water use efficiency. *Crop Science*, **42**, 131–140.

Quarrie, S.A., Stojanović, J. and Pekić, S. (1999) Improving drought resistance in small grained cereals: A case study, progress and prospects. *Plant Growth Regulation*, **29**, 1–21.

Raschke, C. (1986) The influence of the CO_2 content of the ambient air on stomatal conductance and the CO_2 concentration in leaves. In: *Carbon Dioxide Enrichment of Greenhouse Crops* (eds H.Z. Enoch and B.A. Kimball), Volume 2, CRC Press, USA, pp. 87–102.

Read, J.J., Asay, K.H. and Johnson, D.A. (1993) Divergent selection for carbon isotope discrimination in crested wheatgrass. *Canadian Journal of Plant Science*, **73**, 1027–1035.

Rebetzke, G.J., Condon, A.G., Richards, R.A. and Farquhar, G.D. (2003) Gene action for leaf conductance in three wheat crosses. *Australian Journal of Agricultural Research*, **54**, 381–387.

Richards, B.G. and Greacen, E.L. (1986) Mechanical stresses on an expanding cylindrical root analog in antigranulocytes media. *Australian Journal of Soil Research*, **24**, 393–404

Richards, R.A. and Passioura, J.B. (1989) A breeding programme to reduce the diameter of the major xylem vessel in the seminal roots of wheat and its effect on grain-yield in rain fed environments. *Australian Journal of Agricultural Research*, **40**, 943–950.

Richards, R.A., Rebetzke, G.J., Condon, A.G. and von Herwaarden, A.F. (2002) Breeding opportunities for increasing the efficient of water use and crop yield in temperate cereals. *Crop Science*, **42**, 111–121.

Saurer, M., Aellen, K. and Seigwolf, R. (1997) Correlating $\delta^{13}C$ and $\delta^{18}O$ in cellulose of trees. *Plant, Cell and Environment*, **20**, 1543–1550.

Sharp, R.E. and Davies, W.J. (1979) Solute regulation and growth by roots and shoots of water-stressed maize plants. *Planta*, **147**, 43–49.

Sharp, R.E. and Davies, W.J. (1985) Root growth and water uptake by maize plants in drying soil. *Journal of Experimental Botany*, **36**, 1441–1456.

Sharp, R.E., Silk, W.K. and Hsiao, T.C. (1988) Growth of the maize primary root at low water potentials. I. Spatial distribution of expansive growth. *Plant Physiology*, **87**, 50–57.

Sperry, J.S., Stiller, V. and Hacke, U.G. (2002) Soil water uptake and water transport through root systems. In: *Roots: the hidden half* (eds Y. Waisel, A. Eshel and U. Kafkafi) 3rd edn, Marcel Dekker, Inc., USA, pp. 663–681.

Spollen, W.G., LeNoble, M.E., Samuels, T.D., Bernstein, N. and Sharp, R.E. (2000) Abscisic acid accumulation maintains primary root elongation at low water potential by restricting ethylene production. *Plant Physiology*, **122**, 967–976.
Stanhill, G. (1986) Water use efficiency. *Advances in Agronomy*, **39**, 53–85.
Teulat, B., Merah, O., Sirault, X., Borries, C., Waugh, R. and This, D. (2002) QTLs for grain carbon isotope discrimination in field grown barley. *Theoretical and Applied Genetics*, **106**, 118–126.
Turner, N.C. and Jones, M.M. (1980) Turgor maintenance by osmotic adjustment. In: *Adaptations of Plants to Water and High Temperature Stress* (eds N.C. Turner and P. Kramer), John Wiley & Sons, New York, pp. 87–103.
Tyerman, S.D., Bohnert, H.J., Maurel, C., Steudle E. and Smith, J.A.C. (1999) Plant aquaporins: their molecular biology biophysics and significance for plant water relations. *Journal of Experimental Botany*, **50**, 1055–1071.
Tyerman, S.D., Niemietz, C.M. and Bramley, H. (2002) Plant Aquaporins: multifunctional water and solute channels with expanding roles. *Plant, Cell and Environment*, **25**, 173–194.
Webb, A.A.R. and Hetherington, A.M. (1997) Convergence of abscisic acid, CO_2 and extracellular calcium signal transduction pathways in stomatal guard cells. *Plant Physiology*, **114**, 145–161.
Wong, S.C., Cowan, I.R. and Farquhar, G.D. (1978) Leaf conductance in relation to assimilation in *Eucalyptus pauciflora* Sieb. Ex Spreng. Influence of irradiance and partial pressure of carbon dioxide. *Plant Physiology*, **62**, 670–674.
Wright (1977) The relationship between leaf water potential and the levels of abscisic acid and ethylene in excised wheat leaves. *Planta*, **134**, 183–189.
Zhu, Y., Lin, K.R., Huang, Y., Tauer, C.G. and Martin, B. (1998) A cDNA from tomato (*Lycopersicon pennellii*) encoding ribulose 1,5-bisphosphate carboxylase/oxygenase activase (Accession No. AF037361) (PRG98-0053). *Plant Physiology*, **116**, 1603.
Zwieniecki, M.A., Melcher, P.J. and Hobrook, N.M. (2001) Hydrogel control of xylem hydraulic resistance in plants. *Science*, **291**, 1059–1062.

2 What is water use efficiency?

Hamlyn Jones

2.1 Introduction

In their early review of the subject of plant transpiration, Briggs and Shantz (1913) argued strongly that determination of the water requirement of plants must be of great interest for agriculture in dry regions, as the plants capable of expending water most productively 'must evidently be best adapted to regions with a limited water supply'. The same assumption drives much modern research into water use efficiency, with agronomists and plant breeders still having as their 'holy grail' the objective of increasing plant water use efficiency.

It has been known for a century or so that the ratio between the amount of water transpired by plants and their dry matter production varies as a function of environmental conditions, with this ratio being highest for plants grown in dry environments (e.g. Briggs and Shantz, 1913). The ratio between the amount of water transpired by plants and the dry matter accumulated has variously been termed the *transpiration coefficient*, the *transpiration ratio* or the *water requirement of plants*, while the reciprocal quantity was originally termed the *efficiency of transpiration* (see Maximov, 1929). In more recent years the term *water use efficiency* (WUE) has become more usual and is the term that we shall use for preference in this chapter and throughout this book, though as outlined below, the term does have its critics and may be defined in a number of ways depending on the specific purposes of the user.

In this chapter we will restrict ourselves to the study of water use efficiency in plant and crop science, even though there is growing interest, in improving the efficiency of municipal and industrial water use because of the increasing shortage of water worldwide. Some areas of municipal water use efficiency, such as the development of water-efficient landscaping are, however, relevant as they include aspects of plant and soil selection and management (e.g. Robinette, 1984; Vickers, 2001).

2.2 Drought tolerance and the importance of water use efficiency

A crucial step in the colonisation of terrestrial environments by plants has been the evolution of mechanisms that enable plants to control their water loss

Table 2.1 A classification of drought tolerance mechanisms (modified from Jones, 1992), indicating the main types of drought tolerance mechanism that exist and their consequences for plant productivity and competitive ability in natural and agricultural ecosystems.

Mechanism	Potential disadvantages
1. Avoidance of plant water deficits	
(a) *Drought escape* – short growth cycle (e.g. growing on stored water), *dormant* period during dry season	Short season and therefore limited total production
(b) *Water conservation* – small leaves, limited leaf area, stomatal closure, low cuticular conductance, low light absorption	In natural systems may favour competitors which use water faster; in agricultural systems may restrict productivity.
(c) *Maximal water uptake* – good root system	Structural costs reducing photosynthetic area, or seed production
2. Tolerance of plant water deficits	
(a) *Turgor maintenance* – osmotic adaptation, low elastic modulus	Metabolic costs
(b) *Protective solutes, desiccation-tolerant enzymes, etc.*	Metabolic costs
3. Efficiency mechanisms	
(a) *High water use efficiency*	Usually associated with low productivity
(b) *Efficient use of available water*	Low maximum rate thus possibly favouring competitors
(c) *Maximal harvest index*	Unknown

while continuing to fix carbon dioxide in photosynthesis. This step has been so important because the availability of water is probably the key factor determining plant distribution and survival in natural ecosystems, and it is also the most important limiting factor in agricultural production.

All those mechanisms that tend to maintain plant survival or productivity under conditions of limited water supply or high evaporative demand can be described as drought tolerance mechanisms. A convenient classification of these mechanisms is presented in Table 2.1 (following Jones, 1992). From an agricultural point of view at least, it is clear that many of the various drought tolerance mechanisms exhibited by different plants have corresponding 'costs' in terms of either potential dry matter production or competitive ability. Most obvious, and most relevant to the present volume, is the fact that most mechanisms which act to conserve water also tend to reduce photosynthesis and net productivity (see Table 2.1), primarily because the pathway for CO_2 uptake in photosynthesis through the stomata is also the main route for water loss in transpiration. Therefore any stomatal closure acting to conserve water will have a corresponding inhibitory effect on photosynthesis thus reducing potential yield. Nevertheless, there is still significant scope for altering the relationship between photosynthesis and water loss, so this table identifies improvements in the water use efficiency as an important special class of drought tolerance mechanism. Unfortunately high water use efficiency is generally achieved at the expense of productivity.

Different plants exhibit different strategies for survival or growth under drought conditions. The responses found in some species or varieties may be 'conservative' or 'pessimistic' (Jones, 1981) in that they tend to conserve water, while others are more profligate, using any water available rapidly. For this response to be successful the plant must be able to complete its life cycle very rapidly or else it will require further rainfall.

2.3 Definitions of water use efficiency

The term water use efficiency (WUE) is rather an unfortunate misnomer. Firstly, the water taken up by a plant canopy and transpired or lost in evaporation is only 'used' in the very broadest sense – in reality only a very small part of the water taken up is actually utilised in the construction of carbohydrates or even in the composition of plant tissue. Furthermore, as usually defined, it is not even a true 'efficiency', which is a term conventionally reserved for the dimensionless ratio between the output of a quantity and its input. Notwithstanding these objections we will continue to adopt the common usage in what follows.

In the various areas of plant science concerned with drought tolerance a wide range of definitions of WUE have been employed (see Table 2.2), with the basic definition of water use efficiency at a plant scale being the ratio of the rate of biomass production to the rate of plant transpiration. This quantity may be given the symbol, WUE_t, and is otherwise known as the transpiration efficiency. At a crop or vegetation scale, water use efficiency can be expressed as the ratio of production of total biomass (B), shoot biomass (S), or harvested yield (Y) against total evapotranspiration (E_a) or plant transpiration (E_p) (see Loomis and Connor, 1992). Though usually expressed in terms of dry mass (d) these quantities are sometimes expressed in terms of fresh mass (f) or the glucose equivalent of those masses (g). The use of glucose-equivalents is particularly useful for comparing crops with differing chemical composition. Values for different crops depend both on the species and on the nitrogen source used and range from about 1.26 kg glucose-equivalent kg^{-1} biomass for rice to 2.78 kg glucose-equivalent kg^{-1} biomass for the very oily crop peanut (Loomis and Connor, 1992). The closest causal relationships are between biomass production and crop transpiration, since the use of total evapotranspiration can be biased by the soil evaporation, which can dominate water loss especially early in the season before full crop cover, and in many dryland situations.

Table 2.2 Some working definitions of water use efficiency. Within any category, any input can be combined with any output according to the specific requirements.

	Input	Output	Comments
Crop scale:	Seasonal evapo-transpiration (E_a)	Crop yield (Y) – fresh mass (f) – dry mass (d) – glucose equiv. (g)	
	Seasonal transpiration (E_p)	Above ground biomass (B) (f, d, g)	
	Irrigation applied (I)	Total biomass (B) (f, d, g)	
Leaf scale:	Transpiration (E_l) – kg m^{-2} s^{-1} –mol m^{-2} s^{-1}	[a] Net assimilation (A) – g CO_2 m^{-2} s^{-1} – mol m^{-2} s^{-1}	
'intrinsic water use efficiency' (WUE_i):	Stomatal conductance (g_s)	Net assimilation (A)	or the slope of the relationship dA/dg_s
Sundry:	Volume extracted from supply	Volume used (for irrigation)	WUE of irrigation (Burman *et al.*, 1981)

[a] Conversions: Net CO_2 uptake to dry matter = 0.61-0.68 kg DM/kg CO_2.

At a leaf scale WUE_l is often defined as the ratio of instantaneous net CO_2 assimilation rate (*A*) to transpiration (E_l) (A/E_l). Water use efficiencies based on gas exchange can be converted approximately to water use efficiencies defined in terms of dry matter using the conversion: 0.61–0.68 kg dry matter/kg CO_2. For agronomic purposes it is more common to describe water use efficiency in terms of the ratio between seasonal integrals of either biomass production or harvested yield and either total seasonal water use or the total water applied (including rainfall).

A wide range of other variants are used, especially in the agronomic literature: for example agronomists may be primarily concerned with harvestable yield in relation to irrigation applied (I), or may relate yield to total water input during the season, including rainfall.

2.3.1 Intrinsic water use efficiency

The observed value of WUE in any situation depends on the environmental conditions, with, for example, changes in humidity deficit of the atmosphere altering the evaporation rate even if stomatal aperture and hence assimilation rate do not change. It is of particular interest to determine any underlying intrinsic differences in water use efficiency that might exist independent of the specific environmental conditions. Such a measure is particularly useful for those situations where we are interested in improving WUE, for example through plant breeding. Though it is possible to normalise measured WUE on the basis of evaporative demand, as suggested by de Wit (1958) (see below), perhaps the most generally useful approach is to define an 'intrinsic water use efficiency' (WUE_i) as the ratio of A/g_s (e.g. Farquhar *et al.*, 1989). This variable gives a direct measure of the underlying activity of the photosynthetic system normalised to constant stomatal conductance. A further advantage of WUE_i is that the ratio A/g_s is rather constant over quite a range of stomatal conductances, with the relationship being fairly linear except with very wide open stomata where the stomatal limitation to photosynthesis declines (Farquhar *et al.*, 1989). The use of carbon isotope discrimination (see below) also gives a measure of intrinsic water use efficiency, because it primarily measures the ratio A/g (where g is the total gas phase conductance to CO_2).

2.3.2   *Instantaneous versus integral water use efficiency*

Gas exchange studies generally give an estimate of the instantaneous water use efficiency, while harvest analysis or isotopic discrimination methods measured on accumulated plant tissue tend to give a time-integral. Unfortunately it is generally rather difficult to integrate instantaneous measures over longer periods, as diurnal changes in stomatal conductance and environmental conditions mean that the long-term average WUE may only loosely be related to any measured instantaneous value. Significant improvements in WUE can be achieved by the characteristic drought responses of many plants where stomata close at midday (midday depression – Tenhunen *et al.*, 1982) or else open only in the morning. The integral WUE is dominated by the value when stomata are most open.

2.4   Variability in WUE

In the earliest experiments on variation of the transpiration coefficient, it was found under Central European conditions (Hellriegel, 1883; quoted by Maximov, 1929) that for all the crops investigated (barley, oats, wheat, rye, beans, yellow lupin, peas, red clover, buckwheat and *Brassica napus*) the seasonal WUE was approximately 1/300. We now recognise these crops as all being C_3 plant species. When the range of species studied was extended (Schröder, 1895) to include what we now know to be C_4 species it became clear that the transpiration coefficients fell into distinct categories containing the C_3 and the C_4 species, with the latter having much smaller transpiration coefficients (greater WUE). Probably the most detailed set of experiments along these lines were those carried out by Shantz and his co-workers over

Table 2.3 Values of WUE for a range of potted plants grown under screens at Akron, Colorado (from Maximov, 1929, after data of Shantz and Piemeisel, 1927).

	Y_d/E_1 $(\times 10^3)$
C_3 plants:	
Cereals	1.47–2.20
Other *Poaceae*	0.97–1.58
Alfalfa	1.09–1.60
Pulses	1.33–1.76
Sugar beet	2.65
Native plants	0.88–1.73
C_4 plants:	
Cereals	2.63–3.88
Other *Poaceae*	2.96–3.88
Other C_4	2.41–3.85

many years at Akron, Colorado (see e.g. Shantz and Piemeisel, 1927; summarised by Maximov, 1929 and Jones, 1992). Some of these extensive results are summarised in Table 2.3. It should be noted for comparison with other results that these experiments were conducted in a screened enclosure which led to the reported WUEs being about 20% higher than for corresponding plants grown in the open. In addition these experiments only measured the above ground dry matter production, thus underestimating the true WUE. The addition of data for plants having the Crassulacean Acid Metabolism (CAM) pathway of photosynthesis widens the range of WUE observed, with values for *Agave* and pineapple reaching 20×10^{-3} and 35×10^{-3}, respectively (Joshi *et al.*, 1965; Neales *et al.*, 1968) which is up to 10 times the value observed even for C_4 plants.

From the experiments described above, and from many others, it was soon recognised that, when measured in a given environment, WUE is quite stable across a wide range of cultivars or even similar species (e.g. Shantz and Piemeisel, 1927) and also across a wide range of treatments. Analysis of Shantz and Piemeisel's original data on the basis of our more recent knowledge of photosynthetic pathways showed a clear relationship between WUE and photosynthetic pathway. The measured values of Y_d/E for a wide range of C_4 plants fell within a small range (2.41–3.88×10^{-3}), which was consistently higher than the values of Y_d/E measured for a range of C_3 plants (0.88–2.65×10^{-3}). The relative constancy of WUE for different experimental treatments including planting density, water supply, nutrition and even sowing date, when measured in one environment, has been amply confirmed by more recent experiments. In practice, when integrated over the whole life cycle, differences between plants growing in one environment tend to be rather small. There are a number of reasons for this, but an important one is that the overall WUE is dominated by the value when stomata are open which

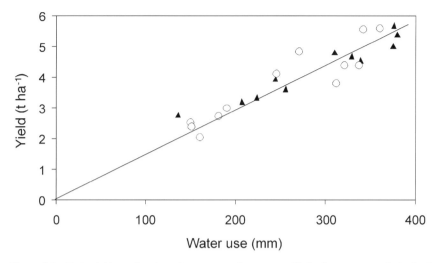

Figure 2.1   Grain yield as a function of water use under a range of irrigation treatments for barley in 1976 (○) and wheat in 1970 (▲) in south-eastern England (data from Day *et al.*, 1978; Innes and Blackwell, 1981). The slope of the line through the origin indicates the WUE for any treatment.

has been referred to above. Some typical examples are illustrated in Figure 2.1.

Most of the above experiments involved comparison of different species growing in a constant environment, but it is well known that the WUE measured in *different* experiments is very sensitive to environmental conditions, especially to the variation in the atmospheric humidity, with the values measured at different sites or, indeed, in different years, varying markedly (e.g. Maximov, 1929). Typical of many studies are the results of N. M. Tulaikov (see Maximov, 1929) with 'Beloturka' wheat in Russia: at one site (the Besenchuk Experiment Station) Y_d/E varied by as much as 90 per cent (from 1.74×10^{-3} to 3.31×10^{-3}) over the years 1911–1917, while in one year (1917) the values ranged from 2.13×10^{-3} at Kostychev to 4.22×10^{-3} in the relatively cool climate at Leningrad (St Petersburg).

The recognition that variation in WUE was primarily related to variation in evaporative demand led de Wit (1958) to propose the normalisation of observed differences in WUE (Y/E) at different sites on the basis of reference evapotranspiration (E_o) according to

$$Y/E = m/E_o \qquad (2.1)$$

where *m* is a plant-specific coefficient that was found to be nearly constant for any species. This relationship only broke down if growth was seriously 'nutrition' limited. Application of this relationship allowed the comparison of data across contrasting climatic regions. An alternative normalisation approach was developed by Sinclair *et al.* (1983), who replaced E_o by $(e^* - e_a) \cong D_a$, where e^* is the vapour pressure within the leaves (actually usually approximated by the saturation vapour pressure at air temperature), e_a is the vapour pressure of the air and D_a is the air vapour pressure deficit.

2.5   Gas exchange

The simple resistance analysis of Bierhuizen and Slatyer (1965) has been widely used to study variation in WUE. Using partial pressures and resistances (rather than conductances to simplify the equations), the assimilation rate (A) is given by

$$A = \frac{p_a - p_\Gamma}{P_a(r_a' + r_s)} \qquad (2.2)$$

where p_a and p_Γ, respectively, are the atmospheric and 'internal' partial pressures of CO_2 (the latter is often assumed equal to the CO_2 compensation partial pressure), P_a is the atmospheric pressure, and r_a', r_s' and r_m', respectively, are the boundary layer, stomatal and intracellular or mesophyll resistances (with the prime referring to CO_2). Similarly for transpiration (E_l), one can write

$$E_l = \frac{e_1 - e_a}{P_a(r_a + r_s)} \qquad (2.3)$$

where e_l and e_a are the saturated water vapour pressure at leaf temperature and the water vapour pressure of the air. Combining equations 2.2 and 2.3 gives

$$\frac{A}{E_l} = \frac{(p_a - p_\Gamma)}{(e_l - e_a)} \cdot \frac{(r_a + r_s)}{r_a' + r_s' + r_m'} \qquad (2.4)$$

If one assumes that p_Γ is constant, the value of WUE in any given environment becomes proportional to $(r_a + r_s)/(r_a' + r_s' + r_m')$. The denominator contains an extra term as compared with the numerator, so it follows that WUE would be expected to increase as stomata close, and this has, for example, been used to justify the extensive effort that has gone into the development of antitranspirants over many years (see e.g. Jones, 1992). This equation also explains the greater WUEs observed for C_4 as compared with C_3 plants, because r_m' (approximately the inverse of the initial slope of an A/p_i response curve) tends to be larger in C_3 plants.

Unfortunately a number of important approximations have been made in this approach:

(a) firstly it assumes that the photosynthetic response to p is linear, but saturation of the A/p_i curve implies that the mesophyll resistance calculated from the simple resistance model is not necessarily constant as stomata open (Jones, 1995);

(b) it assumes that $(e_1 - e_a)$ is constant and equal to D_a in one environment, which is not strictly true as the leaf temperature, and hence D_a, change as stomatal conductance changes;

(c) it is based on instantaneous gas exchange and does not integrate over time (and especially importantly it ignores night-time respiratory losses); and

(d) it ignores soil evaporation. Some consequences of these assumptions are discussed further below.

2.5.1 *Plant-atmosphere coupling*

Unless comparisons are restricted to plants having equal total gas-phase conductances (see Hall *et al.*, 1992), the simple resistance-analogue approach (and the carbon-isotope discrimination method (see below) both depend on the assumption that $(e_1 - e_a)$ can be approximated by D_a. Unfortunately this can lead to quite large errors because $(e_1 - e_a)$ is dependent on the evaporation rate as well as on a range of environmental factors. The magnitude of the error depends on the degree to which the plant or crop is 'coupled' to the environment (Jarvis and McNaughton, 1986; Jones, 1992).

A plant is said to be well coupled when mass and energy exchange between the plant and the bulk atmosphere is efficient, so that leaf temperature closely follows air temperature. In such a situation $D_a \cong (e_1 - e_a)$, and $E \propto 1/r_s$, (where $1/r_s = g_s$ the stomatal conductance) so that equation 2.4 is approximately valid. Where coupling is not perfect, however, as with large areas of aerodynamically smooth crops, the effect of evaporation from plants upwind on the atmospheric humidity, and the fact that leaf temperature diverges from air temperature, both mean that E is not necessarily proportional to the stomatal conductance. In fact, in extreme conditions of very poor coupling, as occur with large boundary layer resistances (e.g. with low wind speeds and large areas of short, smooth crops) E becomes independent of g_s and proportional to incoming energy. Because assimilation is less sensitive to the degree of coupling than is evaporation, changes in stomatal conductance do not necessarily affect the instantaneous WUE as predicted by equation 2.4.

The interactions between boundary layer resistance, radiation and stomatal resistance in their effects on WUE for single leaves and plants have been analysed previously (Jones, 1976). Taking account of the full energy balance resulted in the appearance of an optimal stomatal conductance for maximal WUE (Figure 2.2); this optimal conductance increased as the cuticular water losses increased. It was also pointed out that dark respiratory losses also tended to increase the derived optimal stomatal conductance (Figure 2.2).

It follows from the above that when attempting to test for improvements in WUE, whether achieved by breeding or by agronomic means such as the use of

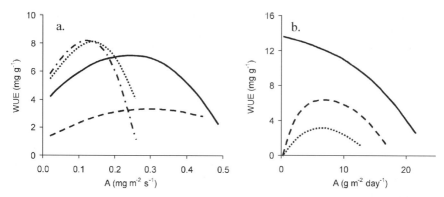

Figure 2.2   (a) Variation in the instantaneous water use efficiency (WUE_i) with photosynthesis rate (which increases as stomata open) calculated for C_3 plant leaves for high (——, - - - , 500 W m^{-2}) and low (–·–·, ····, 50 W m^{-2}) isothermal net radiation and with low (——, –·–·, 10 s m^{-1}) and high (- - - , ····, 100 (s m^{-1}) boundary layer conductances. (b) Variation in daily water use efficiency calculated for C_3 leaves with a net isothermal radiation of 500 W m^{-2} and a boundary layer resistance of 10 s m^{-1} for different likely amounts of night-time respiration (R, mg m^{-2} s^{-1}). ——, R = 0; - - - , R = 0.14 A_d; ····, R = 0.14 A_d + 0.15, where A_d is the daily net assimilation during the light. (Both figures recalculated from Jones, 1976.)

antitranspirants, it is necessary to take account of scale effects. Of necessity, breeders and physiologists usually do their selection on single plants in controlled environments, or on isolated single plants in the field. In each case the plants are well coupled to their environment so that any changes in stomatal conductance would be expected to have a nearly proportional effect on E, with decreases in conductance tending to increase WUE. However, when crops are grown on a large scale in the field, especially aerodynamically smooth short crops such as cereals, coupling is poor. In this case one would not expect the improvements in WUE observed at the small scale to be reflected at the crop scale; indeed one might even expect selection for reduced stomatal conductance to lower WUE (because of the potentially greater sensitivity of A than E to changes in stomatal conductance in decoupled situations), or at least to give variable results (Farquhar *et al.*, 1989).

   In contrast to those situations where selection is based primarily on differences in stomatal conductance, where selection is for differences in assimilation capacity (as occurs when comparing C_3 and C_4 plants, for example) one might expect any differences in WUE to be maintained over a range of scales including the field.

2.5.2   Integration over time

Notwithstanding the relative stability of seasonal WUE, there is much variation diurnally in the leaf-scale WUE as radiation and humidity change diurnally. The seasonal WUE integrates all the short-term changes that occur. The underlying stability of seasonal WUE arises because the seasonal integral is dominated by the value of WUE when stomata are most open. Farquhar *et al.* (1989) expressed the true seasonal WUE in the following terms

$$\text{WUE} = \left(\int A \, dt / \int E_l \, dt\right) \times (1 - \theta_R)/(1 + \theta_W), \tag{2.5}$$

where A is the net photosynthesis, E_l is the leaf transpiration, θ_R is the loss of carbon due to respiration (expressed as a fraction of the carbon fixed) and θ_W is the influence of water lost that is independent of CO_2 uptake (such as cuticular loss from the shoots at night, expressed as a fraction of transpiration).

This integration over daily or longer periods is not straightforward. A particular problem is to estimate the fraction of photosynthesis that is lost in respiration; although photorespiration and any continuing dark respiration of the leaves is included in the measured net assimilation, there are also the losses from stems and non photosynthetic tissues such as roots to be considered, together with night-time losses from the photosynthetic tissues. On average the total respiratory losses through maintenance and growth respiration may be as high as 50 per cent of the daily net assimilation (Jones, 1992).

Integration of E_l is also difficult as the saturation deficit changes diurnally. Tanner (1981) has approximated the daily mean of D_a as 1.45 × the mean of the saturation deficit calculated at minimum and maximum temperatures.

Much effort has been devoted to improvement in the instantaneous value of WUE, but it has been pointed out that maximising WUE over the long term does not necessarily involve continuous maximisation of the instantaneous ratio between assimilation and transpiration (Jones, 1976; Cowan and Farquhar, 1977). Subsequent analyses have been extended to take account of uncertainties in the future environment and to investigate optimal patterns of stomatal behaviour in such circumstances (e.g. Jones, 1981; Cowan, 1982, 1986; Givnish, 1986).

2.6 Methodology

Methods for the study of WUE hardly changed until the 1980s when the use of carbon isotopes as a surrogate measure of water use efficiency was introduced (Farquhar *et al.*, 1982). More recently the oxygen isotopes have also been used to expand the information available. Although stable isotope discrimination can be measured on-line to give estimates of instantaneous WUE (Condon *et al.*, 1990), the major impact of the use of stable isotopes is that their use provides a simple and rapid screening method that integrates WUE over time, because isotope signatures found in tissue carbohydrates depend on the processes involved throughout their formation. Furthermore, the use of isotopic techniques also opens up the possibilities for the study of historic changes in WUE through, for example the analysis of the $\delta^{13}C$ of herbarium specimens (Woodward, 1993) and through the analysis of tree-rings (e.g. Leavitt et al., 2002).

The basis of carbon-isotope discrimination is the fact that photosynthesis tends to discriminate against the heavier isotopes of carbon, so that the ratio of $^{13}C/^{12}C$ in dry matter tends to be somewhat lower than the corresponding ratio for the CO_2 in air. The amount of this discrimination, usually expressed in parts per thousand (‰), is described by Δ, defined by

$$\Delta = \frac{(^{13}C/^{12}C)_{reactants}}{(^{13}C/^{12}C)_{products}} - 1 \qquad (2.6)$$

The value of Δ for plant material is commonly in the range 13 to 28‰ for C_3 plants and between -1 and 7‰ for C_4 plants. Farquhar et al. (1982) showed that the value of Δ would be expected to be dependent on the value of the intercellular space partial pressure of CO_2 (p_i) approximately according to

$$\Delta = 0.0044 + 0.0256 p_i/p_a \qquad (2.7)$$

This prediction has now been confirmed experimentally in a number of cases for C_3 plants (see Farquhar et al., 1988), while Δ is largely insensitive to p_i/p_a in C_4 plants (Evans et al., 1986). Altering equation 2.2 to treat only the gas-phase part of the CO_2 uptake pathway (e.g. Jones, 1992) gives

$$A(r_a' + r_s') = (p_a - p_i)/(P_a) \qquad (2.8a)$$

where p_i is the intercellular space partial pressure of CO_2. This can also be written in terms of the total gas-phase conductance to CO_2 ($g' = 1/(r_a' + r_s')$) as

$$A/g' = p_a(1 - p_i/p_a)/(P_a) \qquad (2.8b)$$

The value of p_i/p_a provides a measure of the ratio between assimilation and the total gas-phase conductance, and hence of WUE, with A/g' (or WUE) increasing as p_i decreases. Where two plants have the same conductance, the assimilation rate, and hence the WUE must be higher for the plant with the lower value of p_i.

Substituting from equation 2.7 and rearranging gives for plants with equal total gas-phase conductance

$$WUE_1/WUE_2 = (0.03 - \Delta_1)/(0.03 - \Delta_2) \qquad (2.9)$$

Unfortunately, as for the other methods for estimating WUE, carbon isotope discrimination neither allows for respiratory loss, nor for evaporation from the soil. Furthermore it also assumes that D_a is a good measure of $(e_1 - e_a)$, though Farquhar *et al.* (1988) have suggested that it may be possible to estimate the vapour pressure difference $(e_1 - e_a)$ weighted by stomatal conductance using information on the stable-isotopic compositions of hydrogen and oxygen in organic material.

Another significant limitation with the use of carbon isotope discrimination in physiological studies is that it only gives a measure of the overall intrinsic water use efficiency, without giving any indication as to whether differences arise as a result of altered photosynthetic activity or as result of altered stomatal conductance. Parallel measurements of oxygen isotope discrimination can give supplementary information on evaporation rate. These measurements are valuable because the ^{18}O signal of leaf organic material is directly related to the rate of evaporation, since the leaf water tends to become enriched in ^{18}O as the lighter isotope (^{16}O) preferentially evaporates (Farquhar *et al.*, 1998; Price *et al.*, 2002).

2.7 Water use efficiency and productivity

In the earliest studies of transpiration ratios it was noted (Kolkunov, 1905, quoted by Maximov, 1929) that 'the more a plant transpires per unit area of dry matter produced, the more it transpires per unit of leaf surface'. In other words, WUE decreases as transpiration rate increases or, because of the close association between transpiration rate and assimilation rate, there is a tendency for WUE to decrease with increasing productivity. This initial observation, however, was somewhat obscured (e.g. Maximov, 1929) by the inclusion of both C_3 and C_4 species in many later studies and was only rediscovered more recently. The importance of this observation is that selection for high water use efficiency in any crop improvement programme may be detrimental for overall productivity whenever water is not limiting. Indeed, there is a widespread misconception that improvement in WUE is all that is required to improve yields under drought conditions, and the related concept that plants have evolved to maximise WUE (e.g. Udayakumar *et al.*, 1998). Because of the fact that in most situations productivity decreases as WUE increases, it follows that simply maximising WUE does not necessarily lead to maximal production.

In consequence the concept of water use efficiency is of only limited applicability to discussion of the fitness of plants for water-limited environments. This is particularly true for natural ecosystems where the dominant factors relating to survival or competitiveness include processes such as the effectiveness of mechanisms (such as stomatal closure, or differences in the exploration of soil by roots) that prevent the occurrence of damaging plant water deficits. Improvements in WUE *per se* are only likely to be of value where there is little competition. It is not much use for a plant to have a high WUE, if its neighbours use up all the available water first! Even in an agricultural or horticultural context, it is still efficiency of use of *available* soil water rather than water use efficiency *per se* that would appear to be important (Jones, 1981).

References

Bierhuizen, J.F. and Slatyer, R.O. (1965) Effect of atmospheric concentration of water vapour and CO_2 in determining transpiration-photosynthesis relationships of cotton leaves. *Agricultural Meteorology* **2**, 259–270.

Burman, R.D., Nixon, P.R., Wright, J.L. and Pruitt, W.O. (1981) Water requirements. In: *Design and Operation Farm Irrigation Systems* (Ed. M.E. Jensen). ASAE, St. Joseph, Michigan.

Condon, A.G., Farquhar, G.D. and Richards, R.A. (1990) Genotypic variation in carbon-isotope discrimination and transpiration efficiency in wheat: leaf gas exchange and whole plant studies. *Australian Journal of Plant Physiology*, **17**, 9–22.

Cowan, I.R. (1982) Water-use and optimization of carbon assimilation. In: *Encyclopedia of Plant Physiology, new series, vol. 12B* (eds O.L. Lange, P.S. Nobel, C.B. Osmond and H. Ziegler). Springer-Verlag, Berlin, pp. 589–613.

Cowan, I.R. (1986) Economics of carbon fixation in higher plants. In: *On the Economy of Plant Form and Function* (ed. T.J. Givnish). Cambridge University Press, Cambridge, pp. 133–170.

Cowan, I.R. and Farquhar, G.D. (1977) Stomatal function in relation to leaf metabolism and environment. *Symposium Society for Experimental. Biology* **31**, 471–505.

Day, W., Legg, B.J., French, B.K., Johnston, A.E., Lawlor, D.W. and Jeffers, W. de C. (1978) A drought experiment using mobile shelters: the effect of drought on barley yield, water use and nutrient uptake. *Journal of Agricultural Science, Cambridge*, **91**, 599–623.

de Wit, C.T. (1958) Transpiration and crop yields. Institute of Biological and Chemical Research on Field Crops and Herbage, Wageningen, Paper 64.6, 88 pp.

Evans, J.R., Sharkey, T.D., Berry, J.A. and Farquhar, G.D. (1986) Carbon-isotope discrimination measured concurrently with gas-exchange to investigate CO_2 diffusion in leaves of higher plants. *Australian Journal of Plant Physiology*, **13**, 281–292.

Farquhar, G.D., O'Leary, M.H. and Berry, J.A. (1982) On the relationship between carbon-isotope discrimination and intercellular carbon dioxide concentration. *Australian Journal of Plant Physiology*, **9**, 121–137.

Farquhar, G.D., Hubick, K.T., Condon, A.G. and Richards, R.A. (1988) Carbon-isotope fractionation and plant water-use efficiency. In: *Stable Isotopes in Ecological Research* (eds. P.W. Rundel, J.R. Ehleringer and K.A. Nagy). Springer-Verlag, New York–Berlin–Heidelberg, pp. 21–40.

Farquhar, G.D., Ehleringer, J.R. and Hubick, K.T. (1989) Carbon-isotope discrimination and photosynthesis. *Annual Review of Plant Physiology and Plant Molecular Biology*, **40**, 503–537.

Farquhar G.D., Barbour M.M. and Henry B.K. (1998) Interpretation of oxygen isotope composition of leaf material. In: *Stable isotopes: integrated of biological, ecological and geochemical processes.* (ed. H. Griffiths) 27–62, BIOS Scientific Publishers, Oxford, pp. 27–62.

Givnish, T.J. (1986) Optimal stomatal conductance, allocation of energy between leaves and roots, and the marginal cost of transpiration. In: *On the Economy of Plant Form and Function* (ed. T.J. Givnish). Cambridge University Press, Cambridge, pp. 171–213.

Hall, A.E., Mutters, R.G. and Farquhar, G.D. (1992) Genotypic and drought-induced differences in carbon-isotope discrimination and gas exchange of cowpea. *Crop Science* **32**, 1–6.

Hellriegel, F. (1883) *Beiträge zu den Naturwiss.* Grundlagen des Ackerbaus, Braunschweig.

Innes, P. and Blackwell, R.D. (1981) The effect of drought on the water use and yield of two spring wheat genotypes. *Journal of Agricultural Science, Cambridge* **96**, 603–610.

Jarvis, P.G. and McNaughton, K.G. (1986) Stomatal control of transpiration: scaling up from leaf to region. *Advances in Ecological Research*, **15**, 1–49.

Jones, H.G. (1976) Crop characteristics and the ratio between assimilation and transpiration. *Journal of Applied Ecology*, **13**, 605–622.

Jones, H.G. (1981) The use of stochastic modelling to study the influence of stomatal behaviour on yield-climate relationships. In: *Mathematics and Plant Physiology* (eds D.A. Charles-Edwards and D.A. Rose, Academic Press, London, pp. 231–244.

Jones, H.G. (1992) *Plants and Microclimate: A Quantitative Approach to Environmental Plant Physiology*. Cambridge University Press, Cambridge.

Jones, H.G. (1995) Partitioning stomatal and non-stomatal limitations to photosynthesis. *Plant, Cell and Environment* **8**, 95–104.

Joshi, M.C., Boyer, J.S. and Kramer, P.J. (1965) CO_2 exchange, transpiration and transpiration ratio of pineapple. *Botanical Gazette*, **126**, 174–179.

Kolkunov, W. (1905) Contributions to the problem of breeding drought resistant plants. I. Anatomical and physiological investigations of the degree of xerophily of certain cereals. *Mém. Polytech. Inst. Kiev*, **5**, no. 4.

Leavitt, S.W., Wright, W.E. and Long, A. (2002) Spatial expression of ENSO, drought, and summer monsoon in seasonal delta C-13 of ponderosa pine tree rings in southern Arizona and New Mexico. *Journal of Geophysical Research – Atmospheres*, **107**, 4349.

Loomis, R.S. and Connor, D.J. (1992) *Crop Ecology: Productivity and Management in Agricultural Ecosystems.* Cambridge University Press, Cambridge.

Maximov, N.A. (1929) *The Plant in Relation to Water.* Allen & Unwin, London.

Neales, T.F., Hartney, V.J. and Patterson, A.A. (1968) Physiological adaptation to drought in the carbon assimilation and water loss of xerophytes. *Nature*, **219**, 469–472.

Price, A.H., Cairns, J.E., Horton, P., Jones, H.G. and Griffiths, H. (2002) Linking drought resistance mechanisms to drought avoidance in upland rice using a QTL approach: progress and new opportunities to integrate stomatal and mesophyll responses. *Journal of Experimental Botany*, **53**, 989–1004.

Robinette, G.O. (1984) *Water Conservation in Landscape Design and Maintenance.* Nostrand Reinhold, New York.

Schröder, M. (1895) The *transpiration* of various crop plants. *Agriculture and Forestry*, 320–336 (in Russian).

Shantz, H.L. and Piemeisel, L.N. (1927) The water requirements of plants at Akron, Colorado. *Journal of Agricultural Research*, **34**, 1093–1190.

Sinclair, T.R., Tanner, C.B. and Bennett, J.M. (1983) Water-use efficiency in crop production. *Bioscience*, **34**, 36–40.

Tanner, C.B. (1981) Transpiration efficiency of potato. *Agronomy Journal*, **73**, 69–54.

Tenhunen, J.D., Lange, O.L. and Jahner, D. (1982) The control by atmospheric factors and water stress of midday stomatal closure in *Arbutus unedo* growing in a natural Macchia. *Oecologia* **55**, 165–169.

Udayakumar M., Sheshshayee M.S., Nataraj K.N., Madhava H.B., Devendra R., Hussain I.S.A., Prasad T.G. (1998) Why has breeding for water use efficiency not been successful? An analysis and alternate approach to exploit this trait for crop improvement. *Current Science, India*, **74**, 994–1000.

Vickers, A. (2001) *Handbook of Water Use and Conservation.* WaterFlow Press, Amhurst, MA., USA.

White, J.W., Castillo, J.A. and Ehleringer, J.R. (1990) Associations between productivity, root growth and carbon-isotope discrimination in *Phaseolus vulgaris* under water deficit. *Australian Journal of Plant Physiology*, **17**, 189–198.

Woodward, F.I. (1993) Plant responses to past concentrations of CO_2. *Vegetatio*, **104**, 105–155.

3 Water use efficiency and photosynthesis

M.M. Chaves, J. Osório and J.S. Pereira

3.1 Introduction

In this chapter we will analyse the relationships between carbon assimilation at the leaf and the whole plant level and water use efficiency (WUE). In particular, we discuss how plants have solved the dilemma of simultaneously assimilating carbon and losing water vapour through the stomatal pores. How can we link WUE with the efficiency of use of other resources, such as light and nitrogen? How can we dissect the links between WUE and drought resistance and relate this to the WUE patterns in the natural vegetation? In general terms, what is the ecological significance of WUE?

Although the expression 'water use efficiency' (WUE) may be misleading, as plants lose water rather than use it as a raw material for the production of biomass (Stanhill, 1986; Monteith, 1993), we shall employ it because of its wide acceptance, and shall define it as the ratio of some measure of carbon assimilation or growth to some measure of the correspondent water loss. We shall also consider it in different spatial and time scales, such as at the level of the leaf or whole plant, and over short-time measurements (instantaneous WUE) or the whole growing season (long-term WUE).

At the leaf level, instantaneous water use efficiency (WUE_i) can be estimated by gas exchange measurements and calculated as the ratio of carbon assimilation to transpiration (A/E, μmol CO_2 mol^{-1} H_2O). As the ratio A/E is largely dependent on the vapour pressure deficit (VPD), the ratio of carbon assimilation to stomatal conductance A/g_s (referred to as *intrinsic water use efficiency*) is often used as a normalised value when comparing instantaneous WUE measured at different VPD.

At the whole plant level, a *season-long water use efficiency* (WUE_{sl}, g DM kg^{-1} H_2O or mmol C mol^{-1} H_2O) can be defined as the ratio of the net gain in dry matter over a given period, divided by the water loss over the same period. However, according to the fields of application, dry matter and water loss can be expressed in several different ways. In the physiological/biological sense, WUE_{sl} is usually regarded as the ratio of total biomass produced per unit of water lost by transpiration, a quantity more appropriately identified as *transpiration efficiency*. Agronomists often use a different definition of WUE_{sl}, known as *crop water use efficiency*, which is a measure of economic yield produced by water lost by transpiration or by evapotranspiration. Another agronomic indicator comparable to WUE_{sl} is *irrigation water use efficiency*, which relates yield to irrigation water applied and is important to monitor and compare the performance of different crop systems. When the objective is to detect differences between plants, it is useful to account for the effect of VPD on WUE_{sl}. Values measured under different climatic conditions may be 'normalised', i.e. multiplied by the average vapour pressure deficit of the air, for better comparisons (Bierhuizen and Slatyer, 1965; Tanner, 1981; Tanner and Sinclair, 1983).

3.2 The carbon compromise

Because carbon dioxide and water vapour share the same stomatal diffusion pathway, and the diffusion gradient that drives water loss is much larger (around 50 times) than that for CO_2 uptake, an increase in leaf stomatal conductance (gs) that enhances CO_2 diffusion (and hence photosynthetic rate, A), inevitably leads to a large increase in transpiration. On the other hand, there is evidence that in C_3 species stomatal conductance is normally such that photosynthesis tends to be co-limited by CO_2 carboxylation and by ribulose-biphosphate regeneration. In other words, the actual rate of net CO_2 assimilation that is dictated by CO_2 availability (and therefore by stomatal conductance), corresponds normally to the intercellular CO_2 partial pressure (p_i) where the balance between the two biochemical limitations to photosynthesis occur. This p_i is called the 'operating point' for photosynthesis. If gs, and hence p_i, increase above the operational point, leaf photosynthetic rate would only marginally increase and WUE will tend to decrease. If, on the other hand, CO_2 supply is strongly limited by stomata closure (e.g. as a result of soil drying), there may be the need for the down-regulation of photosynthesis to avoid photo-inhibitory damage of the chloroplast, and WUE may also decrease.

In summary, it seems that under certain limits, stomata are able to balance the need for the entry of CO_2 to the intercellular spaces – to allow photosynthesis to occur – with the need to avoid dehydration by excessive water loss; in other words, stomata will open to the extent required to provide sufficient CO_2 to meet the requirements for photosynthesis. In fact, there is strong evidence indicating that the ratio A/gs is conservative for a broad range of conditions, a feature that appears to have an important ecological significance (Schulze *et al.*, 1994). This has led to the hypothesis that an optimisation of water use has accompanied plant evolution (Cowan and Farquhar, 1977), so that the partial pressure of CO_2 in the intercellular spaces (p_i) and the instantaneous WUE would remain constant. Cowan and co-authors argued that, within the constraints set by maximal and minimal

Table 3.1 Net photosynthetic rates (A, in μmol m^{-2}s^{-1}) and stomatal conductance (g_s, in μmol m^{-2}s^{-1}) measured in sun leaves of *Quercus suber*, in early summer (during the morning hours and at midday) at 22.5°C and 32.5°C. WUE$_i$ is given by the ratio A/g_s (in mmol CO_2 mol^{-1} H$_2$O).

	Morning		Midday	
	22.5°C	32.5°C	22.5°C	32.5°C
A	11.6	4.8	7.8	3.2
g_s	134.6	57.7	84.6	42.3
A/g_s	86.2	83.2	92.2	75.7

stomatal conductances for a particular leaf type, stomatal response to humidity and temperature would result in maximal daily CO_2 assimilation for a specific daily water use. Indeed, numerous observations showed that $\delta A/\delta E$ remained constant with variations in air humidity and leaf temperature (Table 3.1) or even irradiance (Farquhar *et al.*, 1980, Hall and Schulze, 1980), except when changes are too fast to produce a decrease in A with stomata still open (Wong *et al.*, 1979). The range of values of the ratio of p_i/p_a in healthy C$_3$ plants was shown to collapse in a narrow band (Leuning *et al.*, 1995).

It was further suggested that the capacity of the mesophyll to fix CO_2 could influence g_s (Wong *et al.*, 1979, 1985). The hypothesis of the influence of the mesophyll photosynthesis on stomatal opening, via some unidentified chemical or electrical signal, was supported by the different behaviour of stomata in detached epidermis and in intact leaves (Lee and Bowling, 1995). This hypothesis has been the basis for several models that explain the responses of g_s, p_i and A of well-watered plants to a range of environmental factors (Ball *et al.*, 1987, Leuning *et al.*, 1995). One possibility earlier raised for the mesophyll control of stomata was via the intercellular CO_2 effect on stomatal conductance (Raschke, 1976). This may be an important mechanism in the response of C$_4$ plants to light, but does not seem significant in C$_3$ species (Wong *et al.*, 1979).

As it was stated above, p_i tends to be maintained constant in leaves kept under well-watered conditions. When the water supply is declining, stomata respond to leaf water potential, and both respond to and control the supply and loss of water by the leaves (Leuning *et al.*, 2003). Under these circumstances, intercellular CO_2 could control stomatal opening in response either to the supply of CO_2 to the chloroplast (a function of the diffusion from the air to the site of carboxylation), or to the demand for CO_2 by photosynthesis, governed by chloroplast biochemistry, irradiance or sink strength (Mott, 1988; Assmann, 1999). We usually observe that the decrease in g_s in response to mild water stress leads to a linear decline in transpiration (under constant VPD) and of p_i, because the demand for CO_2 by the chloroplasts stays the same. In this case, because the gradient p_i–p_a increases, photosynthesis either does not decrease or decreases relatively less than g_s and therefore WUE increases (Lambers *et al.*, 1998). In other words, intrinsic WUE commonly increases in response to mild water deficits, because drought-induced stomatal closure restricts water loss more than CO_2 uptake; this is even more evident for C_4 plants, because their CO_2 uptake is less sensitive to the initial decline in g_s than C_3 plants (Long, 1999; Ghannoum *et al.*, 2002).

In long-term water deficits, WUE may be positively or negatively correlated with relative growth rates. For example, in a study with two C_4 Sahelian grasses, we observed that *Dactyloctenium aegyptium,* a drought escaping species, was particularly efficient in using water in the early stages of the life cycle when higher growth rates occurred (with higher intrinsic WUE positively correlated with growth), whereas in the drought resistant species *Schoenefeldia gracilis,* the increased WUE under water deficits was associated with a high allocation to non-photosynthetic tissues (support and vascular). *S gracilis* is a slow grower, with a long growing season; therefore increased WUE was associated with a 'strategy' of conservative use of water resources and survival (Maroco *et al.*, 2000).

3.3 Genetic and environmental constraints to A, g_s, g_w and WUE

It is well known that there are large differences in maximal A (A_{max}, determined by anatomical/diffusional and biochemical characteristics) and in maximal g_s (determined by the density and geometric properties of open stomata) that are genetically governed. This is the case of the species with different pathways of photosynthetic CO_2 fixation, the C_3, C_4 and CAM plants (Schulze and Hall, 1982). The high A/g_s ratios and operational intercellular CO_2 concentration observed in C_4 plants are mainly explained by a CO_2 concentrating mechanism that brings the CO_2 partial pressure in bundle sheath cells to values estimated in 2000 to 3000 bars, enhancing carbon assimilation (Table 3.2). Although the high photosynthetic rates in C_4 plants can be associated with lower stomatal conductance than in C_3 plants (Cowan and Farquhar, 1977; Schulze and Hall, 1982), this is often not the case, mainly when comparing C_3 and C_4 grasses (Körner *et al.*, 1979; Morison and Gifford, 1983). In CAM plants the daytime closure of stomata and the very low stomatal density (ten times lower than in C_3 and C_4 plants) explain their high water use efficiency. Of course, these differences hold for similar conditions of temperature and leaf-air vapour pressure difference and may not apply under natural conditions. For example, a hot habitat for a C_4 versus a cold one for a C_3, that may reverse this trend. It is noteworthy, that the higher intrinsic WUE does not render C_4 plants more tolerant to water deficits. On the other hand, the high WUE of CAM plants is attained at the expense of low productivity, due to limited capacities for storage of malic acid and for CO_2 fixation in the dark.

Table 3.2 Maximal rates of photosynthesis (A_{max}), stomatal conductance at A_{max} (g_s) and the A/g_s, indicative of WUE_i, of two C_4 and two C_3 plants. (Values from Schulze and Hall, 1982.)

	A_{max} (μmol m^{-2}s^{-1})	g_s (mmol m^{-2} s^{-1})	A/g_s (mmol CO_2 mol^{-1} H_2O)
C_4 plants			
Zea	55.8	426.1	131
Sorghum	59.4	365.2	163
C_3 plants			
Fagus	7.3	356.5	20.5
Phaseolus	25.7	473.9	54.2

The ability to optimise net carbon gain for a specific light environment is critical for plant survival. In the summer in temperate or in semi-arid climates, plants are subjected not only to high light but also to high temperatures and VPD. Under these conditions plants exhibit different morphological/ anatomical and biochemical acclimations, enabling them to avoid wilting and to a certain extent, continue growth. Among those, the increase in WUE is critical for the successful acclimation to high light, as shown by Terashima group (Hanba et al., 2002). It was suggested that thick mesophyll cells and low mesophyll porosity, as in sclerophyllous leaves, are responses to water limitation in a high light environment. In shade-adapted plants leaf capacity for CO_2 assimilation is generally lower than in sun plants and is associated with lower stomatal conductance to water vapour (Boardman, 1977; Kubiske et al., 1997). Even though intrinsic WUE can be higher in shade than in sun leaves in some circumstances, shade plants are generally more susceptible to drought (Schulze and Hall, 1982; Kubiske et al., 1997, Niinemets et al., 1998). For example, in studies with natural populations of sun and shade genotypes of the Californian shrub *Heteromeles arbutifolia* in an unusually dry El Niño year, a greater reduction in leaf water potentials (Ψ_1), g_s and A was observed in shade than in sun plants (Valladares and Pearcy, 2002). The increased opportunity for sun plants to rehydrate overnight in the dry months might be responsible for their ability to maintain higher gas exchange rates than shade plants. In addition, because leaves of shade plants tend to be horizontal, their susceptibility to photoinhibition by sunflecks increases during periods of drought, when CO_2 availability is limited by stomatal closure.

Under low light but high air humidity, as mornings or evenings, stomata may be widely open at low photosynthetic rates. The same is true following rapid changes in light, when A decreases instantaneously but stomata remain open, leading to lower intrinsic WUE. On the other hand, the increase in intrinsic WUE at midday in xeric habitats is an important adaptation to water limitation, mainly high VPD. This response may occur even in well-watered

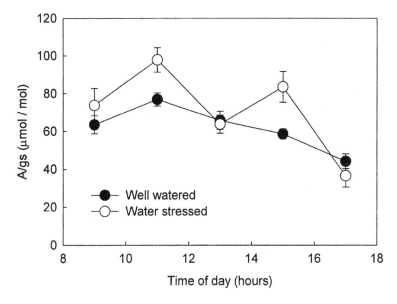

Figure 3.1 Diurnal course of the intrinsic WUE, given by ratio of net photosynthetic rates (A, in μmol $m^{-2}s^{-1}$) and stomatal conductance (g_s, in mmol $m^{-2}s^{-1}$) in sun leaves of nectarines (*Prunus persica*), either well-watered or water-stressed, growing under field conditions in the summer. Values are means of eight measurements per treatment. (From Osório ML, Osório J, Chaves MM unpublished.)

plants, whose leaves may become water-stressed by increased transpiration (Figure 3.1). When midday depression of stomatal conductance becomes much pronounced (e.g. in water stressed plants in Figure 3.1), the daily net carbon assimilation can be dramatically reduced and the low CO_2 availability can induce a decline in carboxylation efficiency, leading to a decrease in intrinsic WUE (Maroco *et al.*, 2002; Valladares and Pearcy, 2002).

In Mediterranean and semi-arid climates there are marked seasonal and diurnal changes in WUE, whose pattern is variable according to the species. In a study with four tree species growing under field conditions side by side, Faria *et al.* (1998) observed that, in two of them (*Quercus suber* and *Q. ilex* ssp. *rotundifolia*) the intrinsic WUE increased by 50 per cent from early July (before drought) to September (end of the drought period) in the morning period, but declined by about 30 per cent in the afternoon (Table 3.3). This type of response indicates that under severe drought the morning period is the most favourable for carbon assimilation. In the afternoon, photosynthesis is restricted not only by stomatal closure but also by photochemical and biochemical limitations. In *Olea europaea* the highest intrinsic WUE (twice that for the *Quercus* species) was reached in July, in the morning hours, and declined by 50 per cent in September. This indicates that this species, with a water saving strategy, had already a low conductance in early July, before the

Table 3.3 Mean values of leaf net photosynthetic rates (A, expressed in μmol m^{-2}s^{-1}), stomatal conductance (g_s, expressed in mmol m^{-2}s^{-1}) measured in July and September, in the morning (MO) and in the afternoon (AF) as well as the ratio of A/g_s and the photosynthetic nitrogen use efficiency (PNUE) of four species *Quercus suber Quercus ilex, Olea europeae* and *Eucalyptus globulus* growing under the field conditions in Portugal (from Faria *et al.*, 1998).

	Q. suber		Q. ilex		O. europeae		E. globulus	
	July	Sept.	July	Sept.	July	Sept.	July	Sept.
A MO	13.9	2.8	13.3	3.3	8.9	1.1	21.1	3.9
AF	4.2	0.9	5.0	1.2	5.0	0.2	4.7	2.2
g_s MO	153.8	18.5	142.5	22.2	41.3	8.9	165.0	37.0
AF	41.3	14.1	52.5	15.5	37.5	7.4	41.2	22.2
A/g_s MO	90.4	151.4	93.3	148.7	215.5	125.8	127.9	105.4
AF	101.7	63.8	95.2	77.4	133.3	27.0	114.1	99.1
PNUE	74.8	14.4	78.6	18.8	49.5	5.5	99.6	23.9

onset of drought. On the other hand, *Eucalyptus globulus*, a fast-growing mesophytic tree, had a fairly constant (and high) intrinsic WUE at all periods, as the result of high photosynthetic rates. A large decline in the nitrogen photosynthetic use efficiency (PNUE) was observed from July to September in all species. Comparing the four species, *O. europaea* had the lowest PNUE as a result of low photosynthetic rates, which were however, associated with the highest intrinsic WUE during the well-watered period. Similar observations were reported by Querejeta *et al.* (2003) when comparing *O. europaea* with a Mediterranean deciduous shrub.

Stomata tend to close in dry air as a direct response to the rate of water loss from the leaf as a consequence of changes in evaporative demand, rather than to air humidity changes (Monteith, 1995; Maroco *et al.*, 1997). It has been suggested that this response is oriented towards photosynthetic optimisation of transpiration, as mentioned above (Cowan and Farquhar, 1977). However, in semi-arid climates, where dry air is not necessarily accompanied by soil water shortage, stomata closure at high evaporative demand might be ecologically disadvantageous (Maroco *et al.*, 1997). If stomata close at high leaf-to-air vapour pressure deficit LAVPD (frequent in semiarid zones between rainfalls) the opportunity for photosynthesis when water in the soil is still available may be lost. This would be particularly detrimental for drought escaping species. On the contrary, this strategy would be useful for 'drought resistant' species, to spare water for further consumption. Indeed, Maroco *et al.* (1997) found out that in drought resistant grasses of the Sahel (*Schoenefeldia gracilis, Ipomoea vagans* and *I. pes-tigridis*) stomata closed with increasing LAVPD, whereas in drought escaping species (*Dactyloctenium aegyptium* and *Eragrostis tremula*) stomatal conductance was independent of LAVPD. The response of intrinsic WUE either increased (*I. pestigridis*), decreased (*S. gracilis*) or did not change significantly with the increasing evaporative demand of the air, as measured by LAVPD.

Under field conditions an increase in temperature tends to accompany the increase in VPD. Therefore it is difficult to separate the effects of the two factors. We found that g_s and A of *Quercus suber* leaves decreased in parallel when temperatures raised from 22.5°C to 32.5°C, so that intrinsic WUE was not substantially changed (Faria *et al.*, 1996) (see Table 3.1). On the other hand, in irrigated *Olea europaea* stomata opened at higher temperatures, leading to a decrease in intrinsic WUE. This response leads to substantial leaf evaporative cooling and suggests an ecological adaptation to high temperature in the absence of soil water limitation (Chaves *et al.*, unpublished). A similar situation was described for Pima cotton (*Gossypium barbadense*) where higher-yielding irrigated cultivars have higher g_s (Radin *et al.*, 1994; Lu *et al.*, 1998). Leaf conductance under well-watered conditions was negatively correlated with foliar temperature because of evaporative cooling. As breeders have selected for high crop productivity, genetic variability for conductance has allowed inadvertent selection for 'heat avoidance' (evaporative cooling) in a hot environment sacrificing WUE, as water was not the limiting factor (Radin *et al.*, 1994).

A strong correlation has been observed between A_{max} and CO_2 transfer conductance from the substomatal cavities to the sites of carboxylation (g_w). This correlation holds constant for several C_3 species ($g_w = 0.012A$) and for both young and old leaves, although photosynthetic capacity and g_w decline in parallel in older leaves (Caemmerer and Evans, 1991; Loreto *et al.*, 1992). A large internal resistance to CO_2 diffusion results in lower partial pressure of CO_2 at the sites of carboxylation (p_c) in the chloroplast (C_3 species) or the cytosol (C_4 species), reducing carbon gain relative to water loss during photosynthesis, i.e., reducing intrinsic WUE. In fact, the CO_2 gradient within the leaves affects the efficiency of Rubisco and the overall N use efficiency. It was estimated that p_c is about 30 per cent lower than p_i for many species in active leaves (Evans and Von Caemmerer, 1996). The minimum g_w will be reached when p_c is reduced to the CO_2 compensation point. Generally, amphistomatous leaves, which possess a lower diffusion path length, have higher g_w as well as greater photosynthetic capacity than hypostomatous leaves (Mott *et al.*, 1982). The lower assimilation rates measured at high irradiance in some sclerophyllous as compared with mesophytic leaves are associated with lower p_c (Evans, 1999); in other sclerophyllous species the larger intercellular air resistance may be compensated by a smaller liquid phase resistance (Evans and Von Caemmerer, 1996). It is noteworthy that p_i/p_a and Δ are linearly related with WUE only if p_c/p_i is constant, therefore an alteration in p_c may alter the predictions of WUE made by p_i/p_a (Evans and Von Caemmerer, 1996).

3.4 WUE in plants growing under elevated CO_2 in the atmosphere

The CO_2 concentration in the atmosphere has been increasing steadily for more than a century and will dominate future scenarios of global environmental change. The positive correlation between A and g_s observed under many circumstances is disrupted in plants growing under elevated CO_2. It has been shown for a large number of species (both C_3 and C_4 plants) that elevated CO_2 leads to a decrease of stomatal conductance by 30–40 per cent on average (Morison, 1987, 1993) although stomata of some forest trees, especially conifers, do not seem to respond to elevated CO_2 (Curtis, 1996). Nevertheless, part of the cases where stomata of woody species failed to respond may have resulted from root growth restrictions (pot effect) or from water deficits (Saxe et al., 1998). In spite of lower leaf g_s under elevated CO_2, photosynthesis is stimulated due to the larger CO_2 concentration difference between the atmosphere and the intercellular spaces. As a consequence, intrinsic WUE increases substantially under elevated CO_2 (up to 2.5 and 3 times), due to the combination of decreased g_s (in C_4 plants) and increased A, in the case of C_3 plants (Drake et al., 1997). The effect of CO_2 on stomatal conductance is generally persistent throughout plant development (Poorter and Perez-Soba, 2001) and in the long-term, stomatal acclimation to CO_2, if it exists, is not as relevant as it can be in CO_2 assimilation rate, where the stimulation of A under 700 ppm CO_2 can be reduced from 40–60 per cent to only 20–30 per cent (Morison, 1998; Lodge et al., 2001).

Intrinsic WUE may increase with plant age in elevated CO_2 grown plants, as was observed in tobacco plants from 3 to 5 weeks old, due to a decrease in g_s while A remained constant (Table 3.4) (Chaves et al., 2001) or WUE may decrease with leaf age in elevated CO_2 grown plants (although maintaining higher values than in ambient CO_2 plants) due to an accelerated decline in photosynthesis with ageing, as was described in Rumex obtusifolius (Pearson and Brooks, 1995).

The effect of elevated CO_2 on intrinsic WUE depends on other environmental factors. For example, we observed that Q suber plants grown

Table 3.4 Mean values of leaf net photosynthetic rates (A, in μmol m^{-2}s^{-1}) and stomatal conductance (g_s, in mmol m^{-2}s^{-1}) measured in tobacco plants (3 and 5 weeks old) grown under two CO_2 concentrations in the atmosphere. WUE$_i$ is given by the ratio A/g_s (from Chaves et al., 2001).

	3 weeks 350 ppm	700 ppm	5 weeks 350 ppm	700 ppm
A	21.5	27.8	22.0	25.8
g_s	285.7	117.9	189.3	67.8
A/g_s	75.3	235.8	116.2	380.5

Table 3.5 Mean values of leaf net photosynthetic rates (A, in μmol m^{-2}s^{-1}), stomatal conductance (g_s, in μmol m^{-2}s^{-1}) measured at three different temperatures (15, 25 and 35°C) and the ratio of A/g_s, indicative of WUE$_i$ of *Quercus suber* plants grown under two CO$_2$ concentrations in the atmosphere (from Faria *et al.*, 1999).

| | 350 ppm | | | 700 ppm | | |
	15°C	25°C	35°C	15°C	25°C	35°C
A	6.8	7.4	8.4	9.6	10.3	11.0
g_s	123.8	153.8	103.1	106.9	105.0	99.4
A/g_s	54.7	48.1	81.5	89.8	98.1	110.7

under elevated CO$_2$ exhibited WUE almost twice that of plants grown in ambient CO$_2$, when measurements were made at 15°C and 25°C as in Table 3.5 (Faria *et al.*, 1999). However, at 35°C the differences between elevated and ambient CO$_2$ plants declined to 30 per cent. This was because stomata of elevated CO$_2$ plants showed little response to temperature, contrary to plants grown at ambient CO$_2$, whose stomata closed when temperature rose, leading to an increase in intrinsic WUE (Chaves *et al.*, 1995). Air humidity and LAVPD also seems to interfere with stomatal response to CO$_2$ (Will and Teskey, 1997; Bunce, 1998). It was shown by Morison and Gifford (1983) that the sensitivity of stomata to LAVPD was lower in elevated than in ambient CO$_2$ grown plants and that the stomatal response to elevated CO$_2$ was reduced at low VPD. The understanding of these complex interactions is fundamental to predict plant water use in response to an increase in the CO$_2$ concentration in the atmosphere.

3.5 Trade-offs in resource use efficiencies – nitrogen, radiation and water

Plant growth and survival depend on how basic resources – water, nutrients and light – are acquired and used. As discussed above, plant nutrition status and atmospheric CO$_2$ concentration, influence WUE. Under natural conditions, plant life is limited by the interplay of the availability and use of several resources. Recently, based on the relative cost of procuring (or using) water versus the cost of procuring (or using) nitrogen for photosynthesis, Wright *et al.* (2003) showed that a standard framework used in microeconomics for optimising the mix of two inputs for a production process could be used to understand the interplay of water and nitrogen in carbon assimilation in plants. This approach integrates the concepts of water- and N-use efficiencies into the wider objective of optimising the input mix for a given situation. The method seems to be an interesting tool for:

(i) Investigating the relative costs of water and nitrogen in contrasting environments.
(ii) Testing hypotheses concerning their interplay in photosynthesis for coexisting species that have different physiology and morphology.
(iii) Identifying shifts in the relationships between leaf traits underlying the carbon fixation strategy of plants.

Although we will not develop here such an approach, we will discuss in this section some of the mechanisms behind the trade-offs between water, nitrogen and light use efficiencies (Monteith, 1984; Wright *et al.*, 2003).

The *photosynthetic nitrogen use efficiency* (PNUE, μmol CO_2 m^{-2}s^{-1}/ mmol N m^{-2} leaf) is defined as the ratio of the rate of photosynthesis to leaf organic nitrogen content per unit area and provides an instantaneous measure of the N cost required to assimilate CO_2 at the tissue level (Poorter and Evans, 1998; Robinson *et al.*, 2001). For ecological, agricultural and forestry purposes a whole-plant, long-term nitrogen use efficiency is frequently more useful than PNUE. A commonly used measure is the *nitrogen use efficiency of dry matter production* (NUE, g DM g^{-1} N), described as the quotient of dry matter production over the incorporated nitrogen. This approach for NUE is merely the inverse of biomass nitrogen concentration and will not provide a good assessment of efficient production at low nutrient input (Gourley *et al.*, 1994). In the agricultural perspective, the term NUE is frequently used to define the agronomic yield per unit of nitrogen fertilizer applied. Assuming that the quotient of total plant carbon to nitrogen reflects the efficiency by which N is distributed and utilised to acquire carbon (Chapin and Van Cleve, 1989), some studies use the C/N ratio as a surrogate NUE (e.g. Patterson *et al.*, 1997; Livingston *et al.*, 1999).

Photosynthetic radiation use efficiency (PRUE) may be defined as the percentage of incoming PAR that is stored in plant dry matter in the form of chemical bonds (Jones, 1992). Assuming a free energy content of sucrose of about 480 kJ mol^{-1} C, an average energy content in the PAR for solar radiation of 220 kJ mol^{-1} C and that the number of photons required to produce one mole of O_2 is eight, a theoretical value of 27 per cent is obtained for the maximal possible efficiency of photosynthesis (PRUE $= (100 \times 480) /(8 \times 220) \approx 27\%$). However, the actual value of gross photosynthesis efficiency in the leaves of C_3 plants was found to be approximately 14 per cent under optimal conditions in the laboratory and between 8 and 10 per cent in a natural environment (Larcher, 1995). Photosynthetic radiation use efficiency of net photosynthesis is one third to one half that of gross photosynthesis (Larcher, 1995).

A long-term approach for radiation use efficiency (ϵ, g MJ^{-1}) is the quotient of dry mass produced by a stand to the amount of light energy absorbed by the foliage canopy in a given time (Monteith, 1972, 1977).

Although theory predicts a maximum value of 2.5 g MJ^1 for ϵ calculated per total radiation (Russell *et al.*, 1989), real estimates for annual crops fall usually in the range of $1.2-1.7$ g MJ^1, due to radiation saturation, environmental stresses, respiratory losses and losses associated with the mortality of leaves and roots.

3.5.1 Trade-offs

In the past couple of decades, relationships among leaf traits underlying the carbon fixation strategy of plants have been extensively studied among and within species (Evans, 1983; Field and Mooney, 1986; Evans, 1989; Reich *et al.*, 1995, 1998, 1999; Niinemets and Tenhunen, 1997; Garnier *et al.*, 1999, Evans *et al.*, 2000; Shipley and Lechowicz, 2000; Mediavilla *et al.*, 2001; Wright *et al.*, 2001). For instance, studies from a large set of diverse C_3 species in six distinct biomes (Reich *et al.*, 1999) found that leaf nitrogen concentration (N_{mass}), specific leaf area (SLA), leaf diffusive conductance (g_s) and light saturated rate of photosynthesis (A_{max}) were positively related to one another and decreased with increasing leaf lifespan. Similar to A_{max}, it was also found that foliar dark respiration rate (R_d) consistently increased with N_{mass} and declined with increasing leaf lifespan.

In general, in broad-leaved species SLA decreases and leaf thickness increases with increasing annual solar radiation and temperature, whereas leaf density relates negatively to precipitation (Niinemets, 2001). As decreases in SLA result normally in higher photosynthesis per unit of leaf area, the suggested adaptive nature of the correlation between leaf density and increased aridity is that WUE and PNUE increase with decreasing SLA. Meziane and Shipley (2001) discussed how SLA, leaf N content, net photosynthesis and stomatal conductance should be ordered in a multivariate model, in order to explain interspecific patterns of direct or indirect co-variation observed in plants subjected to contrasting conditions of irradiance and nutrient supply. The path model that allows the best statistical description of the data pointed out SLA as the forcing variable that directly affects both leaf nitrogen levels and net photosynthetic rates; leaf nitrogen then directly affects net photosynthesis, which in turn affects stomatal conductance. These results draw attention to the importance of SLA as the crossroads to the determination of plant resource-use efficiencies and their trade-offs.

The situation may be complex as differences in SLA result from either changes in leaf thickness or in leaf density and hide other anatomical changes. Each of these components may alter photosynthetic capacity in reverse directions. As pointed out by Niinemets (1999) A_{max} increases with decreased SLA and increased thickness, but increases in leaf density may modify the relationship between A and SLA and tend to decrease A_{max}. In one Mediterranean site Mediavilla *et al.* (2001) studied the effects of SLA and leaf anatomy on CO_2 assimilation, water use efficiency and photosynthetic nitrogen-use efficiency in six Mediterranean woody deciduous and evergreen species with different leaf life spans. Long-lived leaves had lower SLA as well as low assimilation rates and low PNUE. In these species carbon assimilation was reduced by non-stomatal limitations, possibly because of a lower allocation of N to the photosynthetic machinery than in species with high SLA. When leaf thickness increased maintaining a low tissue density, as was the case in intraspecific comparisons, the larger internal air volume in thick leaves, had a positive effect on WUE, probably because it increased the internal CO_2 conductance to the site of carboxylation, suggesting that leaf anatomy was a key factor regulating resource-use efficiency. When comparing different species PNUE changed with different patterns of N partitioning in the leaf, and WUE changed in a more standard way, increasing with leaf thickness.

In a recent review of the role of N in the processes leading to carbon assimilation, Lawlor (2002) provides a detailed explanation of the physiological basis of the A_{max}–leaf N relationship. Basically, photosynthetic capacity and leaf nitrogen content are positively correlated because the bulk of leaf N is found in the photosynthetic complex, in the form of photosynthetic proteins (including large amounts of Rubisco) and, to a lesser extent, of light harvesting complex proteins (Evans, 1983; Field and Mooney, 1986; Evans, 1989; Evans and Seemann, 1989; Lambers et al., 1998). The fact that the A_{max}–leaf N relationship has been used as the conceptual or numerical basis of models to predict photosynthesis over scales ranging from the leaf to the globe (e.g. Farquhar et al., 1980; Woodward and Smith, 1994; Aber et al., 1996) draw attention to its importance and universality.

Despite the consistency of the A_{max}–leaf N relationship among species, functional types and biomes, interspecific variation in PNUE has been repeatedly reported (Field and Mooney, 1986; Sage and Pearcy, 1987; Seemann et al., 1987; Poorter et al., 1990, Reich et al., 1995; Poorter and Evans, 1998). Possible speculative explanations for variation in PNUE (see Poorter and Evans, 1998) are differences between species in one or several of the following features:

(a) Fraction of light absorbed by the leaf at a given irradiance.
(b) Operating intercellular CO_2 partial pressure.
(c) Proportion of organic N allocated to photosynthetic versus non-photosynthetic functions.
(d) Pattern of partitioning of photosynthetic N between light harvesting complexes, electron transport and CO_2 fixation.
(e) Activation state or specific activity of Rubisco.
(f) Rate of respiration in the light.
(g) Light intensity required to saturate photosynthesis.

Some attention has been paid on the trade-offs between PNUE, relative growth rate (RGR) and SLA. In a study on species having contrasting relative growth rates, Poorter *et al.* (1990) concluded that PNUE of fast growing species was higher than that of slow growing ones. Since no differences were found between the two groups of plants in the rate of photosynthesis per unit leaf area, further investigation was made with the aim of testing the hypothesis that fast-growing species have a higher leaf conductance and, consequently, a lower instantaneous WUE (Poorter and Farquhar, 1994). However, no systematic difference was found in the transpiration per unit leaf area, the intercellular CO_2 concentration and the ^{13}C-discrimination. These findings suggest that observed differences in PNUE could not be explained by variation in the operating intercellular CO_2 partial pressure. On the other hand, a compilation of literature sources on a wide range of experiments with interspecific comparisons evidenced a strong positive correlation between SLA and PNUE, with fast-growing plants exhibiting higher SLA than slow growers. This means that, although fast growers have similar photosynthetic rates per unit leaf area than slow growers, the former exhibit higher photosynthetic rates per unit biomass (or per cell). Analysing the relative importance of SLA and of nitrogen partitioning in maximizing carbon gain during the acclimation of plants to their growth irradiance, Evans and Poorter (2001) showed that changes in SLA have in general a greater impact than N partitioning.

Theory predicts a negative relationship between photosynthetic nitrogen use efficiency and photosynthetic water use efficiency. In C_3 plants, leaf p_i is usually not saturating for carbon assimilation, and A tends to increase with increasing g_s (Wong *et al.*, 1979). Such an increase in A does not involve the investment of additional N in photosynthetic enzymes and will therefore result in increased photosynthetic nitrogen use efficiency. On the other hand, any increase in g_s will also lead to a decrease in instantaneous WUE, because of a proportionately greater raise in transpiration rate than in photosynthetic rate. Hence, the contrasting effect of increased g_s on PNUE and WUE_i discloses the negative nature of the trade-off between the two efficiencies. In fact, a negative trade-off between PNUE and WUE_i has been frequently reported in comparisons among species (Field *et al.*, 1983; De Lucia and Schlesinger, 1991), nitrogen and water availabilities (Reich *et al.*, 1989; Fredeen *et al.*, 1991; Van den Boogaard *et al.*, 1995; Livingston *et al.*, 1999), plant age classes (Donovan and Ehleringer, 1992), competitor planting densities (Robinson *et al.*, 2001) and stages of canopy development (Gutiérrez and Meinzer, 1994).

Plants which are exposed to limited availability of any resource required for growth (e.g., water, nutrients or light) usually display some specific responses to overcome the stress condition, such as increased allocation of biomass to the structures involved in resource uptake, increased organ duration to decrease resource losses and increased resource use efficiency (Chapin *et al.*, 1987). In cases where increased use efficiency of the limiting resource is a noticeable response, the physiological and biochemical adjustments implicated in its accomplishment sometimes alter the expected compromises between the use efficiency of the limiting resource and other resources use efficiencies. In this context, the study of the influence of a reduced supply of water and/or nitrogen on the predictable relationship between photosynthetic water and nitrogen use efficiencies seems to be of particular interest.

It is well known that water use efficiency usually increases with reduced water supply (Osório *et al.*, 1998, Ponton *et al.*, 2002) and genotypes with a high WUE are frequently better adapted to water-limited conditions (Lauteri *et al.*, 1997, Silim *et al.*, 2001). However, genotypic or treatment superiority in WUE_i may result from one of the following diverse circumstances:

(a) Higher A and similar E.
(b) Similar A and lower E.
(c) Higher A and E, but proportionally higher A than E.
(d) Lower A and E, but proportionally lower E than A.

It is clear from a review of the literature that the impact of a reduced supply of water and nitrogen on the relationship between the instantaneous measures of water and N-use efficiencies will depend on the circumstances that give rise to the observed variation in WUE. Two examples will be used with the aim of illustrating this point. In a study with a field-grown sunflower cultivar, Fredeen *et al.* (1991) reported that water stress led to increased WUE_i and decreased PNUE relative to the control, while N-stress had the opposite effect. Increased WUE_i observed in water-stressed plants resulted essentially of reduced stomatal conductance, while reduced WUE_i of the nitrogen-stressed ones was ascribed essentially to a non-stomatal limitation of photosynthesis, related to reduced Rubisco activity. Moreover, N-stress reduced both A_{max} and leaf N content, but the decrease was proportionately higher in N content than in A_{max}, thereby enhancing PNUE. According to the authors, sunflower manages the trade-off between nitrogen and water use efficiencies, increasing the use efficiency of the most limiting resource, while decreasing the use efficiency of the other resource. From a quite similar experiment with two wheat cultivars Van den Boogaard *et al.* (1995) concluded that genotypic differences in WUE_i were not consistently related with PNUE. For instance, they found that a high WUE_i was associated with a low PNUE when cultivars were compared at saturating light or when the nitrogen treatments were compared at low-soil water. In these cases, differences in stomatal conductance were the main cause for the differences in WUE_i. However, no trade-off was observed when differences in the rate of photosynthesis, rather than in stomatal conductance, were the main cause for differences in WUE_i, as was the case when cultivars in high water and N conditions were compared at growth (low) irradiance.

Light intensity and nitrogen content are not uniformly distributed between the leaves of a canopy, more N being allocated to the sites under higher irradiance at the top (for a review, see Grindlay, 1997). That is, inside the canopy, the vertical distribution of foliar nitrogen parallels the distribution of photosynthetic photon flux density (PPFD). Moreover, given that processes such as light absorption and regeneration of the activity of enzymes determining CO_2 fixation (e.g. Rubisco) are dependent on both light and leaf nitrogen (Evans, 1989), a matching gradient in leaf photosynthetic capacity (A_{max}) is also observed in the canopy. It has been suggested that N gradients represent a way to maximize carbon assimilation (Hirose and Werger, 1987; Anten et al., 1995), N being optimally allocated when the marginal increase in the assimilation rate with an increase in N content is constant throughout the canopy (Field, 1983).

It is therefore straightforward to postulate that the radiation use efficiency of a crop is largely determined by the patterns of distribution of radiation and leaf N on the canopy (Sinclair and Shiraiwa, 1993). Radiation use efficiency is usually higher under conditions of low light intensities and/or elevated proportions of diffuse radiation (Hammer and Wright, 1994; Bange et al., 1997). For example, data collected by Evans (1993) confirmed that chlorophyll a:b ratio, photosynthetic capacity per unit of chlorophyll and nitrogen content per unit of chlorophyll declined from the top to the bottom of a canopy of lucerne. Accordingly to this, ϵ calculated at the level of the lower leaves of the canopy was not considerably different from that observed at upper levels, mainly due to an increased ability to capture scarce light, related with a higher investment of chlorophyll in light harvesting complexes relative to reaction centres. Also, Dreccer et al. (2000) observed that ϵ increased towards the end of the critical period for grain number definition in wheat and in oilseed rape, essentially because more leaves in the canopy were operating at non-saturating light levels.

The impact of leaf N on radiation use efficiency is expected to be more important in the long rather than in the short term, due to the increased duration of periods of low light intensity (Gastal and Lemaire, 2002). The impact also tends to be more evident in crops with an elevated leaf-area index, as a greater proportion of the leaves are shaded (Gastal and Lemaire, 2002).

3.5.2   A case study

A field experiment installed in March 1986, in central Portugal to evaluate the influence of 'near optimal' water and nutrient supply on the productivity of *Eucalyptus globulus* may be relevant as a case study. The climate is of the Mediterranean type, with summer drought, and the treatments were: irrigation

Table 3.6 Average total biomass (with roots) productivity (NPP; kg.m^{-2} year^{-1}), mean fraction of the solar radiation intercepted by the foliage (f; with Q_o = 4880 MJ m^{-2} year^{-1}), amount of water available (L/m^{-2}) (i.e. 712 mm of total precipitation, plus irrigation water) in the period of June 1988 to June 1989. Leaf area index in 1988 and 1989. The quotient biomass production and radiation intercepted (ϵ) (g MJ^{-1}), the quotient of total biomass produced and evapotranspiration or water use efficiency (WUE) (g L^{-1}) and the nitrogen use efficiency of dry matter production (NUE, g DM mg^{-1} N) calculated as the quotient of dry matter production (NPP) over the mass of nitrogen absorbed. NUE was determined only for the above ground biomass.

Treatment	NPP w. roots (kg m^{-2} year^{-1})	f	Available water (L/m^{-2})	LAI 1988	LAI 1989	ϵ (g MJ^{-1})	Estimated WUE (g L^{-1})	NUE above ground (g mg^{-1} N)
IL	4.6	0.91	1624	3.4	3.8	1.04	3.7	3.0
I	3.6	0.78	1506	2.2	3.0	0.96	3.8	3.7
F	3.3	0.77	712	2.2	2.8	0.88	5.3	2.6
C	3.2	0.71	712	1.6	2.3	0.91	5.1	2.8

between April/May and September to satisfy the evapotranspiration demand in summer, (I), irrigation as in I, plus fertilisers added according to plant needs (IL), and fertilisers added to rainfed plots (F). The control (C) received neither fertilisers (except a small amount at planting time) nor irrigation (Pereira, 1993; Pereira *et al.*, 1994b). The annual production of biomass (above-ground) was measured regularly. The root component was estimated from the root/shoot ratios measured in the site in year 6 assuming that an allometric constant equal to 1 (Madeira *et al.*, 2003).

The treatments were quite effective because, as shown in Table 3.6, net primary productivity (NPP) of IL, I and F trees was 47 per cent, 16 per cent and 5 per cent higher than the control, respectively. Six years after planting, total standing biomass was 81 per cent higher than in the control, and 48 and 27 per cent higher in I and F, respectively (Madeira *et al.*, 2003). But when we consider what happened to the resource use efficiencies, it is obvious that radiation use efficiency (ϵ) increased with abundant water (Pereira *et al.*, 1994b), as IL and I treatments had higher ϵ values than the other treatments. This may have resulted from the expansion of the growth period to the whole summer in the irrigated but not in the non-irrigated treatments (both F and C). The differences in ϵ were proportionally smaller than those in NPP because better access to resources (water and nutrients) led to higher LAI and therefore more absorbed radiation. So, before canopy closure the main investment of the more abundant environmental resources was in new leaves: for example, 80 per cent more leaves produced in IL than in C in 1989 (unpublished). In addition, there was a slight increase in SLA, which was, for example, 13 per cent higher in IL than in C in 1989 (unpublished). A higher SLA permits a greater solar radiation interception for the same leaf biomass.

In the case of the efficiency of water use, however, an extra supply of water led to a decrease in WUE. In Table 3.6, WUE was calculated as the quotient of NPP to annual evapotranspiration in the period between June 1988 and June 1989. The WUE of the irrigated treatments (IL and I) was around 25 per cent lower than in the control. This fact may be explained by a much longer period with open stomata at a time of high evaporative demand – summer – in the irrigated trees, than in rainfed trees. This shows that a slight increase in radiation use efficiency was obtained at the cost of a relatively greater decline in WUE, in these summer-drought conditions. On the other hand, in spite of abundant N, WUE was practically the same in F and C trees (see Table 3.6), the main effect of more N being to increase LAI and photosynthesis rates relative to conductance. Under natural conditions, along a rainfall gradient in Australia, Schulze *et al.* (1998) also found that species with more nitrogen available (and high leaf N concentrations) did not use it to increase photosynthesis, and consequently WUE.

In the present experiment, the addition of nutrients that led to a better NPP had little reward in terms of the efficiency of nitrogen use. In this case, NUE was calculated as the ratio of annual biomass production and the total N. There was a loss in efficiency of 8 per cent in F relative to the control trees. Abundant water however, increased NUE relative to the rainfed treatments. This was especially true when abundant water was combined with low N supplies (e.g., 30 per cent increase in I compared to the control), due to the increases in ϵ and LAI. But abundant water may have not been enough to achieve high NUE in IL plants, because there was a decline in NUE possibly as a result of some 'luxury' consumption.

In the field, before canopy closure, it seems that the extra N available to the plants is used in canopy expansion and less to enrich leaves in N or to increase photosynthetic capacity (Pereira *et al.*, 1994a). Photosynthetic capacity assumed to increase with N concentration per unit leaf area was in fact estimated to be lower in the fast growing irrigated plants than in the slower growing rainfed plants (Pereira *et al.*, 1992). If there are other growth limitations, such as water stress, N concentration per unit leaf mass may increase, as happened in F plants, relative to the control (Pereira *et al.*, 1994b). In terms of carbon assimilation, control rainfed plants closed stomata in the summer and decreased assimilation on a yearly basis, whereas irrigated plants (IL and I) assimilated carbon throughout the summer, with open stomata when evaporative demand was high, therefore leading to a reduction in WUE.

3.6   Patterns of carbon to water balance in natural vegetation

Are there recognisable patterns of WUE at plant community level or is there a generalised WUE at the ecosystem level? Agronomists have used linear relationships between plant yield and transpiration in growth and water-uptake models to estimate yield based on predicted transpiration values (Ben-Gal *et al.*, 2003). In a broader sense it was proposed that large-scale NPP patterns could be estimated through their dependency from actual evapotranspiration (AET), assuming that a given vegetation type has a nearly constant WUE_{sl} (Lieth, 1975). However there are too many cases where AET or water availability are not good estimators of NPP. Alternatively, measurements of stable isotope signatures in plant material are commonly used to investigate both instantaneous and time-integrated plant performance (Ehleringer *et al.*, 1993) because they provide us with mechanistic relationships linking WUE to plant physiology.

3.6.1   Using isotope discrimination to study WUE

Because isotopic fractionation occurs during both photosynthesis and transpiration, the carbon and oxygen stable isotope ratios ($^{13}C/^{12}C$ and $^{18}O/^{16}O$)

can be regarded as surrogate variables integrating these basic physiological processes. The carbon isotope composition ($\delta^{13}C$) or the carbon isotope discrimination (Δ) of plant tissues reflect the variation of the photosynthesis to stomatal conductance ratio over a considerable length of time and under variable environmental conditions, and represents a long-term integrative estimate of the intrinsic water use efficiency (Farquhar et al., 1989). This may be however quite different from long-term WUE$_{sl}$, which changes with climate, especially with the evaporative demand of the air independently from intrinsic WUE.

A few pitfalls and limitations need to be taken into account when using carbon and oxygen stable isotope approaches to investigate plant performance. In additon to the type of tissue sampled, environmental factors influence $\delta^{13}C$ or Δ. Differences in the patterns of temporal and spatial variation in light (Martinelli et al., 1998), in the annual precipitation or in the isotopic composition of source air inside the canopy (Buchmann et al., 1997; Buchmann et al., 1998) are examples of environmental factors to which variation in Δ has been shown in the literature.

Although providing an integrated estimate of intrinsic water use efficiency, $\delta^{13}C$ reflects mostly the carbon assimilation conditions at the time when carbon was integrated into the tissue under analysis. In a study of $\delta^{13}C$ along an aridity gradient in Patagonia, ranging from high annual rainfall to low rainfall grasslands and deserts, Schulze et al. (1996b) found little variation in the average $\delta^{13}C$ of sun leaves (at -27‰), even though there were decreases in mean biomass above and below ground as well as in leaf-area index along the gradient. This suggests that no major differences in A/g_s occurred at the time of leaf construction in the different plant communities. It also suggests that by measuring $\delta^{13}C$ of bulk leaf tissue, we may be measuring mostly the photosynthetic discrimination of the past – when the leaves were produced – rather than the average or current WUE of the vegetation. However, by studying isotope discrimination in different metabolites (e.g. cellulose, starch, soluble sugars) past or recent assimilated carbon can be distinguished (Brugnoli and Farquhar, 2000). The type of information and the corresponding relevance to the objectives of the study will advise us on the tissue, organ or compound to be sampled (see Pate, 2001).

As pointed out by Pate and Arthur (1998), because of improperly defined relationships in space and time between isotope ratios measured in different plant materials, researchers may experience difficulties in understanding the complex patterns of stable isotope discriminations displayed by a whole plant and its variously aged components. One example is that measured leaf $\delta^{13}C$ values of perennial plants are sometimes less negative than would be expected from model-predicted values derived from instantaneous gas exchange measurements (Vitousek et al., 1990; Le Roux-Swarthout et al., 2001). This probably indicates that some ^{13}C-enriched stored carbon was imported from other leaves or from the stem during the early stages of leaf growth. In fact, different leaf cohorts in a canopy may have different $\delta^{13}C$ (Damesin et al., 1998). In long-lived leaves new carbon (with a different ^{13}C signature) may be incorporated in the tissues with time (Hobbie et al., 2002).

Another limitation to the use of plant carbon isotope discrimination as an indicator of WUE is that a higher $\delta^{13}C$ may result either from water saving (lower g_s) or from increased carbon assimilation (higher A_{max},) as with better plant nutrition (e.g. (Holbrook et al., 1995; Gebauer and Ehleringer, 2000). Often, a higher $\delta^{13}C$ (or a lower Δ) results from stomatal closure and reduced carbon assimilation and growth (Donovan and Ehleringer, 1994; Osorio and Pereira, 1994), but this is not always the case. When comparing families of Quercus douglasii, for example, high growth in well watered conditions could be linked to lower p_i and Δ, and higher A/g_s reflecting a greater influence of the maximum assimilation rate compared with stomatal conductance (Matzner et al., 2001).

The combined study of ^{13}C and ^{18}O composition of plant tissues may provide information to separate between stomatal and non-stomatal effects on water use efficiency. The oxygen isotope composition of organic material ($\delta^{18}O$) is related to the ratio of the atmospheric and intercellular vapor pressures (e_a and e_i, respectively) and determined by (i) the isotopic composition of the source or soil water (ii) the enrichment taking place in the leaf water due to the transpiration, resulting in an increased $\delta^{18}O$ of leaf water due to evaporation and (iii) fractionations during biochemical reactions (Dongmann et al., 1974; Scheidegger et al., 2000). The ^{18}O signature of the biomass reflects transpiration rates, as there is no further discrimination for the element during photosynthesis and may represent an integrative measure of stomatal conductance (Farquhar and Lloyd, 1993; Farquhar et al., 1994; Barbour and Farquhar, 2000; Barbour et al., 2000; Scheidegger et al., 2000). Therefore, $\delta^{18}O$ measurements can potentially be used for the interpretation of the $\delta^{13}C$ values, with the purpose of knowing if the observed carbon isotope composition is due to changes in g_s or in A_{max}.

3.6.2   Differences in WUE between plant functional types

WUE in plants change with climate and with plant traits. These include, in addition to carbon metabolism types (see Section above on C_3, C_4 and CAM plants), leaf phenology (deciduous versus evergreen species), leaf morphology, rooting pattern (shallow-rooted versus deeply-rooted species), shade tolerance (tolerant versus heliophilic species), canopy position (understorey versus emergent canopy species), N_2-fixation (N_2 fixing versus non-N_2 fixing species) and pioneer versus late-successional species. Plant age and hydraulic constraints to water flow in the xylem may also influence A/g_s and plant carbon isotope composition.

Inter- and intra-specific differences in leaf $\delta^{13}C$ and WUE may be related to plant functional groups. For example, in boreal forests, Brooks *et al.* (1997) found that the correlations between Δ and life form were related to differences in plant stature and leaf longevity. Deciduous leaves had higher discrimination values than evergreens, indicating that short lived leaves maintained stomata more open than evergreens, which might reflect greater photosynthetic capacity and better nutrition, but a lower A/g_s. When comparing a slow growth Mediterranean evergreen tree *Olea europea*, with a drought deciduous shrub *Rhamnus lycioides*, Querejeta *et al.* (2003) found that *Olea* had a better control over water losses and a higher $\delta^{13}C$, than the deciduous shrub. In addition, they found that the differences between *Olea* and the deciduous shrub *Rahmnus* in leaf $\delta^{13}C$ were larger in mycorrhizal than in non-inoculated plants, suggesting that mycorrhizal infection may contribute to the interspecific differences observed in plant communities. In fact, inoculated arbuscular mycorrhizae plants of *Olea europea* showed improved A_{max}, resulting in higher intrinsic water use efficiency (A_{max}/g_s) than in non-inoculated plants.

In forest canopies shorter plants normally display lower Δ values than taller plants because they receive reduced light intensities. Within the canopy, however, Δ may decrease with increasing average incident radiation, possibly due to more severe water deficits and higher A/g_s in the upper canopy (Niinemets *et al.*, 1999). Changes in the $\delta^{13}C$ of the air due to plant metabolism also contribute to alter plant isotopic composition. Buchmann *et al.* (2002) reported that in dense forests (LAI > 2.5) about 70 per cent of intra-canopy variability in leaf $\delta^{13}C$ was due to changes leaf physiology (Δ) and the remaining 30 per cent resulted from changes in source air isotope composition which might be 0.5 to 1.5‰ in average more negative inside the than outside the canopy. This effect is neggligible though at low LAI.

It is well known that stomatal conductance declines and WUE increases in the foliage of older and larger trees when compared to younger and smaller trees under similar environmental conditions (Yoder *et al.*, 1994; Ryan *et al.*, 2000; Schafer *et al.*, 2000; Fessenden and Ehleringer, 2002). Likewise, pine needles on longer branches (lower hydraulic conductance) had less negative $\delta^{13}C$ (higher A/g_s), reflecting possibly the effect of hydraulic constraint on stomatal conductance (Warren and Adams, 2000). The hypothesis that a decrease in stomatal conductance associated with decreased hydraulic conductance leads to increased CO_2 diffusion limitations in older coniferous trees in the Pacific Northwest of the United States was consistent with the $\delta^{13}C$ data obtained both at the ecosystem and leaf level (Fessenden and Ehleringer, 2002). However, in a review of the reasons of the decline in foliar photosynthetic rates with increasing tree age and size Niinemets (2002) concluded that stomatal limitations alone could not explain the decline in photosynthetic rates in the canopy of older trees. Morphological changes resulting from greater water stress in the foliage of older trees, due to decreases in shoot hydraulic conductance with increasing height, may bring about a decrease in mesophyll conductance and photosynthetic capacity and add to stomatal limitation in bringing about a decrease in A_{max}.

Plant phenology may be quite important in determining water use or the carbon isotope discrimination. Differences in phenology may alter the relationship between bulk leaf tissue Δ and WUE due to changes in the timing of leaf construction. In a study with ponderosa pine (*Pinus ponderosa*) a strong genotype × environment interaction was observed in Δ as a result from geographic location but not from moisture availability within locations (Cregg *et al.*, 2000). It was suggested that the genotype × environment interaction was related to variation in growth phenology among the seed sources, which determined the A/g_s at the time of needle construction.

In the canopies of tropical rainforests contradictory results have been reported about differences in $\delta^{13}C$ between pioneer and late-successional trees. For instance, Huc *et al.* (1994) found more negative $\delta^{13}C$ values in pioneer species whereas Guehl *et al.* (1998) reported a more negative $\delta^{13}C$ for the late-successional trees in French Guiana. On the other hand, foliage of understorey species tend to be more depleted in ^{13}C than that of emergent canopy species, both in dry (Leffler and Enquist, 2002) and in wet (Buchmann *et al.*, 1997) tropical forests. In a study of the functional diversity among tree species of the Amazonian tropical rainforests of French Guiana, Bonal *et al.* (2000) compared different groups based on shade tolerance and position in the canopy. They concluded that the hemi-tolerant species (the potentially emergent species that represented most of the tree biomass) displayed less negative leaf $\delta^{13}C$ values and possibly higher intrinsic water use efficiency than heliophilic (that have a gap occupation strategy) and the shade-tolerant species.

3.6.3   WUE along environmental gradients

As mentioned above, the intrinsic water use efficiency (A/g_s) should increase (or $\delta^{13}C$ become less negative) along a gradient of decreasing moisture availability due to increasing stomatal closure. In fact, the plant community $\delta^{13}C$ measured along a rainfall gradient in eastern Australia (southern Queensland) varied from −25.6‰ in the dry end, with a water deficit (rainfall – pan evaporation), above 2000 mm, to a $\delta^{13}C$ of −31.2‰ in the wetter part of the gradient, with a rainfall excess of 243 mm (Stewart *et al.*, 1995). Likewise, when we compared the $\delta^{13}C$ of 57 clones of eucalyptus planted in Portugal in two sites with a rainfall of 705 mm 1290 mm, respectively, the average difference between the two sites was ca. 2‰ less negative in the dry site (unpublished). When studying the carbon isotope composition of *Pinus pinaster* along a rainfall gradient, Warren and Adams (2000) found a good negative correlation between $\delta^{13}C$ and the index of aridity, precipitation/ potential evapotranspiration, which reflects water availability better than precipitation alone.

The relationship between mean $\delta^{13}C$ and total annual precipitation was studied in a large number of tropical ecosystems, (Von Fisher and Tieszen, 1995; Buchmann et al., 1997; Sobrado and Ehleringer, 1997; Guehl et al., 1998; Martinelli et al., 1998; Schulze et al., 1998; Bonal et al., 2000; Nagy and Proctor, 2000; Leffler and Enquist, 2002). Taking together the results of seventeen study sites throughout the tropics, Leffler and Enquist (2002) were able to identify an overall significant curvilinear relationship (second-order polynomial) between the two variables, suggesting the declining importance of precipitation in determining $\delta^{13}C$ as precipitation increases. Moreover, this curvilinear trend highlighted the apparent existence of a threshold value of annual precipitation, above which additional rainfall has little impact in $\delta^{13}C$.

A similar observation was made along a rainfall gradient in northern Australia in a study to investigate carbon and nitrogen isotope ratios of trees (Schulze et al., 1998). They concluded that community-averaged carbon isotope discrimination (Δ) decreased with decreasing rainfall only in that part of the transect where annual precipitation was between 250 and 450 mm. Where rainfall was from 450 mm to 1800 mm, Δ was roughly constant. Also, Schulze et al. (1996a) found little variation in the $\delta^{13}C$ of the grass flora of Namibia (mostly C_4) along a rainfall gradient (50–600mm). There were however differences in ^{13}C composition between species and functional types and the nearly constant community Δ values (e.g. at medium and high rainfall in Australia) may be the result of replacement of functional types and species as climate changed and possibly denoting no major differences in the A/g_s at the time of leaf construction (Schulze et al., 1996b).

This is in contrast with the results by Stewart et al. (1995), who found that community-averaged values of Δ showed a linear decrease with decreasing rainfall along the total range of annual precipitation of the gradient. According to Schulze et al. (1998), the possible explanations for that could be:

(i)   different composition of the studied communities (in northern Australia only trees, in eastern Australia trees and herbs),
(ii) climatic differences between the two regions (markedly higher seasonal rainfall in northern Australia relative to eastern Australia) or
(iii) a more complex interaction between vegetation and climate in the northern than in eastern Australia.

The timing of rainfall may change drastically both WUE_{sl} and WUE_i. First, primary productivity may be rather different if water is available the year round or if it changes seasonally (Eamus, 2003). The longer the dry season the lower the productivity. Second, WUE_{sl} will tend to be lower in drier than in wet habitats as a consequence of a decrease in overall growth, although A/g_s

and plant $\delta^{13}C$ tend to increase under mild water stress, due to stomatal closure. But the way plants use the water available, i.e. type of root system and phenology, will have a strong impact in the final result. In arid zones, for example, rain may fall in pulses. The relative importance of pulse water versus stored soil water may determine the dominant functional plant types in the environment (Schwinning and Ehleringer, 2001). The phenotypes that maximise pulse water use (small root:shoot ratio, predominantly shallow root system, high leaf conductance with high stomatal sensitivity to plant water status) have a lower carbon isotope discrimination, i.e. a higher A/g_s than the species that use mostly deep soil water (Arndt et al., 2000; Gebauer and Ehleringer, 2000). Also in the tropics, shallow-rooted deciduous- species usually display less negative $\delta^{13}C$ values in comparison with the deeply-rooted evergreen trees in dry (Sobrado and Ehleringer, 1997) as well as in wet (Bonal et al., 2000) environments.

In the Mediterranean sclerophyllous woody plants rely usually on stored water and deep root systems to survive the summer drought (Schenk and Jackson, 2002). This leads normally to a decrease in WUE due to the very evaporative demand of the air in summer as found at the ecosystem level (Reichstein et al., 2002). However these trees and deep-rooted shrubs also have the capability of using pulses of water as the first rains at the end of the summer-autumn. For example, Q. ilex trees in Portugal opened their stomata after a few days of rain in autumn (Pereira et al., 2004). Intense stand carbon assimilation began as soon as stomata opened, while the understorey composed of summer deciduous shrubs and annual plants was still without leaves. This may be important, as the beginning of the rainy season in Mediterranean environments may be erratic, i.e., by pulses. In such an environment, the competing herbaceous and summer deciduous shrubs might waste part of the mild autumn to reconstruct the canopy before effective carbon assimilation. The early start of carbon assimilation after the first autumn rains may be one advantage for evergreen trees because it takes place at a time when active litter and soil organic matter decomposition occurs and nutrients, namely nitrogen, are abundant (Ehleringer et al., 2000; Pereira et al., 2004). Although in arid environments, most species seem to be able to utilise equally well these pulses of water following rainfall events in the dry season, there may be some differences in the nitrogen utilisation patterns, with some species showing capacity to utilise the water pulse, but not the N pulse (Gebauer and Ehleringer, 2000).

References

Aber, I.D., Reich, P.B. and Goulden, M.L. (1996) Extrapolating leaf CO_2 exchange to the canopy: a generalized model of forest photosynthesis validated by eddy correlation. Oecologia, **106**, 257–265.
Anten, N.P.R., Schieving, F. and Werger, M.J.A. (1995) Patterns of light and nitrogen

distribution in relation to whole canopy carbon gain in C3 and C4 mono- and dicotyledonous species. *Oecologia*, **101**, 504–513.

Arndt, S.K., Wanek, W., Clifford, S.C. and Popp, M. (2000) Contrasting adaptations to drought stress in field grown Ziziphus mauritiana Lamk. and Prunus persica L. trees: Water relations, osmotic adaptation and carbon isotope fractionation. *Australian Journal of Plant Physiology*, **27**, 985–996.

Assmann, S.M. (1999) The cellular basis of guard cell sensing of rising CO_2. *Plant, cell and environment*, **22**, 629–637.

Ball, J.T., Woodrow, I.E. and Berry, J.A. (1987) A model predicting stomatal conductance and its contribution to the control of photosynthesis under different environmental conditions. In: *Progress in Photosynthesis Research* (ed. J. Biggins), Martinus Nijhoff, pp. 221–224.

Bange, M.P., Hammer, G.L. and Rickert, K.G. (1997) Effect of specific leaf nitrogen on radiation use efficiency and growth of sunflower. *Crop Science*, **37**, 1201–1207.

Barbour, M.M. and Farquhar, G.D. (2000) Relative humidity- and ABA-induced variation in carbon and oxygen isotope ratios of cotton leaves. *Plant, Cell and Environment*, **23**, 473–485.

Barbour, M.M., Fischer, R.A., Sayre, K.D. and Farquhar, G.D. (2000) Oxygen isotope ratio of leaf and grain material correlates with stomatal conductance and grain yield in irrigated wheat. *Australian Journal of Plant Physiology*, **27**, 625–637.

Ben-Gal, A., Karlberg, L., Jansson, P.-E. and Shani, U. (2003) Temporal robustness of linear relationships between production and transpiration. *Plant and Soil*, **251**, 211–218.

Bierhuizen, J. F. and Slatyer, R. O. (1965) Effect of atmospheric concentration of water vapour and CO_2 in determining transpiration-photosynthesis relationships of cotton leaves. *Agricultural Meteorology*, **2**, 259–270.

Boardman, N. K. (1977) Comparative photosynthesis of sun and shade plants. *Annu Rev Plant Physiol*, 355–377.

Bonal, D., Sabatier, D., Montpied, P., Tremeaux, D. and Guehl, J. M. (2000) Interspecific variability of $\Delta13C$ among canopy trees in rainforests of French Guiana: Functional groups and canopy integration. *Oecologia*, **124**, 454–468.

Brooks, J. R., Flanagan, L. B., Buchmann, N. and Ehleringer, J. R. (1997) Carbon isotope composition of boreal plants: functional grouping of life forms. *Oecologia*, **110**, 301–311.

Brugnoli, E. and Farquhar, G. D. (2000) Photosynthetic fractionation of carbon isotopes. In: *Advances in Photosynthesis. Photosynthesis: Physiology and Metabolism*, Vol. 9 (eds R.C. Leegood, T.D. Sharkey and S.V. Caemmerer), Kluwer Academic, Dordrecht, the Netherlands, pp. 399–434.

Buchmann, N., Brooks, J. R. and Ehleringer, J. R. (2002) Predicting carbon isotope ratios of atmospheric CO_2 within canopies. *Functional Ecology*, **16**, 49–57.

Buchmann, N., Guehl, J. M., Barigah, T. S. and Ehleringer, J. R. (1997) Interseasonal comparison of CO_2 concentrations, isotopic composition, and carbon cycling in an Amazonian rainforest (French Guiana). *Oecologia*, **110**, 120–131.

Buchmann, N., Hinckley, T. M. and Ehleringer, J. R. (1998) Carbon isotope dynamics in *Abies amabilis* stands in the Cascades. *Canadian Journal of Forest Research*, **28**, 808–819.

Bunce, J. A. (1998) Effects of humidity on short term responses of stomatal conductance to an increase in carbondioxide concentration. *Plant, Cell and Environment*, **21**, 115–120.

Caemmerer, S. V. and Evans, J. R. (1991) Determination of the average partial pressure of CO_2 in chloroplasts from leaves of several C3 plants. *Australian Journal of Plant Physiology*, **18**, 287–305.

Chapin, F. S., III and Van Cleve, K. (1989) Approaches to studying nutrient uptake, use and loss in plants. In: *Plant Physiological Ecology: field methods and instrumentation* (eds P.W. Pearcy, J.R. Ehleringer, H.A. Mooney and P.W. Rundel), Chapman and Hall, London, pp. 185–207.

Chapin, F. S., III, Bloom, A. J., Field, C. B. and Waring, R. H. (1987) Plant responses to multiple environmental factors. *BioScience*, **37**, 49–57.

Chaves, M. M., Pantschitz, E. and Schulze, E. D. (2001) Growth and photosynthetic carbon metabolism in tobacco plants under an oscillating CO_2 concentration in the atmosphere. *Plant Biology*, **3**, 1–9.

Chaves, M. M., Pereira, J. S., Cerasoli, S., Clifton-Brown, J., Miglietta, F. and Raschi, A. (1995)

Leaf metabolism during summer drought in Quercus ilex trees with lifetime exposure to elevated CO_2. *Journal of Biogeography*, **22**, 255–259.

Cowan, I. R. and Farquhar, G. D. (1977) Stomatal function in relation to leaf metabolism and environment. *Symposia of the Society for Experimental Biology*, **31**, 471–505.

Cregg, B. M., Olivas-Garcia, J. M. and Hennessey, T. C. (2000) Provenance variation in carbon isotope discrimination of mature ponderosa pine trees at two locations in the Great Plains. *Canadian Journal of Forest Research*, **30**, 428–439.

Curtis, P. S. (1996) A meta-analysis of leaf gas exchange and nitrogen in trees grown under high carbon dioxide. *Plant Cell and Environment*, **19**, 127–137.

Damesin, C., Rambal, S. and Joffre, R. (1998) Seasonal and annual changes in leaf $\delta^{13}C$ in two co-occuring Mediterranean oaks: relations to leaf growths and drought progression. *Functional ecology*, **12**, 778–785.

De Lucia, E. H. and Schlesinger, W. H. (1991) Resource-use efficiency and drought tolerance in adjacent Great Basin and Sierran plants. *Ecology*, **72**, 51–58.

Dongmann, G., Nürnberg, H. W., Förstel, H. and Wagener, K. (1974) On the enrichment of $H^{218}O$ in the leaves of transpiring plants. *Radiation and Environmental Biophysics*, **11**, 41–52.

Donovan, L. A. and Ehleringer, J. R. (1992) Contrasting water-use patterns among size and life history classes of a semiarid shrub. *Functional Ecology*, **6**, 482–488.

Donovan, L. A. and Ehleringer, J. R. (1994) Carbon isotope discrimination, water use efficiency, growth and mortality in a natural shrub population. *Oecologia*, **100**, 347–354.

Drake, B. G., Gonzalez-Meler, M. A. and Long, S. P. (1997) More efficient plants: a consequence of rising atmospheric CO_2? *Annual Review of Plant Physiology and Plant Molecular Biology*, **48**, 609–639.

Dreccer, M. F., Schapendonk, A. H. C. M., van Oijen, M., Pot, C. S. and Rabbinge, R. (2000) Radiation and nitrogen use at the leaf and canopy level by wheat and oilseed rape during the critical period for grain number definition. *Australian Journal of Plant Physiology*, **27**, 899–910.

Eamus, D. (2003) How does ecosystem water balance affect net primary productivity of woody ecosystems? *Functional Plant Biology*, **30**, 187–205.

Ehleringer, J. R., Hall, A. E. and Farquhar, G. D. (eds) (1993) *Stable Isotopes and Plant Carbon-Water Relations*, Academic Press, New York.

Ehleringer, J. R., Roden, J. and Dawson, T. E. (2000) Assessing ecosystem-level water relations through stable isotope ratio analyses. In: *Methods in ecosystem science* (eds Sala, O., Jackson, R., Mooney, H. A. and Howarth, R.) Springer Verlag, New York, pp. 181–198.

Evans, J. R. (1983) Nitrogen and photosynthesis in the flag leaf of wheat (Triticum aestivum L.). *Plant Physiology*, **72**, 297–302.

Evans, J. R. (1989) Photosynthesis and nitrogen relationships in leaves of C3 plants. *Oecologia*, **78**, 9–19.

Evans, J. R. (1993) Photosynthetic acclimation and nitrogen partitioning within a lucerne canopy. I. Canopy characteristics. *Australian Journal of Plant Physiology*, **20**, 55–67.

Evans, J. R. (1999) Leaf anatomy enables more equal access to light and CO_2 between chloroplasts. *New Phytologist*, **143**, 93–104.

Evans, J. R. and Poorter, H. (2001) Photosynthetic acclimation of plants to growth irradiance: The relative importance of SLA and nitrogen partitioning in maximising carbon gain. *Plant, Cell and Environment*, **24**, 755–767.

Evans, J. R. and Seemann, J. R. (1989) The allocation of protein nitrogen in the photosynthetic apparatus: costs, consequences and control. In: *Photosynthesis* (ed Briggs, W. R.) Liss, New York, pp. 183–205.

Evans, J. R. and Von Caemmerer, S. (1996) CO_2 diffusion inside leaves. *Plant physiology*, **110**, 339–346.

Evans, J. R., Schortemeyer, M., McFarlane, N. and Atkin, O. K. (2000) Photosynthetic characteristics of 10 Acacia species grown under ambient and elevated atmospheric CO_2. *Australian Journal of Plant Physiology*, **27**, 13–25.

Faria, T., Garcia-Plazaola, J. I., Abadia, A., Cerasoli, S., Pereira, J. S. and Chaves, M. M. (1996) Diurnal changes in photoprotective mechanisms in leaves of cork oak (Quercus suber L.) during summer. *Tree Physiology*, **16**, 115–123.

Faria, T., Silverio, D., Breia, E., Cabral, R., Abadia, A., Abadia, J., Pereira, J. S. and Chaves,

M. M. (1998) Differences in the response of carbon assimilation to summer stress (water deficits, high light and temperature) in four Mediterranean tree species. *Physiologia plantarum*, **102**, 419–428.

Faria, T., Vaz, M., Schwanz, P., Polle, A., Pereira, J. S. and Chaves, M. M. (1999) Responses of photosynthetic and defence systems to high temperature stress in Quercus suber L seedlings grown under elevated CO_2. *Plant Biology*, **1**, 365–371.

Farquhar, G. D. and Lloyd, J. (1993) Carbon and oxygen isotope effects in the exchange of carbon dioxide between terrestrial plants and the atmosphere. In: *Stable Isotopes and Plant Carbon-Water Relations* (eds Ehleringer, J. R., Hall, A. E. and Farquhar, G. D.) Academic Press, San Diego, pp. 47–70.

Farquhar, G. D., Condon, A. G. and Masle, J. (1994) On the use of carbon and oxygen isotope composition and mineral ash content in breeding for improved rice production under favorable, irrigated conditions. In: *Breaking the yield barrier* (ed Cassman, K. G.) International Rice Research Institute, Manila, pp. 95–101.

Farquhar, G. D., Hubick, K. T., Condon, A. G. and Richards, R. A. (1988) Carbon-isotope fractionation and plant water use efficiency. In: *Stable Isotopes in Ecological Research* (eds. P. W. Rundel, J. R. Ehleringer and K. A. Nagy). Springer-Verlag, New York, pp. 21–40.

Farquhar, G. D., von Caemmerer, S. and Berry, J. A. (1980) A biochemical model of photosynthetic CO_2 assimilation in leaves of C3 species. *Planta*, **149**, 78–90.

Fessenden, J. E. and Ehleringer, J. R. (2002) Age dependent variations in the $\delta^{13}C$ of ecosystem respiration across a coniferous forest chronosequence in the Pacific Northwest. *Tree Physiology*, **22**, 159–167.

Field, C. (1983) Allocating leaf nitrogen for the maximization of carbon gain: leaf age as a control of the allocation program. *Oecologia*, **56**, 341–347.

Field, C. and Mooney, H. A. (1986) The photosynthesis-nitrogen relationship in wild plants. In: *On the economy of plant form and function* (ed Givnish, T. J.) Cambridge University Press, Cambridge, pp. 25–55.

Field, C., Merino, J. and Mooney, H. A. (1983) Compomises between water use efficiency and nitrogen-use efficiency in five species of California evergreens. *Oecologia*, **60**, 384–389.

Fredeen, A. L., Gamon, J. A. and Field, C. B. (1991) Responses of photosynthesis and carbohydrate partitioning to limitations in nutrient and water availability in field-grown sunflower. *Plant, Cell and Environment*, **14**, 963–970.

Garnier, E., Salager, J.-L., Laurent, G. and Sonié, L. (1999) Relationships between photosynthesis, nitrogen and leaf structure in 14 grass species and their dependence on the basis of expression. *New Phytologist*, **143**, 119–129.

Gastal, F. and Lemaire, G. (2002) N uptake and distribution in crops: an agronomical and ecophysiological perspective. *Journal of Experimental Botany*, **53**, 789–799.

Gebauer, R. L. E. and Ehleringer, J. R. (2000) Water and nitrogen uptake patterns following moisture pulses in a cold desert community. *Ecology*, **81**, 1415–1424.

Ghannoum, O., von Caemmerer, S. and Conroy, J. P. (2002) The effect of drought on plant water use efficiency of nine NAD-ME and nine NADP-ME Australian C4 grasses. *Functional Plant Biology*, **29**, 1337–1348.

Gourley, C. J. P., Allan, D. L. and Russelle, M. P. (1994) Plant nutrient efficiency: A comparison of definitions and suggested improvement. *Plant and Soil*, **158**, 29–37.

Grindlay, D. J. C. (1997) Towards an explanation of crop nitrogen demand based on optimization of leaf nitrogen per unit ground area. *Journal of Agricultural Science*, **128**, 377–396.

Guehl, J. M., Domenach, A. M., Bereau, M., Barigah, T. S., Cassabianca, H., Ferhi, A. and Garbaye, J. (1998) Functional diversity in an Amazonian rainforest of French Guyana: a dual isotope approach (δ15N and δ13C). *Oecologia*, **116**, 316–330.

Gutiérrez, M. V. and Meinzer, F. C. (1994) Carbon isotope discrimination and gas exchange in coffee hedgerows during canopy development. *Australian Journal of Plant Physiology*, **21**, 207–219.

Hall, A. E. and Schulze, E. D. (1980) Stomatal responses to environment and a possible interrelation between stomatal effects on transpiration and CO_2 assimilation. *Plant, Cell and Environment*, **3**, 467–474.

Hammer, G. L. and Wright, G. C. (1994) A theoretical analysis of nitrogen and radiation effects on radiation use efficiency in peanut. *Australian Journal of Agricultural Research*, **45**, 575–589.

Hanba, Y. T., Kogami, H. and Terashima, I. (2002) The effect of growth irradiance on leaf anatomy and photosynthesis in Acer species differing in light adaptation. *Plant, Cell and Environment*, **25**, 1021–1030.

Hirose, T. and Werger, M. J. A. (1987) Maximizing daily canopy photosynthesis with respect to the leaf nitrogen allocation pattern in the canopy. *Oecologia*, **72**, 520–526.

Hobbie, E. A., Gregg, J., Olszyk, D. M., Rygiewicz, P. T. and Tingey, D. T. (2002) Effects of climate change on labile and structural carbon in Douglas-fir needles as estimated by 13C and Carea measurements. *Global Change Biology*, **8**, 1072–1084.

Holbrook, N. M., Whitbeck, J. L. and Mooney, H. A. (1995) Drought responses of tropical deciduous forest trees. In: *Seasonally dry tropical forests* (eds Bullock, S. H., Mooney, H. A. and Medina, E.), Cambridge University Press, Cambridge, pp. 243–276.

Huc, R., Ferhi, A. and Guehl, J. M. (1994) Pioneer and late stage tropical rainforest tree species (French Guyana) growing under common conditions differ in leaf gas exchange regulation, carbon isotope discrimination and leaf water potential. *Oecologia*, **99**, 297–305.

Jones, H. G. (1992) *Plants and Microclimate: A Quantitative Approach to Environmental Plant Physiology*, Cambridge University Press, Cambridge.

Korner, C., Scheel, J. A. and Bauer, H. (1979) Maximum leaf diffusive conductance in vascular plants. *Photosynthetica*, **13**, 45–82.

Kubiske, M. E., Abrams, M. D. and Mostoller, S. A. (1997) Stomatal and nonstomatal limitations of photosynthesis in relation to the drought and shade tolerance of tree species in open and understory environments. *Trees: structure and function*, **11**, 76–82.

Lambers, H., Chapin, F. S. and Pons, T. L. (1998) *Plant Physiological Ecology*, Springer, New York.

Larcher, W. (1995) *Physiological Plant Ecology*, Springer-Verlag, New York.

Lauteri, M., Scartazza, A., Guido, M. C. and Brugnoli, E. (1997) Genetic variation in photosynthetic capacity, carbon isotope discrimination and mesophyll conductance in provenances of Castanea sativa adapted to different environments. *Functional Ecology*, **11**, 675–683.

Lawlor, D. W. (2002) Carbon and nitrogen assimilation in relation to yield: mechanisms are the key to understanding production system. *Journal of Experimental Botany*, **53**, 773–787.

Le Roux-Swarthout, D. J., Terwilliger, V. J. and Martin, C. E. (2001) Deviation between $\delta13C$ and leaf intercellular CO_2 concentrations in Salix interior cuttings developing under low light. *International Journal of Plant Science*, **162**, 1017–1024.

Lee, J. S. and Bowling, D. J. F. (1995) Influence of the mesophyll on stomatal opening. *Australian Journal of Plant Physiology*, **22**, 357–363.

Leffler, A. J. and Enquist, B. J. (2002) Carbon isotope composition of tree leaves from dry tropical forests of Guanacaste, Costa Rica: Comparison across tropical ecosystems and tree life history. *Journal of Tropical Ecology*, **18**, 151–159.

Leuning, R., Kelliher, F. M., De Pury, D. G. G. and Schulze, E. D. (1995) Leaf nitrogen, photosynthesis, conductance and transpiration – scaling from leaves to canopies. *Plant, Cell and Environment*, **18**, 1183–1200.

Leuning, R., Tuzet, A. and Perrier, A. (2003) Stomata as part of the soil-plant atmosphere continuum. In: *Forests at the Land-Atmosphere Interface* (eds M. Mencuccini, J. Grace, J. Moncrieff and K. McNaughton), CAB International, Edinburgh, Scotland, pp. in press.

Lieth, H. (1975) Modelling the primary productivity of the world. In: *Primary Productivity of the Biosphere* (eds Lieth, H. and Whittaker, R. H.) Springer-Verlag, New York, pp. 237–263.

Livingston, N. J., Guy, R. D., Sun, Z. J. and Ethier, G. J. (1999) The effects of nitrogen stress on the stable carbon isotope composition, productivity and water use efficiency of white spruce (Picea glauca (Moench) Voss) seedlings. *Plant, Cell and Environment*, **22**, 281–289.

Lodge, R. J., Dijkstra, P., Drake, B. G. and Morison, J. I. L. (2001) Stomatal acclimation to increased CO_2 concentration in a Florida scrub oak species, Quercus myrtifolia Willd. *Plant, Cell and Environment*, **24**, 77–88.

Long, S. P. (1999) C4 photosynthesis – environmental responses. In: *The Biology of C4 Plants* (eds Sage, R. F. and Monson, R. K.), Academic Press, San Diego, pp. 215–249.

Loreto, F., Harley, P. C., Di Marco, G. and Sharkey, T. D. (1992) Estimation of the mesophyll conductance to CO_2 flux by three different methods. *Plant Physiology*, **98**, 1437–1443.

Lu, Z., Percy, R.G., Qualset, C.O. and Zeiger, E. (1998) Stomatal conductance predicts yields in irrigated Pima cotton and bread wheat grown at high temperatures. *Journal of Experimental Botany*, **49**, 543–560.

Madeira, M. V., Fabiao, A., Pereira, J. S., Araujo, M. C. and Ribeiro, C. (2003) Changes in carbon stocks in Eucalyptus globulus Labill. plantations induced by different water and nutrient availability. *Forest Ecology and Management*, **171**, 75–85.

Maroco, J., Rodrigues, M. L., Lopes, C. and Chaves, M. M. (2002) Limitations to leaf photosynthesis in field-grown grapevine under drought – metabolic and modelling approaches. *Functional Plant Biology*, **29**, 451–459.

Maroco, J. P., Pereira, J. S. and Chaves, M. M. (1997) Stomatal responses to leaf-to-air vapour pressure deficit in Sahelian species. *Australian Journal of Plant Physiology*, **24**, 381–387.

Maroco, J. P., Pereira, J. S. and Chaves, M. M. (2000) Growth, photosynthesis and water use efficiency of two C4 Sahelian grasses subjected to water deficits. *Journal of Arid Environments*, **45**, 119–137.

Martinelli, L. A., Almeida, S., Brown, I. F., Moreira, M. Z., Victoria, R. L., Sternberg, L. S. L., Ferreira, C. A. C. and Thomas, W. W. (1998) Stable carbon isotope ratio of tree leaves, boles and fine litter in a tropical forest in Rondônia, Brazil. *Oecologia*, **114**, 170–179.

Matzner, S. L., Rice, K. J. and Richards, J. H. (2001) Factors affecting the relationship between carbon isotope discrimination and transpiration efficiency in blue oak (*Quercus douglasii*). *Australian Journal of Plant Physiology*, **28**, 49–56.

Mediavilla, S., Escudero, A. and Heilmeier, H. (2001) Internal leaf anatomy and photosynthetic resource-use efficiency: interspecific and intraspecific comparisons. *Tree Physiology*, **21**, 251–259.

Meziane, D. and Shipley, B. (2001) Direct and indirect relationships between specific leaf area, leaf nitrogen and leaf gas exchange. Effects of irradiance and nutrient supply. *Annals Botany*, **88**, 915–927.

Monteith, J. L. (1972) Solar radiation and productivity in tropical exosystems. *Journal of Applied Ecology*, **9**, 747–766.

Monteith, J. L. (1984) Consistency and convenience in the choice of units for agricultural science. *Experimental Agriculture*, **20**, 105–117.

Monteith, J. L. (1993) The exchange of water and carbon by crops in a Mediterranean climate. *Irrigation Science*, **14**, 85–91.

Monteith, J. L. (1995) A reinterpretation of the stomatal response to humidity. *Plant, Cell and Environment*, **18**, 357–364.

Monteith, J.L. (1977). Climate and the efficiency of crop production in Britain. *Philosophical Transactions of the Royal Society, Series B*, **281**, 277–294.

Morison, J. I. L. (1987) Intercellular CO_2 concentration and stomatal response to CO_2. In: *Stomatal Function* (E. Zeiger, G. D. Farquhar and I. R. Cowan (eds), Stanford University Press, Stanford, Calif., pp. 229–251.

Morison, J. I. L. (1993) Response of plants to CO_2 under water-limited conditions. *Vegetatio*, **104/105**, 193–209.

Morison, J. I. L. (1998) Stomatal response to increased CO_2 concentration. *Journal of Experimental Botany*, **49**, 443–452.

Morison, J. I. L. and Gifford, R. M. (1983) Stomatal sensitivity to carbon dioxide and humidity: A comparison of two C3 and two C4 grass species. *Plant Physiology*, **71**, 789–796.

Mott, K. A. (1988) Do stomata respond to CO_2 concentrations other than intercellular? *Plant Physiology*, **86**, 200–203.

Mott, K. A., Gibson, A. C. and O'Leary, M. H. (1982) The adaptative significance of amphistomatic leaves. *Plant, Cell and Environment*, **5**, 455–460.

Nagy, L. and Proctor, J. (2000) Leaf $\delta13$ C signatures in heath and lowland evergreen rain forest species from Borneo. *Journal of Tropical Ecology*, **16**, 757–761.

Niinemets, U. (1999) Energy requirement for foliage formation is not constant along canopy light gradients in temperate deciduous trees. *The New Phytologist*, **144**, 35–47.

Niinemets, U. (2001) Climatic controls of leaf dry mass per area, density, and thickness in trees and shrubs at the global scale. *Ecology*, **82**, 453–469.

Niinemets, U. (2002) Stomatal conductance alone does not explain the decline in foliar photosynthetic rates with increasing tree age and size in *Picea abies* and *Pinus sylvestris*. *Tree Physiology*, **22**, 515–535.

WATER USE EFFICIENCY IN PLANT BIOLOGY

Niinemets, U. and Tenhunen, J. D. (1997) A model separating leaf structural and physiological effects on carbon gain along light gradients for the shade-tolerant species *Acer saccharum*. *Plant, Cell and Environment*, **20**, 845–866.

Niinemets, U., Kull, O. and Tenhunen, J. D. (1998) An analysis of light effects on foliar morphology, physiology, and light interception in temperate deciduous woody species of contrasting shade tolerance. *Tree Physiology*, **18**, 681–696.

Niinemets, U., Kull, O. and Tenhunen, J. D. (1999) Variability in leaf morphology and chemical composition as a function of canopy light environment in co-existing trees. *International Journal of Plant Sciences*, **160**, 837–848.

Osório, J., Osório, M. L., Chaves, M. M. and Pereira, J. S. (1998) Effects of water deficits on 13C discrimination and transpiration efficiency of Eucalyptus globulus clones. *Australian Journal of Plant Physiology*, **25**, 645–653.

Osorio, J. and Pereira, J. S. (1994) Genotypic differences in water use efficiency and 13C discrimination in Eucalyptus globulus. *Tree Physiology*, **14**, 871–882.

Pate, J. S. (2001) Carbon isotope discrimination and plant water use efficiency: case scenarios for C3 plants. In: *Stable Isotope Techniques in the Study of Biological Processes and Functioning of Ecosystems* (eds Unkovich, M., Pate, J., McNeill, A. and Gibbs, D. J.) Kluwer, Dordrecht, pp. 19–36.

Pate, J. and Arthur, D. (1998) delta13C analysis of phloem sap carbon: Novel means of evaluating seasonal water stress and interpreting carbon isotope signatures of foliage and trunk wood of Eucalyptus globulus. *Oecologia*, **117**, 301–311.

Patterson, T.B., Guy, R.D. and Dang, Q.L. (1997) Whole-plant nitrogen- and water-relations traits, and their associated trade-offs, in adjacent muskeg and upland boreal spruce species. *Oecologia*, **110**, 160–168.

Pearson, M. and Brooks, G. L. (1995) The influence of elevated CO_2 on age-related changes in leaf gas exchange. *Journal of Experimental Botany*, **46**, 1651–1659.

Pereira, J. S. (1993) Gas exchange and growth. In: *Ecophysiology of photosynthesis*, vol. 100 (eds Schulze, E.-D. and Caldwell, M. M.) Springer-Verlag, Berlin-Heidelberg-New York, pp. 148–181.

Pereira, J. S., Chaves, M. M., Carvalho, P. O., Caldeira, M. C. and Tomé, J. (1994a) Carbon assimilation, growth and nitrogen supply in Eucalyptus globulus plants. In: *Whole-Plant Perspectives of Carbon-Nitrogen Interactions* (eds Roy, J. and Granier, E.), SPB Publish., The Hague, The Netherlands, pp. 79–89.

Pereira, J. S., Chaves, M. M., Fonseca, F., Araujo, M. C. and Torres, F. (1992) Photosynthetic capacity of leaves of Eucalyptus globulus (Labill.) growing in the field with different nutrient and water supplies. *Tree Physiology*, **11**, 381–389.

Pereira, J. S., David, J. S., David, T. S., Caldeira, M. C. and Chaves, M. M. (2004) Carbon and water fluxes in Mediterranean-type ecosystems – constraints and adaptations. In *Progress in Botany*, vol. 65 (eds Esser, K., Lüttge, U., Beyschlag, W. and Murata, J.), Springer-Velag, Berlin-Heidelberg, pp. 467–498.

Pereira, J. S., Madeira, M. V., Linder, S., Ericsson, T., Tomé, M. and Araújo, M. C. (1994b) Biomass production with optimized nutrition in Eucalyptus globulus plantations. In: *Eucalyptus for Biomass Production. The State-of-the-art* (eds Pereira, J. S. and Pereira, H.), Instituto Superior de Agronomia, Lisbon, pp. 13–30.

Ponton, S., Dupouey, J. L., Breda, N. and Dreyer, E. (2002) Comparison of water use efficiency of seedlings from two sympatric oak species: genotype * environment interactions. *Tree Physiology*, **22**, 413–422.

Poorter, H. and Evans, J. R. (1998) Photosynthetic nitrogen-use efficiency of species that differ inherently in specific leaf area. *Oecologia*, **116**, 26–37.

Poorter, H. and Farquhar, G. D. (1994) Transpiration, intercellular carbon dioxide concentration and carbon-isotope discrimination of 24 wild species differing in relative growth rate. *Australian Journal of Plant Physiology*, **21**, 507–516.

Poorter, H. and Perez-Soba, M. (2001) The growth response of plants to elevated CO_2 under non-optimal environmental conditions. *Oecologia*, **129**, 1–20.

Poorter, H., Remkes, C. and Lambers, H. (1990) Carbon and nitrogen economy of 24 wild species differing in relative growth rate. *Plant Physiology*, **94**, 621–627.

Querejeta, J. I., Barea, J. M., Allen, M. F., Caravaca, F. and Roldan, A. (2003) Differential response of δ13 C and water use efficiency to arbuscular mycorrhizal infection in two aridland woody plant species. *Oecologia*, **135**, 510–515.

Radin, J. W., Lu, Z., Percy, R. G. and Zeiger, E. (1994) Genetic variability for stomatal conductance in Pima cotton and its relation to improvements of heat adaptation. *Proceedings of the National Academy of Sciences USA*, **91**, 7217–7221.

Raschke, K. (1976) How stomata resolve the dilemma of opposing priorities [Xanthium Strumarium]. *Philos Trans R Soc Lond, Ser B Biol Sci*, **273**, 551–560.

Reich, P. B., Ellsworth, D. S., Walters, M. B., Vose, J. M., Gresham, C., Volin, J. C. and Bowman, W. D. (1999) Generality of leaf trait relationships: a test across six biomes. *Ecology*, **80**, 1955–1969.

Reich, P. B., Kloeppel, B. D., Ellsworth, D. S. and Walters, M. B. (1995) Different photosynthesis-nitrogen relation in deciduous hardwood and evergreen coniferous tree species. *Oecologia*, **104**, 24–30.

Reich, P. B., Walters, M. B., Ellsworth, D. S., Vose, J. M., Volin, J. C., Gresham, C. and Bowman, W. D. (1998) Relationships of leaf dark respiration to leaf N, SLA, and life-spana test across biomes and functional groups. *Oecologia*, **114**, 471–482.

Reich, P. B., Walters, M. B. and Tabone, T. J. (1989) Response of Ulmus americana seedlings to varying nitrogen and water status. II. Water and nitrogen-use efficiency in photosynthesis. *Tree Physiology*, **5**, 173–184.

Reichstein, M., Tenhunen, J. D., Roupsard, O., Ourcival, J. M., Rambal, S., Miglietta, F., Peressotti, A., Pecchiari, M., Tirone, G. and Valentini, R. (2002) Severe drought effects on ecosystem CO_2 and H_2O fluxes in three Mediterranean evergreen ecosystems: revision of current hypotheses. *Global Change Biology*, **8**, 999–1017.

Robinson, D. E., Wagner, R. G., Bell, F. W. and Swarton, C. J. (2001) Photosynthesis, nitrogen-use efficiency and water use efficiency of jack pine seedlings in competition with four boreal forest plant species. *Canadian Journal of Forest Research*, **31**, 2014–2025.

Russell, G., Jarvis, P. G. and Monteith, J. L. (1989) Absorption of solar radiation and stand growth. In: *Plant Canopies: Their Growth, Form and Function* (eds Russell, G., Marshall, B. and Jarvis, P. G.), Cambridge University Press, Cambridge, pp. 21–39.

Ryan, M. G., Bond, B. J., Law, B. E., Hubbard, R. M., Woodruff, D., Cienciala, E. and Kucera, J. (2000) Transpiration and whole-tree conductance in ponderosa pine trees of different heights. *Oecologia*, **124**, 553–560.

Sage, R. F. and Pearcy, R. W. (1987) The nitrogen use efficiency of C3 and C4 plants. II. Leaf nitrogen effects on the gas exchange characteristics of *Chenopodium album* L. and *Amaranthus retroflexus* L. *Plant Physiology*, **84**, 959–963.

Saxe, H., Ellsworth, D. S. and Heath, J. (1998) Tansley Review No. 98: Tree and forest functioning in an enriched CO_2 atmosphere. *The New Phytologist*, **139**, 395–436.

Schafer, K. V. R., Oren, R. and Tenhunen, J. D. (2000) The effect of tree height on crown level stomatal conductance. *Plant, Cell and Environment*, **23**, 365–375.

Scheidegger, Y., Saurer, M., Bahn, M. and Siegwolf, R. (2000) Linking stable oxygen and carbon isotopes with stomatal conductance and photosynthetic capacity: A conceptual model. *Oecologia*, **125**, 350–357.

Schenk, H. J. and Jackson, R. B. (2002) Rooting depths, lateral root spreads, and belowground/aboveground allometries of plants in water limited ecosystems. *Journal of Ecology*, **90**, 480–494.

Schulze, E. D. and Hall, A. E. (1982) Stomatal control of water loss. In: *Encyclopedia of plant physiology – Physiological plant ecology II*, Vol. 12B (Eds, Lange, O. L., Nobel, P. S., Osmond, C. B. and Ziegler, H.), Springer-Verlag, Berlin, pp. 181–230.

Schulze, E. D., Ellis, R., Schulze, W., Trimborn, P. and Ziegler, H. (1996a) Diversity, metabolic types and δ13C carbon isotope ratios in the grass flora of Namibia in relation to growth form, precipitation and habitat conditions. *Oecologia*, **106**, 352–369.

Schulze, E. D., Kelliher, F. M., Körner, C., Lloyd, J. and Leuning, R. (1994) Relationships between plant nitrogen nutrition, carbon assimilation rate, and maximum stomatal and ecosystem surface conductances for evaporation: A global ecology scaling exercise. *Annual Review of Ecology and Systematics*, **25**, 629–660.

Schulze, E. D., Mooney, H. A., Sala, O. E., Jobbágy, E., Buchmann, N., Bauer, G., Canadell, J., Jackson, R. B., Loreti, J., Oesterheld, M. and Ehleringer, J. R. (1996b) Rooting depth, water availability, and vegetation cover along an aridity gradient in Patagonia. *Oecologia*, **108**, 503–511.

Schulze, E. D., Williams, R. J., Farquhar, G. D., Schulze, W., Langridge, J., Miller, J. M. and Walker, B. H. (1998) Carbon and nitrogen isotope discrimination and nitrogen nutrition of trees along a rainfall gradient in northern Australia. *Australian Journal of Plant Physiology*, **25**, 413–425.

Schwinning, S. and Ehleringer, J. R. (2001) Water use tradeoffs and optimal adaptations to pulse driven arid ecosystems. *The Journal of Ecology*, **89**, 464–480.

Seemann, J. R., Sharkey, T. D., Wang, J. L. and Osmond, C. B. (1987) Environmental effects on photosynthesis, nitrogen-use efficiency, and metabolite pools in leaves of sun and shade plants. *Plant Physiology*, **84**, 796–802.

Shipley, B. and Lechowicz, M. J. (2000) The functional coordination of leaf morphology and gas exchange in 40 wetland species. *Ecoscience*, **7**, 183–194.

Silim, S. N., Guy, R. D., Patterson, T. B. and Livingston, N. J. (2001) Plasticity in water use efficiency of Picea sitchensis, Picea glauca and their natural hybrids. *Oecologia*, **128**, 317–325.

Sinclair, T. R. and Shiraiwa, T. (1993) Soybean radiation-use efficiency as influenced by non-uniform specific leaf nitrogen distribution and diffuse radiation. *Crop Science*, **33**, 808–812.

Sobrado, M. A. and Ehleringer, J. R. (1997) Leaf carbon isotope ratios from a tropical dry forest in Venezuela. *Flora*, **192**, 121–124.

Stanhill. G. (1986) Water-use efficiency. *Advances in Agronomy*, **39**, 53–85.

Stewart, G. R., Turnbull, M. H., Schmidt, S. and Erskine, P. D. (1995) 13C natural abundance in plant communities along a rainfall gradient: a biological integrator of water availability. *Australian Journal of Plant Physiology*, **22**, 51–55.

Tanner, C.B. (1981) Transpiration efficiency of potato. *Agronomy Journal*, **73**, 59–64.

Tanner, C.B. and Sinclair, T.R. (1983) Efficient water use in crop production: research or research? In *Limitations to efficient water use in crop production* (eds H.M. Taylor, W.R. Jordan and T.R. Sinclair), American Society of Agronomy, Madison, USA, pp. 1–27.

Valladares, F. and Pearcy, R. W. (2002) Drought can be more critical in the shade than in the sun: a field study of carbon gain and photo-inhibition in a Californian shrub during a dry EL Nino year. *Plant, Cell & Environment*, **25**, 749–759.

Van den Boogaard, R., Kostadinova, S., Veneklaas, E. and Lambers, H. (1995) Association of water use efficiency and nitrogen use efficiency with photosynthetic characteristics of two wheat cultivars. *Journal of Experimental Botany*, **46**, 1429–1438.

Vitousek, P. M., Field, C. B. and Matson, P. A. (1990) Variation in foliar del 13C in Hawaiian Metrosideros polymorpha: A case of internal resistance? *Oecologia*, **84**, 362–370.

Von Fisher, J. C. and Tieszen, L. L. (1995) Carbon isotope characterization of vegetation and soil organic matter in subtropical forests in luquillo, Puerto Rico. *Biotropica*, **27**, 138–148.

Warren, C. R. and Adams, M. A. (2000) Water availability and branch length determine δ13C in foliage of Pinus pinaster. *Tree Physiology*, **20**, 637–643.

Will, R. E. and Teskey, R. O. (1997) Effect of irradiance and vapour pressure deficit on stomatal response to CO_2 enrichment of four tree species. *Journal of Experimental Botany*, **48**, 2095–2102.

Wong, S. C., Cowan, I. R. and Farquhar, G. D. (1985) Leaf conductance in relation to rate of CO_2 assimilation. II. Effects of short-term exposures to different photon flux densities. *Plant physiology*, **78**, 826–829.

Wong, W. C., Cowan, I. R. and Farquhar, G. D. (1979) Stomatal conductance correlates with photosynthetic capacity. *Nature*, **282**, 424–426.

Woodward, F. L. and Smith, T. M. (1994) Predictions and measurements of the maximum photosynthetic rate at the global scale. In: *Ecophysiology of photosynthesis* (eds Schulze, E.-D. and Caldwell, M. M.), Springer, Berlin, pp. 491–509.

Wright, I. J., Reich, P. B. and Westoby, M. (2001) Strategy-shifts in leaf physiology, structure and nutrient content between species of high and low rainfall, and high and low nutrient habitats. *Functional Ecology*, **15**, 423–434.

Wright, I. J., Reich, P. B. and Westoby, M. (2003) Least-cost input mixtures of water and nitrogen for photosynthesis. *The American Naturalist*, **161**, 98–111.

Yoder, B. J., Ryan, M. G., Waring, R. H., Schoettle, A. W. and Kaufmann, M. R. (1994) Evidence of reduced photosynthetic rates in old trees. *Forest Science*, **40**, 513–527.

4 Water use efficiency and chemical signalling

Sally Wilkinson

4.1 Introduction

4.1.1 Definitions

As described in Chapter 2, there are several definitions of water use efficiency (WUE). At the leaf level, WUE is a ratio of net assimilation to net transpiration (moles or grams of carbon, or CO_2 fixed per mole or kg of water transpired) – also termed 'instantaneous transpiration efficiency'. At a community or crop level WUE, is usually referred to as a ratio of dry matter production or grain yield to cumulative water use. In this chapter, we shall examine how plants prevent water loss or use the water available to them more efficiently as various types of stress develop, and thereby how they improve their WUE at both the leaf and the crop yield level.

Plants require access to the atmosphere, via open stomatal pores in the leaf, to take up gaseous CO_2 for photosynthesis. However, this means that they also risk dehydration via unavoidable transpirational water loss through the same pores, especially when water deficits are developing in the environment around them. Plants will inevitably 'lose water to fix carbon'. To meet the opposing requirements of access to CO_2 while at the same time retaining sufficient water, plants have adapted to allow a degree of control over water loss. The most important means by which this control is exerted at the leaf level is via modulation of stomatal aperture, controlled by the pair of guard cells surrounding thousands of stomatal pores per leaf. Guard cells are highly sensitive to both the internal and the external environment, and their turgor, and therefore pore size, is adjusted as a result of changes in both of these environments. As stomata close there is a non-linear correlation between the decrease in water loss that is achieved and the decrease in carbon assimilation that unavoidably occurs. Water loss is restricted by stomatal closure sooner and to a greater extent than the reduction in CO_2 uptake, due to the differences between the vapour pressure/concentration gradients for water and CO_2 between the inside and the outside of the plant. This phenomenon means that most plants tend to show an increase in WUE as stomata begin to close, e.g. as a soil or air water deficit begins to develop around the plant, because rates of photosynthesis remain high while water loss is restricted (Ehrlinger and Cooper,

1988; Jones, 1993). It is believed that plant evolution has occurred to optimise this trade-off between carbon uptake and water loss (e.g. Raven, 2002).

In this chapter we shall examine the means by which stomatal aperture is reduced by changes (mainly in water availability) in the external environment, and thereby how leaf WUE is improved and dehydration is avoided as the plant encounters stress. The mechanisms by which rates of photosynthesis are maintained (and thereby WUE is improved) in plants with reduced stomatal apertures or in otherwise drought-adapted plants are described in Chapter 3. The effect of stomatal closure to improve WUE occurs ostensibly during the early stages of water stress, but as severe stress develops, WUE usually decreases again, as leaf mesophyll cells dehydrate and photosynthesis becomes severely inhibited.

Stomatal closure is not the only means, however, whereby WUE is sustained or improved, and dehydration is avoided in plants experiencing water stress, and we will also examine the ways in which some of these other determinants of improved WUE are generated by stress in plants. At the leaf level, these include leaf rolling, dense leaf trichome layers and steep leaf angles (epinasty). These changes minimise water loss from the canopy by reducing light absorbance/increasing light reflection by/from leaves (see Chaves *et al.*, 2002, 2003). In addition many of the same, or similar signalling systems that plants utilise to control stomatal aperture when they sense stress (see below), are also used to induce reductions in canopy leaf area through either reduced leaf and stem growth or shedding of older leaves. This reduces the surface area from which water can be lost, and redirects or reallocates nutrients to the stem, roots, younger leaves, or fruits/grains. Increases in the root/shoot ratio are a common response to water deficit. Not only is shoot growth reduced, but root growth may be sustained or even increased under stress (Munns and Sharp, 1993). This maintains or enhances the ability of plants to scavenge moisture from a larger volume of soil, while minimising water loss from the aerial parts of the plant. While a reduction in the area of a single leaf does not improve its WUE (as water loss and photosynthesis will be restricted to the same extent), a general reduction in the growth of aerial vegetative tissues (and an increase in root growth) means that more of the water (and nutrients) available to the plant in the soil will be used to maintain the growth of the reproductive parts in favour of the vegetative parts (see Chapters 5 and 8). Thus, plants use the water available to them in the soil more efficiently by reducing vegetative vigour, while maintaining reproductive vigour, such that yield WUE, as opposed to leaf WUE, is improved. Evidence for the involvement of chemical signals in maintaining reproductive vigour at the expense of vegetative vigour under stress comes from studies using partial root drying (PRD – see below), and is described in Chapter 5. Evidence for improved partitioning of assimilates to developing fruits and grains when water deficits are perceived is described in Chapter 8.

In this chapter we shall examine the chemical signals that lead to reduced leaf growth rates and to the maintenance or improvement of root growth rates, which may in turn lead to improvements in yield WUE. In addition it has been shown that while soil-water deficits do not always influence grain/fruit yield, under some circumstances fruit growth is reduced alongside that of the vegetative plant parts, although not always to the same extent (see section 4.2.4). We will also explore the chemical signals that might be used by plants to control fruit growth in response to stress, as we can argue that 'fruit WUE' might have a role in plant adaptation to stress.

4.1.2   Background to chemical signalling

Stomatal closure and leaf growth inhibition are among the earliest responses to drought, and they can occur in response to very mild soil drying, even when moisture tensions are low enough that a supply of water is still freely available (e.g. Henson *et al.*, 1989). Traditional explanations for the mechanisms whereby environmental water deficit regulates these processes (and thereby WUE) have centred around a role for the often co-occurring decline in shoot water status (i.e. hydraulic signals). While leaf-water deficit can clearly develop and influence the physiology and growth of leaves (e.g. Comstock and Mencuccini, 1998), this often accompanies more severe water deficits, while physiological change has frequently been shown to occur under mild deficit before the development of any change in shoot water status (Gowing *et al.*, 1990; Henson *et al.*, 1989). It is now accepted that many plants will regulate stomatal aperture and leaf growth rates (and thereby often maintain shoot water status) independently of hydraulic signals. In many cases this has been found to result from the action of chemical regulators generated by interactions between the root or the leaves and the drying soil or air (e.g. Zhang and Davies, 1989). By pressurising the soil around the root systems of intact plants growing in dry soil, to restore shoot water relations to control levels, chemical signals have been shown to act independently of hydraulic signals, in that stomata still close and growth is still reduced (e.g. Gollan *et al.*, 1986). In addition, the PRD (partial root drying) or 'split-root' system (see Chapter 5) has demonstrated the importance of chemical signalling in the absence of changes in leaf-water deficit, for the control of stomatal aperture and of leaf and fruit growth (Gowing *et al.*, 1990; Stoll *et al.*, 2000; Mingo *et al.*, 2003). In this system the roots are trained into two pots, only one of which is watered. The roots in the wet soil provide enough water to the shoot to maintain water potential. Despite this, stomata still close and leaf (and sometimes fruit) growth rates can be reduced because the roots in the dry pot are still generating chemical messages such as an increased synthesis and export to the shoot of the plant hormone, abscisic acid (ABA). These chemical signals that become involved in the control of WUE as soil or air water

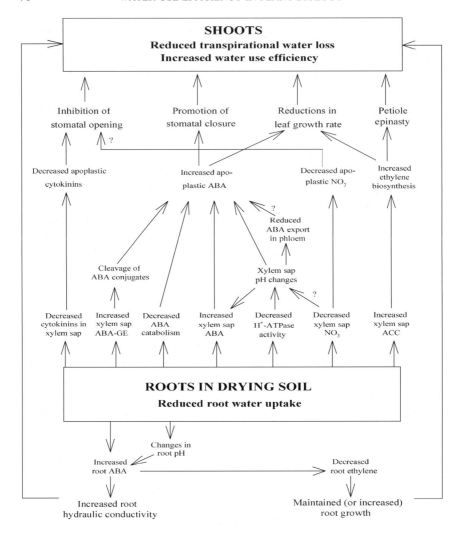

Figure 4.1 Schematic diagram of the proposed signalling networks in a model plant experiencing a soil-water deficit. Question marks denote hypothetical relationships for which there is, as yet, no direct evidence.

deficits (or other stresses) develop are the focus of this chapter, and those discussed here are summarised in Figure 4.1.

Chemical signals can involve changes in the transport of plant hormones and other chemical species between organs, usually via the xylem. For example, soil drying can increase the synthesis of the plant hormone ABA in the root, and its transport from the root to the leaf via the xylem vessels (e.g. Zhang and Davies, 1989; see section 4.2). In the leaf, ABA accumulates to a high enough concentration to act on its target cells to close stomata (e.g. Loveys, 1984) and/or to reduce leaf growth (e.g. Zhang and Davies, 1990; Dodd and Davies, 1996). There is much evidence for the mechanism whereby ABA closes stomata (e.g. Assmann and Shimazaki, 1999), but very little for the mechanism of its action on growing cells. There is circumstantial evidence for both extracellular and intracellular ABA receptors in stomatal guard cells, but ABA receptors on/in growing cells have yet to be identified. Once bound to receptors on the guard cell membrane, ABA induces a signal transduction cascade inside the guard cell (see Assmann and Shimazaki, 1999). This involves an increase in cytoplasmic calcium via increased calcium uptake from the apoplast and calcium release from internal stores. This calcium increase, and/or an increase in cyoplasmic pH, depolarises the guard cell plasma membrane, causing outward K^+ and Cl^- channels to open, bringing about a massive efflux of these ions. Thus, the osmotic potential of the cell is reduced, causing water loss, reduced turgor and stomatal closure.

ABA synthesis in, and transport from the root can occur before the decreasing water content of the soil causes any measurable change in the water status of the leaves, i.e. before the generation of hydraulic signals. Thus, this movement of ABA is defined as a chemical signal to the shoot that provides information about the environment around the root. This signal alerts the shoot to take appropriate measures to conserve the water that it already contains and any that it will continue to receive. Other chemical signalling molecules also act in plants, although not all are discussed in this Chapter. For example, soil drying also reduces the transport of stomatal-opening, growth-promoting cytokinins from the root to the shoot (section 4.4.3; Fusseder *et al.*, 1992). Soil drying and flooding increase the generation of leaf-growth-reducing ethylene from shoots (section 4.4.1), by up-regulating the synthesis and root to shoot transport of the ethylene precursor ACC (1-amino-cyclopropane-1-carboxylic acid, e.g. Else and Jackson 1998). Soil drying, changes in the aerial environment, soil flooding and soil nutrient deficiencies can induce changes in xylem sap pH (section 4.2), which can influence stomatal aperture by preferentially compartmentalising ABA within the apoplast adjacent to its sites of action on the stomatal guard cell membrane (Wilkinson and Davies. 1997, 2002). These and several other signalling mechanisms are discussed further below.

Chemical signals and the responses that they induce often occur earlier in a soil drying/inundation cycle than hydraulic signals (e.g. Henson *et al.*, 1989; Bahrun *et al.*, 2002; Jackson *et al.*, 2003). Unpublished results from Lancaster University laboratory (W. Sobeih, personal communication) have shown that the earliest changes occurring in tomato plants in response to very gradual soil drying are chemical. The first change measured was an increase in xylem pH, which arose after 24 h, prior to the reduction in stomatal aperture. Increases in ethylene generation by leaves occurred slightly later (days 3 to 4), but preceded reductions in leaf growth rate. Xylem (ABA) increased on day 7. All these changes occurred prior to any change in shoot water relations, indicating that limitations in conductance and growth can occur (perhaps mediated respectively by the pH change and the increase in ethylene – see below) in the absence of hydraulic signalling. Chemical signalling in response to changes in soil and air water availability (and to changes in nutrient availability, soil compaction and temperature), in the control of stomatal aperture and leaf (and fruit and root) growth rate is discussed in detail below.

Figure 4.1 depicts a summary of the chemical signalling networks proposed to exist in a model plant with its roots exposed to drying soil. Question marks indicate hypotheses for which there is as yet no direct evidence.

4.2 Abscisic acid and xylem sap pH as signals of soil water and nutrient availability

4.2.1 Effects of ABA in plants

Abscisic acid plays primary regulatory roles in the initiation and maintenance of seed and bud dormancy and in the plant's response to stress, especially water stress. It also interacts with auxin, cytokinin (section 4.4.3), gibberellin and ethylene (section 4.4.1), usually as an antagonist, to influence many aspects of development and the response to stress. Levels of ABA fluctuate dramatically in specific tissues during development or in response to the changing environment, whether or not this is 'stressful'. Almost all cells can synthesise ABA (Cutler and Krochko, 1999), and it is transported around the plant via both the xylem and the phloem. As soil dries around roots these synthesise ABA. ABA in stressed roots not only maintains their growth rates such that they can continue to access soil water (Munns and Sharp, 1993), but it also increases root hydraulic conductivity (Glinka and Reinhold, 1971) by modifying root membrane properties such that they take up more water per unit surface area. Both effects of ABA in the root can lead to an increase in plant WUE. Some of the ABA synthesised in the root is transported to the shoot with the xylem to accumulate in the leaves (see section 4.1.2). ABA can also be synthesised in leaves experiencing soil drying- or air-drying-induced changes in water status. This can accumulate in the leaf and/or be transported down to the root in the phloem, and back up to the shoot via the xylem.

In the leaf, ABA induces stomatal closure, which, as described above, is one of the most important factors controlling transpirational water loss and WUE under stressful conditions. Mutants that lack the ability to synthesise ABA, exhibit permanent wilting because of their inability to close their stomata. These have been termed wilty mutants, and examples include *flacca* in tomato and Az34 in barley. Applications of exogenous ABA to such mutants restores stomatal closure and shoot water status to wild-type levels (Imber and Tal, 1970). As described above (section 4.1.2) ABA also induces reductions in leaf and stem growth rates, reducing the transpirable leaf surface area and preserving water. ABA also greatly accelerates the senescence of leaves (Zacarias and Reid, 1990), while ethylene induces their abscission. Both processes reduce the leaf surface area over which water can be lost, thereby improving WUE. ABA can also reduce the xylem osmotic potential to increase water flux across the root or stem (e.g. Glinka, 1980). This helps to prevent xylem embolism in stressed plants, thereby maintaining water flux to all parts of the plant. ABA is arguably the most important stress response hormone involved in the modulation of WUE in plants.

4.2.2 How the ABA/pH signal works

As described above, an increase in xylem sap and/or bulk leaf ABA concentration is often associated with soil and/or air water deficit and found to be required for the water-retentive response that occurs in the shoot (stomatal closure or reduced leaf growth). However in reality the ABA-based chemical signalling system is much more complex than a simple up-regulation of root (or shoot) ABA biosynthesis and of xylem ABA transport to the leaf. Decreases in stomatal aperture and leaf growth rate frequently occur before any increase in the total, or 'bulk' leaf ABA content can be measured (e.g. Cornish and Zeevaart, 1985), and even before increases in xylem sap, ABA can be observed (see below). Despite this the responses observed are still dependent on the presence of ABA in the plant. This is because the strength of the 'ABA signal' perceived at its final site of action (taking the stomatal guard-cell membrane as a particular example) is influenced in many ways along its journey from root to shoot (and shoot to root), and does not always reflect coarser measurements of ABA concentration. This means that arbitrary measurements of, for example, xylem sap ABA concentrations, are often not indicative of the ABA concentration at its ultimate site of action, i.e. of that to which the stomata respond. This has been elegantly demonstrated by Zhang and Outlaw (2001a), who were able to measure the ABA concentration in the tiny volume of liquid-filled space around a single guard cell pair (the guard-cell apoplast) as a separate sub-compartment of the bulk-leaf apoplastic ABA concentration (this being a sub-component of the total bulk tissue ABA concentration). These authors found that mildly stressing *Vicia faba* L. roots could increase the apoplastic guard-cell ABA concentration in the absence of a change in total bulk leaf or xylem sap ABA concentration, and even in the absence of a change in the bulk apoplastic ABA concentration. The apoplastic guard-cell ABA concentration correlated with changes in stomatal aperture most effectively. Such localised changes in ABA are the result of factors that affect the extent to which the cells of the leaf are able to 'filter out' a portion of the ABA that arrives in the leaf before it reaches the stomata in the epidermis (see below, and for detailed reviews see Wilkinson and Davies, 2002; Davies *et al.*, 2002). The effectiveness of this symplastic filtration at removing ABA from the apoplast can be changed via alterations in the rate of cellular catabolic ABA degradation, or via changes in the pH of the leaf apoplast. There is evidence that soil drying influences both of these processes. This is described in detail below.

A concentration of approximately 0.1 μM ABA is found in the xylem sap of a well-watered plant, but if all this ABA reached the guard cells at an unadjusted concentration, the stomata would be permanently closed and the plant would not survive (Trejo *et al.*, 1993). Stomata in isolated epidermal tissue fragments directly exposed to buffers containing a range of concentrations of ABA begin to close at concentrations as low as 1nM, and have reached full closure at 0.1μM (see Wilkinson *et al.*, 2001). So, how is the ABA concentration of the xylem sap adjusted so that it is in the appropriate range to sensitively control stomatal aperture by the time it reaches the guard cells?

Much of the ABA that enters the leaf apoplast at the xylem vessel endings is taken up and stored or metabolised by the mesophyll (and epidermal) cells of the leaf as the sap moves through the apoplast and past these cells towards the stomatal pores (e.g. Trejo *et al.*, 1993; Daeter and Hartung, 1995). In fact most of the ABA transported from the roots of well-watered plants is destined to be catabolised and destroyed in the cells of the leaf. This has been termed ABA filtration as described above, and the pH of the in-coming xylem sap and of the leaf apoplast itself can determine the extent to which the leaf cells are able to filter out and remove the ABA entering the leaf apoplast via the xylem (Wilkinson and Davies, 1997; Figure 4.2). This is because the distribution of ABA between the cells and compartments of the leaf follows the 'anion trap' concept (see Wilkinson and Davies, 2002). ABA accumulates in the most alkaline compartments to an extent determined by the steepness of the pH gradient across the membrane separating two compartments (e.g. the apoplast and the mesophyll cell cytoplasm). The steeper the pH gradient, the more ABA is taken up by the cell. In well-watered plants the xylem sap (and thus the apoplastic) pH is often acidic (approx. pH 6.0 – see Wilkinson and Davies, 2002). The cytoplasm of the leaf cells is approximately pH 7.4. There is thus, a steep pH gradient between the apoplast and the cell cytoplasm so that the ABA entering the leaf is efficiently taken up by these cells (and filtered out of the apoplast). A relatively low ABA concentration is all that remains in the transpiration stream by the time it reaches its final destination, the apoplast immediately adjacent to the guard cells surrounding the stomatal pores. Under these circumstances, the ABA concentration is low enough that stomata remain open (Figure 4.2A). However, during the early stages of water stress the pH of the xylem sap and thereby of the leaf apoplast becomes more

A - Well-watered leaf loses water
ABA taken up by symplast

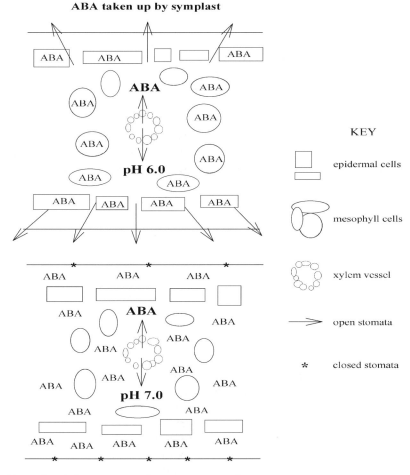

KEY

epidermal cells

mesophyll cells

xylem vessel

open stomata

closed stomata

B - Leaf with stressed roots retains water
ABA remains in apoplast

Figure 4.2 Schematic diagram depicting the effect of xylem/apoplastic sap pH on ABA penetration to the guard cells in the epidermis of a stylised leaf, in a well-watered plant (A), in a plant exposed to a soil water deficit (B).

alkaline (approx. pH 7.0 – see section 4.2.3). Thus, the ABA carried into the leaf by the xylem vessels is less readily filtered out of the apoplast by the cells of the leaf that it encounters on its passage towards the guard cells. The concentration of ABA that remains in the transpiration stream by the time it reaches the guard cells is less affected, and remains high enough to induce stomatal closure (Figure 4.2B). All this can occur without an increase in ABA input anywhere in the system (i.e. in the absence of *de novo* ABA synthesis or greater xylem ABA concentrations/flux rates). Thus, the pH change that occurs as soil (or air) dries (see section 4.3) can function as a root (or leaf) sourced chemical signal that alerts the shoot of the need to conserve water and increase its WUE, and provides the means by which to do so. As stress develops further, a combination of increased xylem ABA and sap alkalinisation will provide an even stronger signal to the shoot than either signal could have done alone.

As well as increasing the residence time of xylem-sourced ABA in the leaf apoplast (by reducing the ability of leaf cells to remove ABA from the transpiration stream), a pH change can increase the local (ABA) at the guard cell apoplast in other ways. A change in the pH of the xylem sap can actually increase the ABA concentration in the xylem sap before it reaches the leaf, in the absence of *de novo* ABA synthesis. This comes about because the increase in sap alkalinity induces ABA stored in the stem parenchyma to be released into the xylem lumen (Sauter and Hartung, 2002). It may also be the case that a greater proportion of the ABA synthesised in the root (or transported to the root from the shoot) will enter the xylem vessels when the pH of the xylem sap is higher. There is some indirect evidence to support this hypothesis (Slovik *et al.*, 1995). In addition root dehydration has been shown to increase the ability of root cells to release the ABA that they contain/synthesise, by causing cytoplasmic acidification (Daeter *et al.*, 1993). This may change the pH gradient over the root cell membrane in favour of ABA release to the root apoplast, from where the ABA can more easily enter the xylem vessels. There is also some evidence to suggest that an increase in leaf apoplastic alkalinity will reduce the export of ABA out of the leaf via the phloem, as loading of ABA into this transport conduit will also be dependent on pH gradients (Jia and Zhang, 1997; Wilkinson and Davies, 2002). Some work provides evidence to suggest that the increased pH of the leaf apoplast actually causes the release of ABA from stores in leaf mesophyll cells (Hartung *et al.*, 1983), to enrich the ABA in the apoplast sourced from the root, the stem and the xylem. However this may only occur under more severe stress when leaf-water deficits occur and leaf cell cytoplasm becomes acidified (Kaiser and Hartung, 1981). This may be one way in which chemical and hydraulic signals interact. The net result of all these pH-dependent processes is an accumulation of ABA to physiologically active concentrations in the leaf apoplast adjacent to the guard cells, often in the absence of measurable increases in bulk tissue or even of xylem sap (ABA), and before shoot water status is affected by the stress. In addition, non-pH mediated stress-induced increases in the ABA signal all contribute to the raised guard cell apoplastic concentration. The obvious inclusion here is *de novo* ABA synthesis in the root (or leaf), but there is also some evidence for a stress-induced reduction in the ability of the cells of both the root and the leaf to catabolise ABA (Liang *et al.*, 1997; Jia and

Zhang, 1997). In addition, ABA-conjugates can also be synthesised in stressed roots, transported to the shoot via the xylem, and cleaved in the leaf apoplast to release free ABA to this compartment (section 4.2.5). When stress becomes more severe, chemical signals are accompanied by hydraulic changes that induce ABA synthesis in the leaf, and ABA release from stores in the cells of the leaf (see above). In all, there are many ways in which the basic ABA signal can be enriched and strengthened in a plant experiencing stress of varying magnitudes (for a detailed review see Wilkinson and Davies, 2002). To accurately measure the true 'ABA signal' to which the plant is responding, the ABA concentration in the tiny apoplastic micro-compartment surrounding the guard cell pair (or the growing cells) should be sampled (Zhang and Outlaw, 2001a,b), although this is obviously not always possible.

4.2.3   Evidence for the pH signal

Changes in xylem sap pH were originally investigated in relation to their role as chemical signals of soil drying (e.g. Schurr et al., 1992; Wilkinson and Davies, 1997). As evidence for this hypothesis accumulated, we also suggested that a pH change could be a signal of other types of perturbation at the roots of plants, namely flooding (Wilkinson, 1999) and nutrient deficiency (Wilkinson and Davies, 2002). There is now much evidence for the occurrence of very early increases in xylem sap pH as soil dries (see below). In addition, it has also very recently been confirmed that, as predicted, xylem sap pH becomes more alkaline as soil floods (Jackson et al., 2003), or becomes nutrient deficient (e.g. Dodd et al., 2003). It has been shown that such changes in pH can induce stomatal closure and/or reduce growth rates by affecting ABA compart-mentation (Wilkinson and Davies, 1997, 2002; Wilkinson et al., 1998; Bacon et al., 1998; see below). Several reviews can be referred to (Wilkinson, 1999; Wilkinson and Davies, 2002; Davies et al., 2002).

Xylem/apoplastic sap pH has been shown in several species to increase from approximately 5.5/6.5 in well-watered plants, to approximately pH 7.0 in droughted plants (e.g. Wilkinson et al., 1998; Mingo et al., 2003), and from approximately 5.6 to 6.6 as tomato plants become flooded (Jackson et al., 2003). When plants are transferred to nitrate deficient soil the sap pH changes from 5.6 to 7.3 (Kirkby and Armstrong, 1980 – castor oil) or from 5.75 to 6.1 (Dodd et al., 2003 – pepper – root exudation). These changes can occur before the shoot water status of the plant is affected by perturbation at the roots (Schurr et al., 1992, Stoll et al., 2000; Dodd et al., 2003). They can occur within 1–2 days (Bahrun et al., 2002; Mingo et al., 2003; Dodd et al., 2003) or even within a few hours (Jackson et al., 2003), prior to, or coincident with the changes in plant physiology that they are believed to induce. The pH change is sometimes not evident however, until after the change in stomatal conductance has been induced by the imposed stress (Liu et al., 2003 –

soybean). In this particular case the whole pot pressurisation technique used to express xylem sap may have masked all but the greatest *in planta* pH changes (Megat-Wahab, personal communication). In addition stress-induced pH changes may be more pronounced when leaf apoplastic sap is sampled, as opposed to xylem sap sourced from closer to the root system (Hoffmann and Kosegarten, 1995). Changes in apoplastic sap pH mirror changes in tomato stomatal conductance more closely than changes in xylem sap pH (Dodd, personal communication). Liu *et al.* (2003) sampled sap from de-topped roots. Measurements of xylem sap pH in earlier studies in which de-topped roots were sampled may also have resulted in under-estimations of the stress-induced pH change (e.g. Bacon *et al.*, 1998).

Three areas of research provided evidence that led us to investigate the possibility that an increase in xylem pH could be a novel root-sourced signal of soil-drying that closes stomata and reduces leaf growth. First, the literature reported non-ABA anti-transpirant activity of unknown origin (e.g. Munns and Sharp, 1993). Second, xylem alkalisation correlated with increases in the apparent sensitivity of stomata to xylem (ABA) (Schurr *et al.*, 1992; Gollan *et al.*, 1992). Finally very detailed studies from Hartung's laboratory demonstrated that increasing the pH of media bathing isolated leaf tissues such as epidermal strips reduced the uptake of ABA by the tissue from the solution (e.g. Kaiser and Hartung, 1981). Computer simulations based on these data predicted that even well-watered plants contain enough ABA to accumulate at guard cells to a high enough concentration to close stomata if the sap pH is high enough to prevent the symplast from mopping up this ABA as it passes through the leaf (Slovik and Hartung, 1992). It was confirmed that a pH change can function as a chemical signal (Wilkinson and Davies, 1997). It has now been demonstrated directly that an increase in xylem pH can be a signal that reduces rates of water loss from detached leaves of *C. communis* L., tomato, and pepper (Wilkinson and Davies, 1997; Thompson *et al.*, 1997; Wilkinson *et al.*, 1998; Hartung *et al.*, 1998; Dodd *et al.*, 2003), and reduces rates of leaf growth in de-rooted seedlings of barley (Bacon *et al.*, 1998). It was determined, using detached leaves from which the ABA had been removed, or using ABA-deficient mutants, that these effects were dependent on the relatively low endogenous concentration of ABA normally found in the xylem (Wilkinson and Davies, 1997; Thompson *et al.*, 1997; Wilkinson *et al.*, 1998; Bacon *et al.*, 1998). Only when ABA was re-supplied to the ABA-deficient shoots was the wild-type response to an increase in xylem sap pH seen, namely a reduction in stomatal conductance or leaf growth. The hypothesis that the pH signal works by reducing symplastic filtration/sequestration and thereby increasing the penetration probability of ABA to the guard cells in the epidermis (Figure 4.2) was also confirmed and extended (Wilkinson and Davies, 1997; Hartung *et al.*, 1998). These findings greatly reduce the need to search for as yet unknown signalling compounds, and provide evidence that a pH change could be one of the earliest signals of soil drying, flooding (see section 4.2.5) and nutrient deficiency, that induces an increase in WUE. In addition we have recently determined that pH changes can occur in response to changes in the aerial environment around the leaf (see section 4.3), as well as in response to perturbations at the root.

4.2.4   pH as a signal controlling fruit growth

As described above (section 4.1.1) stress-induced reductions in leaf growth rate can improve WUE yield because they reduce the surface area from which water can be lost by the plant, such that fruit/grain yield is less compromised by the stress had vegetative vigour been maintained. Stress-induced reductions in fruit growth, that often occur in some species alongside reductions in vegetative vigour, obviously do not improve yield WUE (although again, fruit growth is often reduced by stress to a lesser extent than vegetative growth such that effects of stress on yield and WUE yield are minimsed). However it can be argued that because fruits also lose water via transpiration (Hetherington et al., 1998), but do not photosynthesise, a decrease in fruit size could increase fruit water use efficiency. While the importance of such 'fruit WUE' to stress adaptation is debatable, it is worth examining the chemical signals that could influence this parameter, as evidence is accumulating that they may be similar to those used to control leaf growth under stress.

As described above (and see section 4.4.1) chemical signals have been shown to have a substantial role in soil drying-induced reductions in leaf and stem growth rates, often in the absence of a hydraulic influence. Recently chemical signals have also been implicated in stress-induced limitations of tomato fruit growth (Davies et al., 2000; Mingo et al., 2003). This type of control over fruit growth was originally dismissed as unlikely, as the xylem connection between the shoot and the tomato fruit is lost as fruits develop (e.g. Davies et al., 2000), and fruit access to water is limited to the phloem. Nevertheless, Davies et al. (2000) were able to show that soil-water deficits could still limit fruit expansion in the absence of changes in shoot water status (under PRD), implicating a role for root-sourced chemical signals and the free transmission of these from the shoot into the fruit. It has since been suggested that the xylem connection to the fruit is maintained until the later stages of fruit development in some cultivars, and only then will chemical signals cease to be transmitted (see Mingo et al., 2003). On the other hand grape berries seem to become isolated from the xylem (and the root-sourced chemical signals carried via this route) early on in their development, as PRD reduces leaf growth and stomatal conductance in the absence of a change in berry size (see Stoll et al., 2000). It is likely that fruits of different species will rely on chemical signals to control growth in relation to perturbations at the root to differing extents, partly depending on the extent of the xylem connection between the shoot and the fruit. Thus, fruit size of some species is particularly insensitive to soil drying, especially under PRD. In some species, fruit and seed development may even be accelerated when stress is perceived (see Chapter 8), and chemical signals may also be involved in this response.

Mingo et al. (2003) have shown that tomato fruit exposed to drying soil under PRD (whereby the shoot water status is maintained) exhibited reduced growth rates in the absence of measurable changes in fruit mesocarp cell turgor. These authors also showed, for the first time, that fruit sub-epidermal pH can increase when a soil-water deficit is imposed (Figure 4.3). The pH change occurred concomitantly with the change in growth rate, three days after stress imposition, again in the absence of a change in fruit (and shoot) cell turgor. When a full soil-water deficit was imposed (instead of splitting the roots between a watered and a drying pot as under PRD), the fruit sub-epidermal pH change was evident within a single day. Under PRD the increase in fruit apoplastic pH was mirrored by similar pH increases in the apoplast of the adjacent leaves (Figure 4.3). Preliminary experiments have demonstrated that feeding buffers to detached fruit trusses, of a pH equivalent to that of the fruit apoplast of plants exposed to a soil-water deficit, can directly reduce fruit growth rates (Mingo, personal communication). These data provide circumstantial evidence to suggest that chemical signals, namely a change in pH, have a key regulatory role in controlling fruit cell expansion during soil-water deficit, and can freely pass from the shoot into the fruit.

It is not known, however, whether the pH signal affects fruit growth in the same way as it does leaf growth. We have shown that the effect of high pH to reduce barley leaf growth is dependent on ABA (Bacon et al., 1998). It is also tempting to suggest that the increase in pH in leaves and fruits re-distributes and directs incoming ABA towards the active sites for growth restriction in fruits (Mingo et al., 2003). Alternatively in vitro studies have shown that the pH imposed on tomato fruit cell walls can directly modify cell wall extensibility, which decreases as pH increases above an optimum of 5.0 (Thompson, 2001), possibly via pH-induced changes in the activity of enzymes such as expansins. The range of pH values recorded in the fruit apoplast by Mingo et al. (2003) lies close to both those used to elicit in vitro changes in tomato fruit cell wall extensibility (Thompson, 2001), and those that directly induced ABA-dependent changes in barley leaf growth (Bacon et al., 1998).

4.2.5   pH as a signal of soil flooding

Paradoxically, the most immediate crisis facing plants following soil flooding is a decreased ability to take up water (this occurs within 2 h). The reduction in root hydraulic conductance may be caused by the rapid depletion of oxygen and/or accumulation of CO_2 in the inundated rhizosphere (Else et al., 1995). Shoot dehydration and death quickly follow unless water loss from leaves is

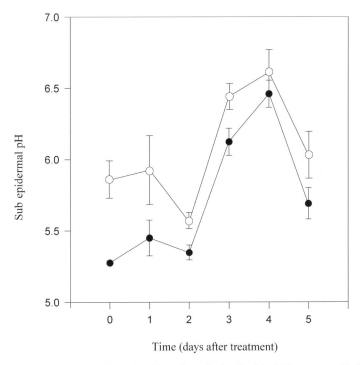

Figure 4.3 Mean fruit (O) and leaf (●) sub-epidermal apoplastic pH of the PRD treatment. Each point represents the mean (n=3) ± SE (r²=0.9 and P<0.01). From Mingo *et al.* (2003) – reprinted by permission of the *Journal of Experimental Botany* and Oxford University Press, UK.

minimised, i.e. unless WUE is improved. Although many differences exist between plants exposed to flooding and those affected by other edaphic stresses such as soil drying, it has been proposed that, as for soil drying, a pH-mediated redistribution of existing ABA into the shoot apoplast may regulate stomatal apertures and improve WUE in flooded plants (Wilkinson, 1999; Jackson *et al.*, 2003). This has been proposed to occur despite the fact that ABA delivery to the shoot from the root was reduced by 75 per cent within the first 4 h of flooding in tomato (Else *et al.*, 1996) as a result of flooding-induced reductions in the oxygen that is required for ABA biosynthesis in the root. However the involvement of shoot ABA is still implicated in the stomatal response, as grafting experiments with ABA-deficient mutants showed that wild-type concentrations of shoot-sourced ABA were necessary to invoke complete stomatal closure in flooded tomato and pea plants (Jackson, 1991).

Jackson *et al.* (2003) have shown that increases in xylem sap pH do indeed occur in flooded tomato plants within 3 h, preceding any changes in shoot physiology or shoot water relations. Evidence exists to propose that flooding-induced xylem sap alkalinisation could constitute a root-sourced ABA-based chemical signal that modifies stomatal behaviour and WUE via redistribution of the existing shoot ABA. As described above pH buffers in the range encompassing those described for flooded tomatoes by Jackson *et al.* (2003) have previously been shown to reduce transpiration rates of detached tomato leaves when applied directly to the cut petiole, in an ABA-dependent manner (Wilkinson *et al.*, 1998). This could not be repeated by Jackson *et al.* (2003) however. In addition Sauter and Hartung (2002) have shown that artificial xylem sap perfused through maize stems becomes enriched by stores of ABA from the stem parenchyma when its pH is buffered to more alkaline values (7.0 as opposed to 6.0). It is envisaged that such pH-induced ABA enrichment will also contribute to the anti-transpirant potential of sap extracted from flooded tomato plants. This potential may not be fully realised until the sap reaches the target cells (the stomatal guard cells) in the leaf, as the sap will go on accumulating ABA from the whole length of the stem and possibly from the cells of the leaf (Hartung *et al.*, 1983) as it flows through the plant. Such hypotheses in flooded plants, which have limited root-sourced ABA supplies, are supported by recent evidence from grafting studies with ABA-deficient tomato mutants. These revealed that the stomatal response to drying soil can be strongly influenced by the capacity of the shoot, and not the roots, to synthesise ABA (Holbrook *et al.*, 2002).

Of course, the effect of pH on ABA compartmentation and stomatal aperture is not the only potential chemical signal to stomata of soil inundation thought to exist in tomato. It has been proposed that a novel non-pH, non-ABA anti-transpirant exists in the sap of flooded tomato plants (Else *et al.*, 1996), and it is also possible that root-sourced ABA conjugates (see 4.2.7) may have a role, or even explain the novel anti-transpirant activity. Hydraulic signals will undoubtedly sustain stomatal closure, but these do not arise until after closure has already occurred (Else *et al.*, 1995). Other chemical signals induced by soil flooding include an up-regulation of xylem ACC and shoot ethylene production (see section 4.4.1). These have no effect on stomata, but induce petiole epinasty and possibly reductions in leaf and stem growth rates, which also impact on plant WUE.

4.2.6 *Proposed mechanisms for the pH change*

Given that changes in xylem, apoplastic and symplastic pH seem to be fundamental responses to soil drying and flooding (and other types of stress) that grossly change the basic 'ABA signal', it seems remiss that we still know very little about how these pH changes arise. Changes in xylem and symplastic pH may be linked to effects of cellular (root or leaf) dehydration on the activity of H^+-ATPases associated with xylem parenchyma, leaf mesophyll and root cortical cells (Hartung and Radin, 1989). Reductions in

H^+-ATPase activity will also be prompted by the rapidly decreasing levels of oxygen in the rhizosphere of flooded plants (Ratcliffe, 1997).

It has also been suggested that soil drying-induced perturbations in the ionic composition of xylem sap may influence its pH (Schurr *et al.*, 1992; Wilkinson and Davies, 2002). We also propose that soil inundation may induce a similar response. A reduction in xylem (nitrate) of at least 50 per cent is one of the earliest and most sensitive chemical changes that occurs in response to perturbation at the roots, that precedes associated increases in xylem sap pH and reductions in stomatal aperture (Schurr *et al.*, 1992; Bahrun *et al.*, 2002; Jackson *et al.*, 2003). Imposed N-deprivation is also associated with an increase in xylem sap pH in the absence of soil-water deficit (Kirkby and Armstrong, 1980; Dodd *et al.*, 2003). We have proposed (Wilkinson and Davies, 2002) that soil-drying (flooding and nutrient-deficiency) induced xylem alkalisation may result from a switch in nitrate reductase activity (NRA) from shoots to roots. This switch can be prompted by the reduced supply of nitrate from the root to the leaf (Andrews, 1986; Lips, 1997). Organic N products of root NRA, such as malate or citrate, are then loaded into the xylem in preference to inorganic nitrate (e.g. Andersen *et al.*, 1995) where they exert an effect that increases pH (Kirkby and Armstrong, 1980; Patonnier *et al.*, 1999). Patonnier *et al.* (1999) fed malate and related organic products to the xylem of detached ash leaves at concentrations equivalent to those that they had detected in intact droughted plants (that were elevated in comparison to well-watered concentrations), and found that these closed stomata in a manner dependent on the negative charge of the compound. We have shown that the nitrate reductase inhibitor sodium tungstate prevented soil-drying-induced sap alkalisation in tomato seedlings (Wilkinson, unpublished results). Presumably tungstate was inhibiting the soil-drying-induced production of organic acids in the root and therefore their export to the shoot in the xylem. Instead sap became more acidic as the soil dried. This was probably due to the direct and opposing effect on pH of the decreased xylem (nitrate) (e.g. Kosegarten *et al.*, 1999) that accompanies soil drying, in the absence of the concomitant increase in xylem [malate]. Elements of this hypothesis, that reduced xylem nitrate concentrations increase the penetration probability of xylem ABA to the guard cells by influencing xylem sap pH, have already been proposed by Raven and Smith (1976), and McDonald and Davies (1996).

It must be noted however, that drought- and flooding-induced reductions in xylem sap nitrate are not a universal phenomenon in the plant world. It appears that it is mainly fast-growing herbaceous species, that commonly assimilate a high percentage of their nitrate in the shoot (and which therefore transport large amounts of nitrate from roots to shoots under control well-watered conditions), that are susceptible to drought- (and presumably flooding-) induced reductions in xylem sap (nitrate) (Lips, 1997; Andrews,

1986). Examples of such species are *Helianthus* (Schurr *et al.*, 1992) and tomato (Jackson *et al.*, 2003). It is likely therefore, that it will only be species from this group that exhibit drought-induced increases in xylem sap pH. Slower growing (often deciduous tree) species, that assimilate most of their nitrate in the root, never transport a significant amount of N as nitrate within the xylem to the shoot, even under well-watered conditions. These species transport the organic products of root nitrate assimilation (or of re-mobilised stored N) in the xylem instead of nitrate, these being amino acids or organic acids. Therefore, soil-water deficit or soil inundation are unlikely to change the xylem sap (nitrate) of these species, and sap pH will remain unchanged. In support of this it has recently been shown that the xylem sap pH of deciduous woody species remained unchanged or became acidified in response to soil drying (Thomas and Eamus, 2002; *Forsythia* – Cameron *et al.*, 2004). In addition *Ricinus communis*, that also exhibits sap acidification rather than alkalisation as soil dries, does not exhibit reduced xylem nitrate concentrations in response to soil drying (Schurr and Schulze, 1996).

There is also some evidence in the literature that changes in the cation/anion ratio, or the strong ion difference (SID) of the xylem sap may be induced by perturbations at the root. This has been suggested to be the cause of the concomitant pH change measured by Gollan *et al.* (1992) in sunflower as the soil around the roots dried. In this case the major change in xylem sap composition was in the nitrate concentration, which was reduced as described above, although organic acids were also increased. Similar changes in xylem (nitrate) imposed by other means were shown to correlate with pH changes in *Ricinus communis* in the absence of concomitant changes in (malate) (Gerendas and Schurr, 1999). However, Bahrun *et al.* (2002) found that the large reductions in xylem nitrate concentration that they observed in maize as soil dried were not paralleled by changes in the cation/anion ratio, while the xylem sap pH was nevertheless increased. Organic acid concentrations were not measured in the xylem sap of this study. Changes in xylem biochemistry that lead to SID-induced xylem alkalisation may well be involved in the soil drying response, and perhaps in the flooding response of some species. This may occur in addition to or instead of changes in the xylem organic acid content, that clearly influence stomatal aperture in a manner dependent on the effect that they exert on sap pH (Patonnier *et al.*, 1999). Common to both hypotheses for the mechanism of pH change is the reduction in nitrate availability. Indeed it may be the case that the drought-induced increases in xylem malate (or citrate) measured by several authors are involved in charge balancing SIDs generated by the loss of nitrate (which alone tends to cause an acidification of sap pH – Kosegarten *et al.*, 1999). The two hypotheses are therefore by no means mutually exclusive.

4.2.7   ABA-conjugates

The 'basic' ABA signal can also be influenced by soil drying via synthesis of ABA conjugates such as ABA-glucose ester (ABA-GE), and their transport in the xylem to leaves (see Sauter *et al.*, 2002 for a comprehensive review, and for references to statements made in this section where not otherwise referenced). Under stress conditions, their concentration can rise substantially, even to above that of free ABA in the xylem. They are formed under stress in root cells and transported symplastically to xylem parenchyma cells, which release them to the xylem vessels for transport to the shoot. Unlike free ABA, ABA-GE is lipophobic and unable to cross the lipid cell membranes between the xylem and its surrounding tissues, and will thus be transported to the shoot without loss to the surrounding stem parenchyma. This means that ABA-GE is potentially a much more efficient means of transporting ABA from the root to the shoot, especially under conditions which cause the xylem sap to become too acidic to retain free ABA (as stated above ABA tends to accumulate in the most alkaline compartments). Free ABA that is translocated over long distances in the xylem through stems may be redistributed to the stem parenchyma, significantly weakening the ABA signal (Jokhan *et al.*, 1999). This may be particularly important when xylem sap remains acidic, increasing the tendency for free ABA to move over the xylem vessel membrane into the more alkaline cytoplasm of the adjacent xylem parenchyma cells. For example, grapevine xylem sap pH has been shown to be particularly acidic (4.0 in well-watered plants, 4.6 under PRD – Stoll *et al.*, 2000), and some species e.g. *Ricinus communis* (Schurr and Schulze 1996) and *Forsythia* x *intermedia* (Cameron *et al.*, 2004) tend to exhibit xylem sap acidification as soil dries (rather than the more usual sap alkalisation discussed above and below), and it may be the case that in such species the major transport form of ABA within the plant is ABA-GE.

After arrival in the leaf apoplast, apoplastic esterases, hydrolases and/or β-glucosidases cleave the conjugate and release free ABA to the target cells and tissues (Dietz *et al.*, 2000). The activity of such enzymes has been shown to increase when barley plants are subjected to salt stress, but it is not yet known whether water deficit or flooding stress has the same effect. However, interestingly an increase in apoplastic pH over the range of the increase detected in leaves of plants experiencing soil-water deficit stress has also been shown to enhance extracellular β-glucosidase activity. So far these enzymes have only been found, appropriately, to occur in the root cortical apoplast or the leaf apoplast, and not within the xylem lumen. The conjugates themselves are unable to affect stomatal aperture or plant growth, and the ABA that they 'carry' only becomes active upon its release from the conjugate by enzyme cleavage. Since ABA-GE remains undetectable in the usual experimental assays employed for ABA detection, simple measurements of xylem ABA do not always reflect the amount of anti-transpirant potentially present in the xylem stream. Indeed under some circumstances xylem ABA-GE may be the major method of translocation of the 'ABA signal' from the root to the shoot, and is therefore an important determinant of plant WUE in some species.

4.3 Abscisic acid and xylem sap pH as signals of changes in the aerial environment

Circumstantial evidence has recently been provided that short-term climatic changes around leaves can chemically modulate the strength of the basic ABA

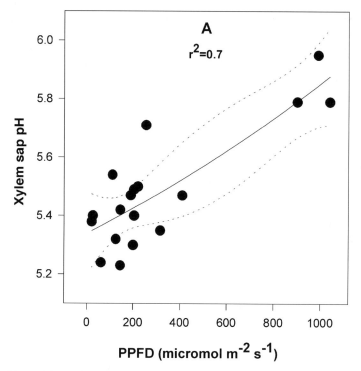

Figure 4.4 Correlations between photosynthetic photon flux density (PPFD) incident on the first fully expanded leaf of *Forsythia x intermedia* (cv Lynwood) plants of approx. 100cm in height (3–4 branches) growing in 3l pots at full irrigation capacity, and A the pH of xylem sap expressed immediately afterwards from the top 10cm of the main shoot at up to 2 bars over-pressure. In B the correlation between PPFD and the stomatal conductance of the leaves is shown; with C showing the correlation between xylem sap pH and stomatal conductance. Data for A and C are from the same set of plants, whilst that for B is a separate experiment. The points on the graphs represent data from individual plants, and 2nd order regressions are shown, along with 95% confidence intervals (dotted lines) and r^2 curve coefficients, as calculated on Sigmaplot version 1.02. From Davies *et al.* (2002) reprinted by permission of the *New Phytologist* and Blackwell Publishing Ltd., UK.

signal sent from roots. We have shown that changes in pH sourced from within the leaf also occur in response to climatic fluctuations, in the absence of, or along-side leaf-water deficits (Wilkinson and Davies, 2002 – *Hydrangea macrophylla* cv Bluewave; Davies *et al.*, 2002 – *Forsythia* x *intermedia* cv Lynwood). In pot-grown *Forsythia* exposed to natural variations in atmospheric conditions, high PPFD (photosynthetic photon flux density – a measure of light intensity) (and by association high temperature and VPD – vapour pressure deficit) was associated with high xylem sap pH (Figure 4.4A) and low stomatal conductance (Figure 4.4B), such that pH was negatively correlated with stomatal conductance (Figure 4.4C). We propose, therefore, that when the climate becomes potentially stressful, increases in xylem and leaf apoplastic pH increase the penetration probability of ABA to the guard cells as described above, to control stomatal aperture. This may include an effect of pH to inhibit the removal of ABA from the leaf in the phloem (see Jia and Zhang, 1997), as we determined that the leaf apoplast was the most likely source of the pH change (Davies *et al.*, 2002). These pH changes do not act directly to close stomata in other species (rather the opposite – see Wilkinson and Davies, 1997; Wilkinson *et al.*, 1998), and require ABA to do so. In

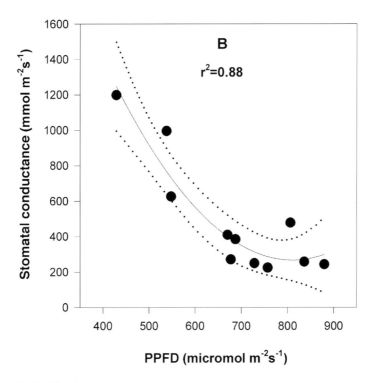

Figure 4.4   Continued

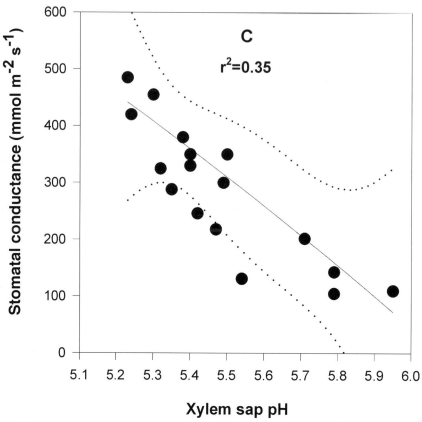

Figure 4.4 Continued

Forsythia, but not *Hydrangea*, the climate-induced increase in pH, and stomatal closure, correlated with increased bulk leaf (but not xylem sap) ABA. It has frequently been demonstrated that increasing VPD (and temperature) increase the delivery rate (flux) of ABA to leaves, as the transpiration stream flows more rapidly to, through and from the leaf (see Trejo *et al.*, 1995), and in the past this has been suggested to be the basis of VPD- and/or temperature-induced increases in stomatal sensitivity to xylem ABA. However stomata do not always respond to VPD – or temperature-induced increases in bulk leaf ABA (Trejo *et al.*, 1995), especially in well-watered plants, due to its symplastic removal from the apoplast and its accumulation in the mesophyll 'reservoir' (Wilkinson *et al.*, 2001; Wilkinson and Davies, 2002). In some cases, there is no increase in bulk leaf ABA and stomata still respond to the change in VPD with an increasing sensitivity to xylem ABA in the absence of gross hydraulic signals (Tardieu and Davies, 1992, 1993). To explain these data it is suggested that VPD (or PPFD or temperature) induced increases in pH, paralleled or not by increases in ABA flux, could create a very powerful ABA signal to the guard cells in the epidermis. Of course the signal would be more powerful if the increase in ABA flux and the increase in pH were concurrent. Tardieu and Davies (1992, 1993) on the other hand, suggested that subtle (unmeasurable) hydraulic signals induced by increasing VPD might act directly on guard cells to increase their sensitivity to a given dose of ABA. However Zhang and Outlaw (2001b) demonstrated VPD-induced local increases in ABA around guard cells that paralleled the heightened sensitivity of these to xylem ABA. These authors suggested that localised water loss at the stomatal pore was the reason for the concentration of the hormone (and other apoplastic solutes) in these microsites. It is likely that in different species and under different conditions/growth stages, different signals will modulate the stomatal response to VPD, or combinations of several of these might operate at once (see below).

The mechanism of high PPFD/VPD- (or temperature-) induced increases in apoplastic sap pH is as yet unknown. Wilkinson and Davies (2002) have speculated that effects on H^+-ATPases might be involved (Hartung and Radin, 1989), or that increased removal of CO_2 and/or nitrate locally from the leaf apoplast (during increased photosynthesis) might increase its pH (Stahlberg *et al.*, 2001; Mengel *et al.*, 1994).

Wilkinson and Davies (2002) have proposed that potent ABA-based pH signals will be influential in a wide range of environments. In addition to soil-water deficits and changes in the leaf microclimate, flooding, nutrient availability, fungal infection, and time of day and season also affect xylem and leaf and/or fruit apoplastic pH (see section 4.2 and Wilkinson *et al.*, 1998). Environmental factors that affect aerial parts of the plant (e.g. PPFD, temperature, VPD, season or fungal infection) will interact with factors that affect underground parts (drought, flooding) and/or the whole plant (water/ nutrient availability) to achieve a final apoplastic pH which reflects the total plant environment, that governs the ABA concentration arriving at the guard cells or the growing cells. The plant may therefore respond simultaneously to multiple environmental stimuli through its final combined apoplastic pH.

Such interactions (and see section 4.5) could explain the hydraulic homeo-stasis observed in shoots under fluctuating soil water and VPD, among other environmental variables (Tardieu and Davies, 1992, 1993), in many species including trees. They also suggest that root-sourced chemical signalling can still be important for stomatal regulation and WUE in large woody species, despite long transport times and distances: the root-sourced ABA con-centration may be modified locally in the leaf by climate-induced changes in apoplastic pH.

It must be noted, however, that there is plenty of evidence that not all species modulate stomatal responses to VPD via chemical changes that increase the strength of the ABA signal. Assmann *et al.* (2000) found that increasing VPD closed stomata in both ABA-deficient and ABA-insensitive mutants of *Arabidopsis thaliana*. Indeed the more traditional view is that dry air has a more direct effect on guard cell and/or epidermal cell turgor. However, it is likely that responses to VPD (like those to soil drying) differ between species, some employing hydraulic signals (e.g. Saliendra *et al.*, 1995), others chemical (e.g. Tardieu and Davies, 1993; Wilkinson and Davies, 2002; Davies *et al.*, 2002), and others both (e.g. Fuchs and Livingstone, 1996).

4.4   Other chemical regulators (cytokinins, ethylene and nitrate)

4.4.1   ACC and ethylene

Increases in gaseous ethylene production from shoots are a feature of the developmental cues that induce leaf abscission, flower senescence and fruit ripening. Stresses such as wounding, disease, chilling, flooding, soil compaction, high temperature and drought also increase its biosynthesis (for an overview see Taiz and Zeiger, 1998). Stress ethylene is involved in stress responses such as abscission, senescence, wound healing, the proliferation of adventitious roots and root hairs, stem and root swelling (that increase the tensile strength of the growing organ for example, increased soil penetration) and leaf epinasty (a downward curvature of the leaf that minimises evaporative demand), some of which lead to an improvement in WUE (see section 4.1.1). More recently stress ethylene has also been found to be involved in decreasing leaf expansion rates (e.g. Lee and Reid, 1997; Sharp *et al.*, 2000). This makes ethylene an important chemical regulator of WUE under stresses such as flooding (Else and Jackson, 1998) and soil compaction (Hussain *et al.*, 2000). It has also been shown to increase as soil dries (Abeles, 1992; see section 4.1.2). One school of thought maintains that ethylene reduces leaf growth by triggering ABA biosynthesis (e.g. Grossmann and Hansen. 2001; also see below).

Increased xylem fluxes of root-sourced ACC (the immediate ethylene precursor) to shoots can occur within 4 h of soil flooding, and these induce increases in shoot ethylene production 8 h later, which trigger petiole epinasty and leaf growth reductions in tomato seedlings (Else and Jackson, 1998). ACC is the intermediate root-sourced chemical signal of soil flooding in this case. The ACC in the roots cannot be converted to ethylene until it reaches the shoot as oxygen is required for this interaction, and flooded roots are anaerobic. On arrival in shoot tissues, ACC conversion to ethylene is facilitated by increased cytoplasmic ACC oxidase (ACO) activity, but the signal that prompts the increase in ACO is unknown.

As well as direct responses to increases in ethylene, stress-induced *reductions* in ethylene biosynthesis are important chemical signals that can lead to improvements in plant WUE. Evidence has been provided to show that ABA is required for a maintenance of and/or increase in root extension in water stressed roots, by preventing excess ethylene production, thereby inhibiting its root growth-reducing activity (e.g. Spollen *et al.*, 2000). This occurs despite the fact that in well-watered roots additions of ABA reduce, rather than increase, growth rates. Controversially, this group have also provided evidence that ABA is required to restore leaf growth to wild-type rates in ABA-deficient *flacca* plants (Sharp *et al.*, 2000), again by suppressing excessive ethylene evolution (which is double wild-type levels in the *flacca* mutant). Again this is despite the wealth of evidence demonstrating ABA's role as a shoot and leaf growth inhibitor (see section 4.1.2). This group have proposed, on the basis of their work, that ethylene, and not ABA, is the major regulator of leaf growth rates in wild-type plants under stress. They propose that the role of ABA is to restore or maintain leaf growth in plants exposed to drying, flooded or compacted soil by preventing excessive ethylene production, instead of, as traditionally accepted, to directly inhibit leaf growth (Sharp, 2002). This is supported by findings of Roberts *et al.* (2002) that increased ABA concentrations help to maintain rather than inhibit the shoot growth of tomato plants subjected to soil compaction.

It is difficult to reconcile this set of findings with the large body of work showing that increases in ABA reduce leaf growth rates (see section 4.1.2). Sharp points out that most previous evidence for ABA's role as a shoot (or root) growth suppresser was determined in shoots (or roots) in which the water status was not affected (see Sharp, 2002). The effect of ABA on growth seems to be inhibitory in plants in which the shoot water status is as yet unaffected by stress at the roots, but promotive in shoots in which water stress has developed, by suppressing ethylene synthesis. The restoration of shoot growth rates by ABA in plants that have accumulated ethylene, and that initially exhibit reduced growth rates, may have functional importance in allowing plants to continue to grow once other detrimental effects of the particular stress being experienced by the plant have been overcome. For example, flooded plants initially close stomata, reduce leaf growth rates and exhibit leaf epinasty, which prevent excessive water loss. These early adaptations to reduced water uptake at the root (mediated by increases in xylem sap pH and increased shoot ethylene emmissions – section 4.2.5) may enable the plant to tolerate the flooding stress enough to resume leaf growth, once ABA starts to be produced in the leaf after a few days of stress. Once ABA has started to accumulate in shoot tissues, a degree of osmotic adjustment may be induced such that cells are more able to tolerate low water potentials that might be generated by a resumption of growth. More experimentation is required to test this hypothesis. It is important to note that the earlier a plant can resume growth during or after stress, the greater is the advantage that it can acquire over its competitors.

It seems likely that ethylene is the main suppresser of shoot and leaf growth under stress conditions in which the production of ABA is restricted, for whatever reason. For example, under both flooding stress and soil compaction stress ethylene production in the shoot is high, and in the case of flooding stress, ABA synthesis in the root and transport to the shoot is decreased. Thus, ethylene rather than ABA probably controls shoot growth and WUE during the initial stages of flooding stress. However if Sharp is correct, the increase in ABA a few days later may restore shoot and leaf growth rates to control levels. Is this likely to also be the case in plants experiencing soil-water deficit stress? This seems unlikely given that soil-water deficits sensitively increase the ABA signal (see section 4.2) and reduce growth in the shoots before these exhibit a reduced water status. In addition growth seems to remain suppressed as the stress develops and shoots become water stressed, often despite the combined presence of both ethylene and ABA. Indeed high ethylene and high ABA are coincident in many plant tissues (Abeles *et al.*, 1992), indicating that it is certainly not always the case that ABA suppresses ethylene biosynthesis. Detailed studies need to be conducted to determine the true sequence of events in plants exposed to soil-water deficits. Work with tomato in the laboratory at Lancaster University has shown that ethylene increases before xylem ABA in the shoots of plants experiencing a gradual soil-drying cycle, and that the reduction in growth coincides with the increase in ethylene (see section 4.1.2). However it is not known whether the subsequent increase in ABA combined with a reduction in shoot water status would sustain slower growth or restore this to approaching well-watered rates. The former seems more likely (and more adaptive), given several lines of evidence that reductions in shoot water status actually increase the sensitivity of both stomatal closure and reductions in leaf growth to xylem ABA (e.g. Tardieu and Davies, 1992, 1993; Dodd and Davies, 1996). In addition preliminary work from our laboratory (Figure 4.5) shows that the ABA synthesis inhibitor sodium tungstate can increase growth rates in *Forsythia* x *intermedia* plants experiencing a range of shoot water potentials (induced by the natural variations in VPD being experienced by the plants). Leaf ABA concentrations were reduced on most sampling occasions, although this was only significant in two out of eight occasions. Ethylene levels were not tested in these experiments, but high concentrations seem to be precluded by the rapid growth rates of these plants. The work provides preliminary evidence to support the fact that ABA is a growth inhibitor rather than a growth promoter in fully-developed light-grown shoots experiencing a range of shoot water potentials (from -0.35 to -1.0 MPa) as soil dries. Direct injection of ABA into the stems of intact well-watered plants of this species reduced leaf growth rates as expected (data not shown).

It may well be the case that ABA promotes shoot growth by suppressing ethylene biosynthesis under some stress conditions, especially when increases in shoot ABA are initially restricted (e.g. during soil flooding), or when shoot

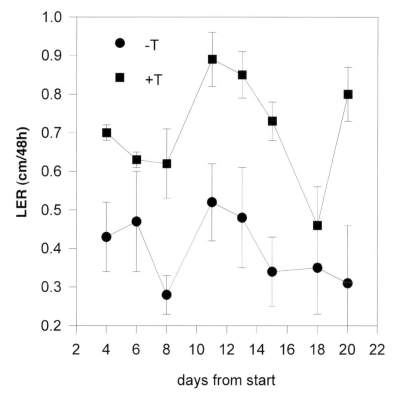

Figure 4.5 The effect of 0.1mM sodium tungstate (T) on the leaf extension rate (LER) of the 3rd leaf from the apex of *Forsythia* x *intermedia* (cv Lynwood) plants of approximately 100cm in height growing in 3l pots. The experiment was conducted in a poly tunnel such that plants were experiencing natural variations in PPFD/VPD, and hence experienced variable shoot water potentials over the course of the experiment (from −0.35 to −0.1 Mpa). Tungstate was applied daily at 5pm with the irrigation supply. Each point represents a mean (n=3) ± SE.

ethylene production is particularly high (e.g. during soil compaction – Hussain *et al.*, 2000). However, it seems more probable that ABA suppresses rather than restores growth in the shoots of mature plants experiencing soil-water deficits, both before and after shoot water deficits are generated (Figure 4.5). Work from Sharp's laboratory showing that ABA restores growth rates in water-stressed shoots may have been a special case which is not applicable to plants growing in the field. The plants were genetically modified to be ABA-deficient, and/or were grown under very artifical conditions (darkness and 100% humidity) in which ABA transport to and within the shoot and/or ABA perception/cross-talk with ethylene signal transduction pathways may have been changed/impaired. Nevertheless, the finding that normal levels of ABA are required to relieve an effect of excess ethylene to reduce growth is important as it demonstrates a novel function for ABA that may be important during the later stages of flooding and compaction stress. The promotion of root growth by ABA-induced ethylene suppression in water stressed roots is likely to be a more universal phenomenon, and to figure significantly in maintaining root growth under soil-water deficit and other stresses.

4.4.2   Nitrate

Stress-induced changes in plant nitrate concentrations have traditionally been considered to be secondary stress-response mechanisms to other signals, that relate to the ability of nitrate to change concentrations of osmotica, or on the basis that its metabolism provides the amino acid building blocks for protein synthesis and growth and development (see McDonald and Davies, 1996). However there is a growing body of evidence that nitrate can act as a signalling molecule in its own right (McDonald and Davies, 1996; Zhang and Forde, 2000). We have already described above how changes in xylem nitrate availability (in response to soil drying, flooding, compaction or fertilisation) may be able to affect stomatal aperture (and leaf growth) by changing the pH balance of the plant (section 4.2). Reduced xylem nitrate concentrations correlate with increases in xylem and apoplastic pH. The pH increase can close stomata and reduce leaf growth rates by preferentially compartmentalis-ing ABA adjacent to its sites of action on guard cells, and by inference on growing cells. Soil drying, soil flooding, soil compaction and soil nitrate deficiencies all sensitively reduce xylem nitrate concentrations, and all but soil compaction induce sensitive increases in xylem pH that pre-empt stomatal closure (section 4.2). We propose that stress-induced changes in nitrate availability can thereby act as primary root-sourced chemical signals to the shoot, that control WUE, via effects on pH and ABA.

However, very recently evidence has been provided that nitrate molecules have a more direct effect on stomatal guard cells than previously realised. Guo *et al.* (2003) showed that *Arabidopsis* guard cells took up nitrate as stomata opened in the light, and that this initiated the well-known cascade of signalling events that leads to stomatal opening (see Assmann and Shimazaki, 1999). Previously chloride or malate were believed to be the only anions that were used by plants to balance potassium uptake and increase the osmolarity of guard cells, thereby causing them to swell and stomata to open. Mutants, deficient in guard cell nitrate uptake transporters exhibited reduced nitrate accumulation and reduced stomatal apertures in the light, and had an enhanced tolerance to drought. It seems, therefore, that nitrate can support stomatal opening, and its absence in guard cells increases the sensitivity of the plant stress response. Because a reduction of root nitrate uptake and transport in the xylem to leaves is a sensitive response to soil drying, flooding and compaction in some species, we propose that nitrate can act directly as a long-distance signal of environmental stress, that affects transpirational water loss and plant WUE. It is proposed that root perturbation-induced reductions in nitrate uptake and transport from roots to shoots act as negative chemical messages that cause direct reductions in stomatal aperture, and reductions in the sensitivity of guard cells to other stomatal effectors (e.g. ABA). It is also proposed that nitrate fertilisation (of crops in the field) will reduce the extent to which stresses such as soil drying can impact on the nitrate content of the root, xylem and guard cell, such that the negative message induced is not strong enough to reduce stomatal aperture. These hypotheses would be fairly simple to test.

Thus, both direct and indirect effects of nitrate anions on stomata may be the basis of the often-reported increase in sensitivity of stomata to ABA in nitrate deficient or droughted plants (Schurr *et al.*, 1992; see McDonald and Davies, 1996; Wilkinson and Davies, 2002).

4.4.3   Cytokinins

Cytokinins regulate many cellular processes, but their affect to increase cell division and the effects of this on the promotion of plant growth and development is considered their most traditionally known role. For an overview see Taiz and Zeiger (1998). Defining effects of a cytokinin (the most common of which is *trans*-zeatin) are the induction of cell division in callus cells in the presence of auxin, the promotion of bud or root formation from callus cultures in an appropriate molar ratio to auxin, the delaying of leaf senescence, and the promotion of the expansion of dicot cotyledons (Cleland, 1996). Cytokinins also promote the mobilisation of nutrients into leaves from other parts of the plant. They inhibit cell elongation in stems and roots (at the same concentrations that promote cell division and expansion in dicot leaves), possibly by promoting ethylene biosynthesis. More recently cytokinins have been shown to promote stomatal opening (Incoll and Jewer, 1987a,b).

Root apical meristems are major sites of synthesis of free cytokinins. These move through the xylem into the shoot, usually in the conjugated form of zeatin riboside, and the bioactive free zeatin or its glucoside is released in the leaf (although some may be compartmentalised and stored away from sites of action). Cytokinin synthesis and transport to the shoot occurs continuously at a high rate in well-watered plants. When the soil around the root dries, the cytokinin content of the xylem sap is reduced (Itai and Vaadia, 1971), presumably as a result of reduced synthesis. The reduction in growth-promoting and stomatal-opening cytokinin in the xylem constitutes a negative chemical message to the shoot, that reduces growth and stomatal opening and may induce senescence, thereby impinging on the WUE of the plant. The reduced cytokinin concentration in the root also stimulates root cell expansion (possibly by down-regulating root ethylene production).

Ethylene is not the only plant hormone that interacts with cytokinin and its effects on plant physiology. As described above, the promotion of cell synthesis and cell cycle effects of cytokinins require the additional presence of auxin. The ratio of auxin to cytokinin determines the differentiation of cultured plant tissues into either roots or buds. High ratios promote roots, and low ratios promote buds. Auxin accumulation in and transport from the shoot apex to the root makes the shoot apex a sink for cytokinin from the root, and this may be one of the factors involved in apical dominance. In addition ratios of abscisic acid to cytokinin have been found to be important in controlling plant growth and functioning. Cytokinins in the xylem decrease stomatal sensitivity to xylem ABA in, for example, cotton (Radin et al., 1982) and field-grown almond trees (Fusseder et al., 1992). Where the concentration of cytokinin itself in the xylem does not dramatically change as soil dries, a change in the ABA/cytokinin ratio may be sufficient to regulate plant development and physiology (Radin et al., 1982). More recently, Stoll et al. (2000) found that the reduction in xylem cytokinin concentration induced by soil drying could occur in the absence of a shoot water deficit in grapevines, as the response was also induced under PRD (see section 4.1.2). In this situation stomatal conductance was reduced, while cytokinin concentrations in roots, shoots and xylem sap decreased and xylem ABA concentrations and xylem sap pH increased. Addition of exogenous synthetic cytokinins to the leaves of the PRD plants re-opened the stomata and restored lateral shoot development, indicating that the reduction of cytokinin concentration in the xylem was a primary (although negative) chemical signal to stomata of drying soil in this case.

4.5   Factors affecting stomatal responses to chemical signals

Once a chemical signal has arrived at its target cell, e.g., the stomatal guard cell in the leaf epidermis, the ability of this signal to control stomatal aperture (and thereby WUE) can still be modulated. This is because changes in both the external and in planta environments can directly change guard cell biochemistry, and cause these cells to exist in various states of susceptibility to the in-coming chemical signal (see e.g. Assmann and Shimazaki, 1999). There is much evidence to suggest that the sensitivity of a plant cell to a hormone is at least partially established by the interplay of several hormones (e.g. Trewavas, 1992). For example, a high leaf apoplastic cytokinin or nitrate concentration will make the guard cell less likely to close in response to an ABA signal from the root, as both of these molecules induce responses such as increased potassium concentrations inside guard cells that increase their turgor and oppose the effect of ABA (section 4.4). On the other hand a high CO_2 concentration in the atmosphere will cause the guard cell to be more receptive to an ABA signal from the root, as the two molecules can induce some of the same responses in the guard cell that lead to stomatal closure (Zeiger and Zhu, 1998; see Wilkinson and Davies, 1992). As described above increasing the VPD around the leaves may directly affect guard cell biochemistry, as well as affecting stomatal aperture via an increase in the strength of the ABA signal (section 4.3). The next paragraph describes in detail one particular example of how the external environment can modulate stomatal sensitivity to an incoming ABA signal, and indeed how it can control stomatal aperture (and thereby WUE) independently of this hormone.

At low, chilling temperatures, stomata in isolated strips of *C. communis* epidermis exhibit a reduced sensitivity to ABA, by closing less readily in its presence (Wilkinson *et al.*, 2001). It is assumed that as temperatures are decreased, ABA binding to its receptors in the guard cell plasma membrane (or inside the cell) decreases, and/or the activity of the ion transporters involved in the ABA signal response chain becomes reduced. This explains several lines of evidence in the literature that chill sensitive species exhibit excessively wide stomatal apertures as temperatures decrease below approximately 8°C, and stomata can become 'locked open'. Initially this was somewhat paradoxical because the same low temperatures induced stomatal closure in intact leaves of chill-tolerant species, including *C. communis*, despite the lack of response of stomata to ABA. This is an important response in plants as low temperatures reduce water uptake at the root such that stomatal closure is necessary to maintain shoot water potential and WUE. So if the sensitivity of guard cells to ABA is reduced, how do chilling temperatures induce stomatal closure?

We have provided evidence that calcium is involved in the closing response to low temperatures (Wilkinson and Davies, 2002). Although *C. communis* stomata in isolated epidermal strips floating on pre-chilled media remained open at low temperatures, the closing response to low temperatures seen in the intact plant could be restored by including calcium in the medium (Figure 4.6), at the concentration estimated to exist in the leaf apoplast at room temperature (10^{-5}M – De Silva *et al.*, 1996). When this concentration was supplied to the strips at room temperature, the stomata remained open. Thus, chill-induced stomatal closure occurs via a sensitisation of guard cells to the resting concentration of calcium in the apoplast (Figure 4.6; Wilkinson and Davies, 2001). In other cell types calcium uptake from external solutions is enhanced at chilling temperatures (e.g. Knight *et al.*, 1991), and as we already know (section 4.2.1), an influx of calcium to the guard cell apoplast is an integral part of the stomatal closing mechanism.

Despite guard-cell desensitisation to apoplastic ABA, and the fact that calcium causes the initial stomatal closure when the chilling stress is first sensed, ABA may still have a role in controlling stomatal aperture (and WUE) as the chilling stress progresses (a few hours later). Many studies demonstrate

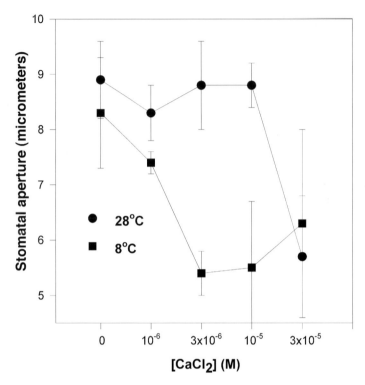

Figure 4.6 The effect of ($CaCl_2$) on stomatal aperture in abaxial epidermal strips of *Commelina communis* L. at two different temperatures. Strips (approx. 1.0×1.5 cm^2 – 35–40 in total) peeled from detached leaves were preincubated on 25cm^3 70mM KCl at room temperature for 3 h, then transferred to a second KCl solution (4.0 cm^3) in 4.0 cm^2 glass watch glasses with 2cm-diameter indentations. These solutions also contained the appropriate concentration of $CaCl_2$, and had been pre-chilled where necessary. Incubations were carried out at the appropriate temperature under a PPFD of 350 μmol m^{-2} s^{-1} for 1 h. 25 stomatal apertures in each of four strips per treatment were measured under a projection microscope immediately after removal from the treatment solution. Each point represents the mean for all four strips ±SE. From Wilkinson *et al.* (2001) this material is copyrighted and reprinted by permission of *Plant Physiology* and the American Society of Plant Biologists, USA.

that ABA protects plant tissues against chilling stress, and show that this hormone can be up-regulated in plants by low temperatures, although this usually occurs after stomata have already been induced to close (see Wilkinson and Davies, 2001). However stomata seem to become attenuated to the chill-induced calcium-sensitisation response after a few hours, so it may be the case that calcium provides the initial means for stomata to close such that plants can conserve water, and the slower build up of ABA serves to maintain these in a closed state in the longer term.

4.5 Conclusion

Water use efficiency can and must be increased in many ways when plants experience stress, and in particular stomatal closure and reductions in leaf growth rates are important regulators of water loss from leaves. Different stresses generate different sets of signalling mechanisms that regulate stomatal aperture and/or leaf growth, and the chemical signalling mechanisms involved in these responses have been examined in this chapter. These include up-regulations of ABA biosynthesis, changes in xylem and apoplastic sap pH that increase the penetration probability of root-, stem- and leaf-sourced ABA to the guard cells in the epidermis and/or the growing cells, synthesis of ABA-conjugates in roots and their transport to shoots where free ABA is released to the leaf apoplast, changes in the synthesis and transport of cytokinins, ACC, ethylene and nitrate to the leaves, and changes in the ability of the target cells to respond to the dose of the hormone that it finally perceives. Chemical signals interact with each other, and with hydraulic signals, and signals sent from the roots and those that arise in the shoot interact, to provide the plant with a unified signal to which to respond. The plant's water use efficiency is thereby a product of multiple chemical signals, which integrate the requirements of both the shoot and the root, reflecting the status of the total environment around the plant.

References

Abeles, F.B., Morgan, P.W. and Saltveit, M.E. Jr. (1992) *Ethylene in Plant Biology*, 2nd edn., Academic Press Inc., San Diego, CA.

Andersen, P.C., Brodbeck, B.V. and Mizell, R.F. (1995) Water stress and nutrient solution-mediated changes in water relations and amino-acids, organic-acids, and sugars in xylem fluid of *Prunus-salicina* and *Lagerstroemia-indica*. *Journal of the American Society for Horticultural Science*, **120**, 36–42.

Andrews, M. (1986) The partitioning of nitrate assimilation between root and shoot of higher plants. *Plant, Cell and Environment*, **9**, 511–519.

Assmann, S.M. and Shimazaki, K-L. (1999) The multisensory guard cell, stomatal responses to blue light and abscisic acid. *Plant Physiology*, **119**, 809–816.

Assmann, S.M., Snyder, J.A. and Lee, Y.R.J. (2000) ABA-deficient (aba1) and ABA-insensitive (abi1–1, abi2–1) mutants of *Arabidopsis* have a wild-type stomatal response to humidity. *Plant, Cell and Environment*, **23**, 387–395.

Bacon, M.A., Wilkinson, S. and Davies, W.J. (1998) pH-regulated leaf cell expansion in droughted plants is abscisic acid dependent. *Plant Physiology*, **118**, 1507–1515.

Bahrun, A., Jensen, C.R., Asch, F. and Mogensen, V.O. (2002) Drought-induced changes in xylem pH, ionic composition, and ABA concentration act as early signals in field-grown maize (*Zea mays* L.). *Journal of Experimental Botany*, **53**, 251–263.

Baier, M. and Hartung, W. (1991) Movement of abscisic acid across the plasma membrane of phloem elements of *Plantago major*. *Journal of Plant Physiology*, **137**, 297–300.

Cameron, R.W.F., Wilkinson, S., Davies, W.J., Harrison-Murray, R.S., Dunstan, D. and Burgess, C. (2004) Regulation of plant growth in container-grown ornamentals through the use of controlled irrigation. *Acta Horticulturae*, **630**, 305–312.

Chaves, M.M., Maroco, J.P. and Pereira, J.S. (2003) Understanding plant responses to drought – from genes to the whole plant. *Functional Plant Biology*, **30**, 239–264.

Chaves, M.M., Pereira, J.S., Maroco, J., Rodrigues, M.L., Ricardo, C.P.P., Osorio, M.L., Carvalho, I., Faria, T. and Pinheiro, C. (2002) How plants cope with water stress in the field. Photosynthesis and growth. *Annals of Botany*, **89**, 907–916.

Cleland, R.E. (1996) Growth substances. In: *Units, Symbols and Terminology for Plant Physiology* (ed F.B. Salisbury), pp. 126–128, Oxford University Press, New York.

Comstock, J. and Mencuccini, M. (1998) Control of stomatal conductance by leaf water potential in *Hymenoclea salsola* (T. & G.), a desert subshrub. *Plant, Cell and Environment*, **21**, 1029–1038.

Cornish, K. and Zeevaart, J.A.D. (1985) Movement of abscisic acid into the apoplast in response to water stress in *Xanthium strumarium*. *Plant Physiology*, **78**, 623–626.

Cutler, A.J. and Krochko, J.E. (1999) Formation and breakdown of ABA. *Trends in Plant Science*, **4**, 472–478.

Daeter, W. and Hartung, W. (1995) Stress-dependent redistribution of abscisic acid (ABA) in *Hordeum vulgare* L. leaves: the role of epidermal ABA metabolism, the tonoplastic transport and the cuticle. *Plant, Cell and Environment*, **18**, 1367–1376.

Daeter, W., Slovik, S. and Hartung, W. (1993) The pH-gradients in the root system and the abscisic acid concentration in xylem and apoplastic saps. *Philosophical Transactions of the Royal Society of London* B, **341**, 49–56.

Davies, W.J. and Zhang, J.H. (1991) Root signals and the regulation of growth and development of plants in drying soil. *Annual Review of Plant Physiology and Plant Molecular Biology*, **42**, 55–76.

Davies, W.J., Bacon, M.A., Thompson, D.S., Sobeih, W. and Gonzalez Rodriguez, L. (2000) Regulation of leaf and fruit growth on plants in drying soil: exploitation of the plants' chemical signalling system and hydraulic architecture to increase the efficiency of water use in agriculture. *Journal of Experimental Botany*, **51**, 1617–1637.

Davies, W.J., Wilkinson, S. and Loveys, B.R. (2002) Stomatal control by chemical signalling and the exploitation of this mechanism to increase water use efficiency in agriculture. *New Phytologist*, **153**, 449–460.

De Silva, D.L.R., Honour, S.J. and Mansfield, T.A. (1996) Estimations of apoplastic concentrations of K^+ and Ca^{2+} in the vicinity of stomatal guard cells. *New Phytologist*, **134**, 463–469.

Dietz, K-J, Sauter, A., Wichert, K., Messdaghi, D., and Hartung, W. (2000) Extracellular β-glucosidase activity in barley involved in the hydrolysis of ABA glucose conjugate in leaves. *Journal of Experimental Botany*, **51**, 937–944.

Dodd, I.C. and Davies, W.J. (1996) The relationship between leaf growth and ABA accumulation in the grass leaf elongation zone. *Plant, Cell and Environment*, **19**, 1047–1056.

Dodd, I.C., Tan, L.P. and He, J. (2003) Do increases in xylem sap pH and/or ABA concentration mediate stomatal closure following nitrate deprivation? *Journal of Experimental Botany*, **54**, 1281–1288.

Ehrlinger, J.R. and Cooper, T.A. (1988) Correlations between carbon isotope ratio and microhabitat in desert plants. *Oecologia*, **76**, 562–566.

Else M.A., Davies, W.J., Croker, S. and Jackson, M.B. (1996) Stomatal closure in flooded tomato plants involves abscisic acid and a chemically unidentified anti-transpirant in xylem sap. *Plant Physiology*, **112**, 239–247.

Else, M.A., Davies, W.J., Malone, M. and Jackson, M.B. (1995) A negative hydraulic message from oxygen-deficient roots of tomato plants? Influence of soil flooding on leaf water potential, leaf expansion and the synchrony between stomatal conductance and root hydraulic conductivity. *Plant Physiology*, **109**, 1017–1024.

Else, M.A. and Jackson, M.B. (1998) Transport of 1–aminocyclopropane-1–carboxylic acid in the transpiration stream of tomato (*Lycopersicon esculentum*) in relation to foliar ethylene production and petiole epinasty. *Australian Journal of Plant Physiology*, **25**, 453–458.

Fuchs, E.E. and Livingstone, N.J. (1996) Hydraulic control of stomatal conductance in Douglas fir [*Pseudotsuga menziesii* (Mirb) Franco] and alder [*Alnus rubra* (Bong)] seedlings. *Plant, Cell and Environment*, **19**, 1091–1098.

Fusseder, A., Wartinger, A., Hartung, W., Schulze, E-D. and Heilmeier, H. (1992) Cytokinins in the xylem sap of desert grown almond (*Prunus dulcis*) trees: Daily courses and their possible interactions with abscisic acid and leaf conductance. *New Phytologist*, **122**, 45–53.

Gerendas, J. and Schurr, U. (1999) Physicochemical aspects of ion relations and pH regulation in plants – a quantitative approach. *Journal of Experimental Botany*, **50**, 1101–1114.

Glinka, Z. (1980) Abscisic acid promotes both volume flow and ion release to the xylem in sunflower roots. *Plant Physiology*, **65**, 537–540.

Glinka, Z. and Reinhold L. (1971) Abscisic acid raises the permeability of plant cells to water. *Plant Physiology*, **48**, 103–105.

Gollan, T., Passioura, J.B. and Munns, R. (1986) Soil-water status affects the stomatal conductance of fully turgid wheat and sunflower leaves. *Australian Journal of Plant Physiology*, **13**, 459–464.

Gollan, T., Schurr, U. and Schulze, E.D. (1992) Stomatal response to drying soil in relation to changes in xylem sap composition of *Helianthus annuus*. I. The concentration of cations, anions, amino acids in, and pH of, the xylem sap. *Plant, Cell and Environment*, **15**, 551–559.

Gowing, D.J., Davies, W.J. and Jones, H.G. (1990) A positive root-sourced signal as an indicator of soil drying in apple, *Malus* x *domestica* Borkh. *Journal of Experimental Botany*, **41**, 1535–1540.

Grossmann, K. and Hansen, H. (2001) Ethylene-triggered abscisic acid: a principle in plant growth regulation? *Physiologia Plantarum*, **113**, 9–14.

Guo, F-Q., Young, J. and Crawford, N.M. (2003) The nitrate transporter AtNRT1.1 (CHL1) functions in stomatal opening and contributes to drought susceptibility in *Arabidopsis*. *The Plant cell*, **15**, 107–117.

Hartung, W. (1983) The site of action of abscisic acid at the guard cell plasmalemma of *Valerianella locusta*. *Plant, Cell and Environment*, **6**, 427–428.

Hartung, W. and Radin, J.W. (1989) Abscisic acid in the mesophyll apoplast and in the root xylem sap of water-stressed plants: The significance of pH gradients. *Current Topics In Plant Biochemistry and Physiology*, **8**, 110–124.

Hartung, W., Kaiser, W.M. and Burschka, C. (1983) Release of ABA from leaf strips under osmotic stress. *Zeitschrift für Pflanzenphysiologie*, **112**, S131–S138.

Hartung, W., Radin, J.W. and Hendrix, D.L. (1988) Abscisic acid movement into the apoplastic solution of water stressed cotton leaves. Role of apoplastic pH. *Plant Physiology*, **86**, 908–913.

Hartung, W., Wilkinson, S. and Davies, W.J. (1998) Factors that regulate abscisic acid and its concentration in the xylem. *Journal of Experimental Botany*, **49**, 361–367.

Henson, I.E., Jenson, C.R. and Turner, N.C. (1989) Leaf gas exchange and water relations of lupins and wheat. I. Shoot responses to soil water deficits. *Australian Journal of Plant Physiology*, **16**, 401–413.

Hetherington, S.E., Smillie, R.M. and Davies, W.J. (1998) Photosynthetic activities of vegetative and fruiting tissues of tomato. *Journal of Experimental Botany*, **49**, 1173–1181.

Hoffmann, B. and Kosegarten, H. (1995) FITC-dextran for measuring apoplast pH and apoplastic pH gradients between various cell types in sunflower leaves. *Physiologia Plantarum*, **95**, 327–335.

Holbrook, N.M., Shashidhar, V.R., James, R.A. and Munns, R. (2002) Stomatal control in tomato with ABA-deficient roots: response of grafted plants to soil drying. *Journal of Experimental Botany*, **53**, 1503–1514.

Hussain, A., Black, C.R., Taylor, I.B. and Roberts, J.A. (2000) Does an antagonistic relationship between ABA and ethylene mediate shoot growth when tomato plants encounter compacted soil? *Plant, Cell and Environment*, **23**, 1217–1226.

Imber, D. and Tal, M. (1970) Phenotypic reversion of flacca, a wilty mutant of tomato, by abscisic acid. *Science*, **169**, 592–593.

Incoll, L.D. and Jewer, P.C. (1987a) Cytokinins and stomata. In: *Stomatal Function* (eds E. Zeiger, G.D. Farquhar and I.R. Cowan), pp. 281–292, Stanford, Stanford University Press.

Incoll, L.D. and Jewer, P.C. (1987b) Cytokinins and the water relations of whole plants. In: *Cytokinins: Plant Hormones in Search of a Role, Monograph 14* (eds R. Horgan and B. Jeffcoat), pp. 85–97, British Plant Growth Regulator Group, Bristol.

Itai, C. and Vaadia, Y. (1971) Cytokinin activity in water-stressed shoots. *Plant Physiology*, **47**, 87–90.

Jackson, M.B. (1993) Are plant hormones involved in root to shoot communication? *Advances in Botanical Research*, **19**, 103–187.

Jackson, M.B. (1991) Regulation of water relationships in flooded plants by ABA from leaves, roots and xylem sap. In: *Abscisic acid: Physiology and Biochemistry* (eds W.J. Davies and H.G. Jones), pp. 217–226, Bios Scientific, Oxford.

Jackson, M.B., Saker, L.R., Crisp, C.M., Else, M.A. and Janowiak, F. (2003) Ionic signalling from roots to shoots of flooded tomato plants in relation to stomatal closure. *Plant and Soil*, **253**, 103–113.

Jia, W.S. and Zhang, J.H. (1997) Comparison of exportation and metabolism of xylem-delivered ABA in maize leaves at different water status and xylem sap pH. *Plant Growth Regulation*, **21**, 43–49.

Jokhan, A.D., Harink, R.J. and Jackson, M.B. (1999) Concentration and delivery of abscisic acid in xylem sap are greater at the shoot base than at a target leaf nearer to the shoot apex. *Plant Biology*, **1**, 253–260.

Jones, H.G. (1993) Drought tolerance and water use efficiency. In: Water Deficits: Plant Responses from Cell to Community (eds J.A.C. Smith and H. Griffiths), pp. 193–203, BIOS Scientific, Oxford.

Kaiser, W.M. and Hartung, W. (1981) Uptake and release of abscisic acid by isolated photoautotrophic mesophyll cells, depending on pH gradients. *Plant Physiology*, **68**, 202–206.

Kirkby, E.A. and Armstrong, M.J. (1980) Nitrate uptake by roots as regulated by nitrate assimilation in the shoot of castor oil plants. *Plant Physiology*, **65**, 286–290.

Knight, M.R., Campbell, A.K., Smith, S.M. and Trewavas, A.J. (1991) Transgenic plant aequorin reports the effects of touch and cold-shock and elicitors on cytoplasmic calcium. *Nature*, **352**, 524–526.

Kosegarten, H.U., Hoffmann, B. and Mengel, K. (1999) Apoplastic pH and Fe^{3+} reduction in intact sunflower leaves. *Plant Physiology*, **121**, 1069–1079.

Lee, S.H. and Reid, D.M. (1997) The role of endogenous ethylene in the expansion of *Helianthus annuus* leaves. *Canadian Journal of Botany*, **75**, 501–508.

Liang, J., Zhang, J. and Wong, M.H. (1997) How do roots control xylem sap ABA concentration in response to soil drying? *Plant Cell Physiology*, **38**, 10–16.

Lips, S.H. (1997) The role of inorganic nitrogen ions in plant adaptation processes. *Russian Journal of Plant Physiology*, **44**, 421–431.

Liu, F.L., Jensen. C.R. and Andersen, M.N. (2003) Hydraulic and chemical signals in the control of leaf expansion and stomatal conductance in soybean exposed to drought stress. *Functional Plant Biology,* **30**, 65–73

Loveys, B.R. (1984) Diurnal changes in water relations and abscisic acid in field grown *Vitis vinifera* cultivars III. The influence of xylem-derived abscisic acid on leaf gas exchange. *New Phytologist*, **98**, 563–573.

McAinsh, M.R., Brownlee, C. and Hetherington, A.M. (1997) Calcium ions as second messengers in guard cell signal transduction. *Physiologia Plantarum*, **100**, 16–29.

McDonald, A.J.S. and Davies, W.J. (1996) Keeping in touch: responses of the whole plant to deficits in water and nitrogen supply. *Advances in Botanical Research*, **22**, 229–300.

Mengel, K., Planker, R. and Hoffmann, B. (1994) Relationship between leaf apoplast pH and iron chlorosis of sunflower (*Helianthus annuus* L.). *Journal of Plant Nutrition*, **17**, 1053–1065.

Mingo, D.M., Bacon, M.A. and Davies, W.J. (2003) Non-hydraulic regulation of fruit growth in tomato plants (*Lycopersicon esculentum* cv. Solairo) growing in drying soil. *Journal of Experimental Botany*, **54**, 1205–1212.

Munns, R and Sharp, R.E. (1993) Involvement of ABA in controlling plant growth in soils at low water potential. *Australian Journal of Plant Physiology*, **20**, 425–437.

Patonnier, M.P., Peltier, J.P. and Marigo, G. (1999) Drought-induced increase in xylem malate and mannitol concentrations and closure of *Fraxinus excelsior* L. stomata. *Journal of Experimental Botany*, **50**, 1223–1229.

Radin, J.W., Parker, L.L. and Guinn, G. (1982) Water relations of cotton plants under nitrogen deficiency. V. Environmental control of abscisic acid accumulation and stomatal sensitivity to abscisic acid. *Plant Physiology*, **70**, 1066–1070.

Ratcliffe, R.G. (1997) In vivo NMR studies of the metabolic response of plant tissue to anoxia. *Annals of Botany*, 79, 39–48.

Raven, J.A. (2002) Selection pressures on stomatal evolution. *New Phytol*, **153**, 371–386.

Raven, J.A. and Smith, F.A. (1976) Nitrogen assimilation and transport in vascular land plants in relation to intrcellular pH regulation. *New Phytol*, **76**, 415–431.

Roberts, J.A., Hussain, A., Taylor, I.B. and Black C.R. (2002) Use of mutants to study long-distance signalling in response to compacted soil. *Journal of Experimental Botany*, **53**, 45–50.

Saliendra, N.Z., Sperry, J.S. and Comstock, J.P. (1995) Influence of leaf water status on stomatal response to humidity, hydraulic conductance, and soil drought in *Betula occidentalis*. *Planta*, **196**, 357–366.

Sauter, A. and Hartung, W. (2000) Radial transport of abscisic acid conjugates in maize roots: its implication for long distance stress signals. *Journal of Experimental Botany*, **51**, 929–935.

Sauter, A. and Hartung, W. (2002) The contribution of internode and mesocotyl tissues to root to shoot signalling of abscisic acid. *Journal of Experimental Botany*, **53**, 297–302.

Sauter, A., Dietz, K-J. and Hartung, W. (2002) A possible stress physiological role of abscisic acid conjugates in root-to-shoot signalling. *Plant, Cell and Environment*, **25**, 223–228.

Schurr, U., Gollan, T. and Schulze, E.D. (1992) Stomatal response to drying soil in relation to changes in the xylem sap composition of *Helianthus annuus* 2. Stomatal sensitivity to abscisic acid imported from the xylem sap. *Plant Cell and Environment*, **15**, 561–567.

Schurr, U. and Schulze, E-D. (1996) Effects of drought on nutrient and ABA transport in *Ricinus communis*. *Plant, Cell and Environment*, **19**, 665–674.

Sharp, R.E. (2002) Interaction with ethylene: changing views on the role of abscisic acid in root and shoot growth responses to water stress. *Plant, Cell and Environment*, **25**, 211–222.

Sharp, R.E., LeNoble M.E., Else, M.A., Thorne, E.T. and Gherardi, F. (2000) Endogenous ABA maintains shoot growth in tomato independently of effects on plant water balance. Evidence for the involvement of ethylene. *Journal of Experimental Botany*, **51**, 1575–1584.

Slovik, S., Daeter, W. and Hartung, W. (1995) Compartmental redistribution and long distance transport of abscisic acid (ABA) in plants as influenced by environmental changes in the rhizosphere. A biomathematical model. *Journal of Experimental Botany*, **46**, 881–894.

Slovik, S. and Hartung, W. (1992) Compartmental distribution and redistribution of abscisic acid in intact leaves. II. Model analysis. *Planta*, **187**, 26–36.

Spollen, W.G., LeNoble, M.E., Samuels, T.D., Bernstein, N. and Sharp, R.E. (2000) Abscisic acid accumulation maintains maize primary root elongation at low water potentials by restricting ethylene production. *Plant Physiology*, **122**, 967–976.

Stahlberg, R., Van Volkenburgh, E. and Cleland, R.E. (2001) Long-distance signalling with *Coleus* x *hybrides* leaves; mediated by changes in intra-leaf CO_2? *Planta*, **213**, 342–351.

Stoll, M., Loveys, B. and Dry, P. (2000) Hormonal changes induced by partial rootzone drying of irrigated grapevine. *Journal of Experimental Botany*, **51**, 1627–1634.

Taiz, L. and Zeiger, E. (1998) *Plant Physiology*, 2nd edn, Sinauer Associates Inc., USA.

Tardieu, F. and Davies, W.J. (1992) Stomatal response to abscisic acid is a function of current plant water status. *Plant Physiology*, **98**, 540–545.

Tardieu, F. and Davies, W.J. (1993) Integration of hydraulic and chemical signalling in the control of stomatal conductance and water status of droughted plants. *Plant, Cell and Environment*, **16**, 341–349.

Thomas, D.S. and Eamus, D. (2002) Seasonal patterns of xylem sap pH, xylem [ABA], leaf water potential and stomatal conductance of 6 evergreen and deciduous Australian savanna tree species. *Australian Journal of Botany*, **50**, 229–236.

Thompson, D.S. (2001) Extensiometric determination of the rheological properties of the epidermis of growing tomato fruit. *Journal of Experimental Botany*, **52**, 1291–1301.

Thompson, D.S., Wilkinson, S., Bacon, M.A. and Davies, W.J. (1997) Multiple signals and mechanisms that regulate leaf growth and stomatal behaviour during water deficit. *Physiologia Plantarum*, **100**, 303–313.

Trejo, C.L., Clephan, A.L. and Davies, W.J. (1995) How do stomata read abscisic acid signals? *Plant Physiology*, **109**, 803–811.

Trejo, C.L., Davies, W.J. and Ruiz, L.D.P. (1993) Sensitivity of stomata to abscisic acid an effect of the mesophyll. *Plant Physiology*, **102**, 497–502.

Trewavas, A.J. (1992) Growth-substances in context a decade of sensitivity. *Biochemical Society Transactions*, **20**, 102–108.

Wilkinson, S. (1999) pH as a stress signal. *Plant Growth Regulation*, **29**, 87–99.

Wilkinson, S. and Davies, W.J. (1997) Xylem sap pH increase: a drought signal received at the apoplastic face of the guard cell that involves the suppression of saturable abscisic acid uptake by the epidermal symplast. *Plant Physiology*, **113**, 559–573.

Wilkinson, S. and Davies, W.J. (2002) ABA-based chemical signalling: the co-ordination of responses to stress in plants. *Plant, Cell and Environment*, **25**, 195–210.

Wilkinson, S., Clephan, A.L. and Davies, W.J. (2001) Rapid low temperature-induced stomatal closure occurs in cold-tolerant *Commelina communis* L. leaves but not in cold-sensitive *Nicotiana rutica* L. leaves, via a mechanism that involves apoplasic calcium but not abscisic acid. *Plant Physiology*, **126**, 1566–1578.

Wilkinson, S., Corlett, J.E., Oger, L. and Davies, W.J. (1998) Effects of xylem pH on transpiration from wild-type and *flacca* tomato leaves: a vital role for abscisic acid in preventing excessive water loss even from well-watered plants. *Plant Physiology*, **117**, 703–709.

Zeiger, E. and Zhu, J. (1998) Role of zeaxanthin in blue light photoreception and the modulation of light CO_2 interactions in guard cells. *Journal of Experimental Botany*, **49**, 433–442.

Zacarias, L. and Reid, M.S. (1990) Role of growth regulators in the senescence of *Arabidopsis thaliana* leaves. *Physiologia Plantarum*, **80**, 549–554.

Zhang, H.M. and Forde, B.G. (2000) Regulation of *Arabidopsis* root development by nitrate availability. *Journal of Experimental Botany*, **51**, 51–59.

Zhang, J. and Davies, W.J. (1989) Abscisic acid produced in dehydrating roots may enable the plant to measure the water status of the soil. *Plant, Cell and Environment*, **12**, 73–81.

Zhang, J. and Davies, W.J. (1990) Does ABA in the xylem control the rate of leaf growth in soil dried maize and sunflower plants? *Journal of Experimental Botany*, **41**, 1125–1132.

Zhang, S.Q. and Outlaw, W.H. Jr. (2001a) The guard-cell apoplast as a site of abscisic acid redistribution in *Vicia faba* L. *Plant, Cell and Environment*, **24**, 347–356.

Zhang, S.Q. and Outlaw, W.H. Jr. (2001b) Abscisic acid introduced into the transpiration stream accumulates in the guard-cell apoplast and causes stomatal closure. *Plant, Cell and Environment*, **24**, 1045–1054.

5 Physiological approaches to enhance water use efficiency in agriculture: exploiting plant signalling in novel irrigation practice

B.R. Loveys, M. Stoll and W.J. Davies

5.1 Introduction

More than 70% of freshwater used throughout the world is for agriculture. In 1900, some 600 km^3 of water was used for agriculture and by 2000 this figure had increased to 3100 km^3. The bulk of this water is used for irrigating approximately 240 million ha of crops. On average, irrigated agricultural production is two and a half times as productive as rain-fed. Water for irrigation can come from a variety of sources, including established water courses, rainwater catchments or from groundwater. Changes in these sources suggest that current irrigation practices are often unsustainable. For example, flows in many of the world's major rivers such as the Nile, Ganges, Yellow River, Colorado, Murray/Darling are at times reduced to zero as a direct result of irrigation diversions. The situation is even worse with ground water supplies, with significant falls in ground water reserves being recorded in most major irrigation areas. There is therefore significant pressure on irrigators to bring about improvements in the efficiency of water use as current irrigation practices are clearly unsustainable.

Water use efficiency (WUE) of crop plants may be defined as the amount of water taken into a plant in order to produce a unit of output, where output may be total biomass or biomass assigned to a harvestable crop. This definition recognizes the inevitable link between transpiration and carbon fixation and it is this nexus that may be subject to manipulation through changes to the physiology of the plant that may result from altered stomatal conductance or carbon fixation. Assigning actual values to WUE as defined above can be quite difficult as the determination of how much water passes through the plant is not straightforward, relying on direct measurement of transpiration with sap flow sensors (Granier, 1985; Green and Clothier, 1988), extrapolation from single leaf gas exchange analysis, whole canopy gas exchange measurement or perhaps through a subtractive process, having determined other components of the cropping system water balance such as rainfall, irrigation, drainage, runoff, evaporation and storage. These measurements may be further complicated by diurnal and seasonal fluctuations that may be difficult to assess. Using weighing lysimeters Meyer *et al.* (1987) and Meyer *et al.* (1990) determined that it takes between 715 and 750 l of water to produce a kilogram of dry wheat or soybean grain. For field production the efficiency falls considerably due to losses associated with irrigation and harvest, and for paddy rice Meyer (1994) estimates that approximately 1550 l of water are required for each kg of dry grain produced. This type of analysis suggests another definition of WUE that is more readily measurable and meaningful to irrigators. That is units of product produced for a unit of irrigation water (t/ML). This definition may be modified to include rainfall if this is a significant contributor to water input during the growing season. If we move further back in the irrigation water supply chain efficiencies fall further due to evaporation and transmission losses, so that often less than 30% of stored water is transmitted to the crop (Wolters, 1992).

Thus, growing plants for food requires large quantities of water and improvements in the efficiency of use of this water may come in a number of ways. Engineering measures may improve water delivery systems, while attention to the nature of the crop plants may improve WUE as initially defined. Water costs will ultimately be the most important driver of reform in irrigation industries that have often been established with significant government subsidy. In Australia, for example, it has been estimated that agriculture, which uses more than 70% of all water supplied, and accounts for some 20% of supply costs, paid only 5% of the water industry's revenue in 1992–93 (Anon., 1999). This culture of low water cost has not been conducive to improvements in efficiency of water use and within any particular irrigation district there can be huge differences in water use and crop yields, resulting in large differences in WUE, expressed as irrigation water use per unit of crop produced. For example, Skewes and Meissner (1997) showed WUE varied ten-fold for winegrape production in the Riverland of South Australia, from 1.07 to 10.15 t/ML. Similar differences were identified for citrus production in the same district. As the cost of irrigation water increases and availability decreases due to over exploitation or the need to return water to river systems for the maintenance of environmental flows, pressures will increase for improvements in efficiency.

Crop plant WUE can be influenced by a number of environmental and management variables including temperature, humidity, water, nutrition, canopy structure, genetic makeup of rootstocks and scions. This chapter will focus on how an understanding of the physiological mechanisms by which some of these variables impact on plant function can be used to improve crop WUE.

5.2   Understanding the ways in which WUE can be influenced

5.2.1   Stomatal regulation of gas exchange

Terrestrial plants inhabit a fundamentally hostile environment where the leaves are subjected to a high evaporative demand, particularly during the middle of the day. This can generate very high rates of water loss and unless these are regulated, the plant can lose its entire fresh weight in only a few hours. Reduction of water content of more that 20% or so will be catastrophic for most plants, as they cannot recover from these deficits without substantial lesions developing which will potentially reduce growth, functioning and economic value. Fortunately, plants can exert some control over water loss from leaves via the narrowing of stomatal apertures, and this is a commonly observed plant response to reduction in water availability.

One important question for those interested in artificial manipulation of water loss via modification in stomatal behaviour is whether an effective stomatal closing agent will actually save the plant (and the farmer!) any water. There are many naturally-occurring and artificial stomatal-active compounds available for use. For example, we have known for many years that the plant stress hormone abscisic acid (ABA) can control stomatal behaviour when applied to leaves at concentrations as low as 10^{-10} M. This kind of plant response should reduce water loss and this may be an effective water-saving strategy as the great majority of water lost as transpiration is simply an inevitable consequence of evaporation from the moist cell surfaces surrounding the substomatal cavities, and this water flux contributes nothing to the growth or functioning of the plant. The general consensus is that transpiration could be cut to only a few percent of the potential value determined by the plant's microclimate, without affecting plant functioning. Assuming that such a manipulation is possible, one impact will almost certainly be the heating up of the leaf. In relatively still air, leaf temperatures of non-transpiring leaves may be as much as 20°C above ambient air temperature and much early writing suggested that one of the 'purposes' of transpiration was to cool leaves under hot and dry conditions to avoid metabolic damage by high temperatures. While there is little doubt that high leaf temperatures can enhance the possibility of oxidative stress, particularly in droughted plants, many plants will routinely experience high leaf temperatures as a result of mid-day stomatal closure. Patterns of stomatal behaviour seem to have evolved to prevent potential desiccation injury during periods of high evaporative demand, rather than to prevent high temperature stress. The plant has developed a range of protective measures against high temperature-induced oxidative stress and these often result in a reversible protective down-regulation of leaf functioning rather than damage *per se*.

It seems then that large unregulated fluxes of water are not essential to plant functioning and that without deleterious consequences, water could be saved by manipulation of stomatal behaviour. However, narrowing of stomatal aperture will eventually reduce the uptake of CO_2 by the leaf and this is likely to impact on the yield of many crops where yield is carbon limited. Theory suggests that partial closure of stomata can initially restrict water loss without a significant restriction in the rate of carbon gain, but in reality transpiration will often be driven at the equilibrium rate determined by the radiation load on the leaf (Jones, 1985). This is because narrowing of stomatal aperture may initially reduce water loss but this will result in an increase in leaf temperature which itself will drive transpiration harder to cancel out the impact of reduced conductance (Jarvis and McNaughton, 1985). This will be the case particularly with short, smooth crops which are uncoupled from the turbulence of the bulk atmosphere. Better stomatal control of water loss can be achieved with taller and rougher vegetation which is better coupled to the atmosphere.

The successful use of artificial antitranspirants in the field is dependent upon the degree of coupling between vegetation and atmosphere. The apparent lack of effect in the field of several stomatal-active compounds under test in the 1970s and 1980s has subsequently been attributed to poor coupling between the atmosphere and short vegetation. Breeding for reduced stomatal size and number in grass species as a means of reducing water loss and increasing water use efficiency has been largely ineffective for similar reasons. Both of these endeavours had proved to be quite successful when tested with individual plants that were well coupled to the turbulent atmosphere of the controlled environment chamber and results of this kind highlight the importance of conducting experiments with crops under realistic climatic conditions and preferably in the field.

5.2.2 Regulation of plant development and functioning

While application of stomatal-regulating compounds under appropriate climatic conditions can increase water use efficiency, to achieve a worthwhile saving of water, use of these methods will require repeated application of chemicals throughout the season. Alternatives to this kind of management are to select genotypes with high water use efficiency (see Bacon, this volume) or even select genotypes that accumulate high concentrations of stomatal regu-lating chemicals. Quarrie and co-workers (Quarrie, 1991) have investigated the water use and growth of lines selected for their capacity to accumulate high concentrations of ABA. They have stressed the importance of doing these selections in material growing in the field since high-ABA lines selected in the glasshouse may have the capacity to produce large amounts of ABA, but they may not always do this due to compensating responses which are apparent under field conditions. Even in plants expressing high-ABA accumulation in the field, impact on water use efficiency is not always predictable. This suggests that if endogenous ABA balance does regulate water use by droughted plants, its activity may not depend on substantial accumulation of the hormone. We know now (see below) that subtle redistribution of ABA between different compartments of the plant can generate substantial and important changes in plant growth and functioning without the necessity for the accumulation of high concentrations of ABA and that substantial accumulations may only be necessary to promote the development of desiccation resistance mechanisms which may be important in generating resistance to very severe plant water deficits.

A considerable amount of work over the last 30 years or so has revealed much of the complexity of the plant's regulation of development, growth and functioning under drought. It is now clear that an important component of this regulation is the chemical control of a multitude of plant processes and we have come to view these chemicals as signals of changing resource availability to key organs of the plant. Such changes can have important local effects but they will also have implications for functioning of plant parts which are remote from the site of stress imposition, and it is clear that the plant has evolved signalling mechanisms to communicate important information to remote sites. These signals may be hydraulic, e.g. the well-reported reduction in water transport through the plant as the soil dries, but it has also become very clear that both local and long-distance chemical signalling can be potent means of regulation in plants experiencing environmental stress. It follows then that some understanding of these signalling processes should open the way to their artificial manipulation in agriculture.

It is now well known that many soil perturbations will modify long-distance signalling mechanisms that co-ordinate shoot adaptive responses. For example, soil compaction will directly affect root growth and functioning. Flooding of the soil affects root metabolism and function, often as a result of impeded gas exchange that imposes oxygen shortage and accumulations of gases such as CO_2 and ethylene in the rhizosphere. Soil drying exerts a range of influences on roots, some of which may be a result of a change in soil strength, modified nutrient availability or modification to the gaseous environment, but mostly through changes in the water relations of the cortical cells and the cells in the stele. The consequences of these changes are not restricted to roots. Their influence soon extends to aerial parts as root functions upon which shoots depend become increasingly modified. Plant responses to soil compaction and soil drying can be extremely finely tuned to changes in soil conditions, with changes in signal intensity apparent when only a relatively few roots are encountering modification in soil properties. The sensitivity of the chemical signalling response to soil drying will presumably be a function of genotype, although there is little clear information on this point. It seems very likely then that the capacity of different rootstocks to synthesise hormones will impact on signalling and that this may be a way for horticulturalists to exploit the power of signalling to increase the efficiency of

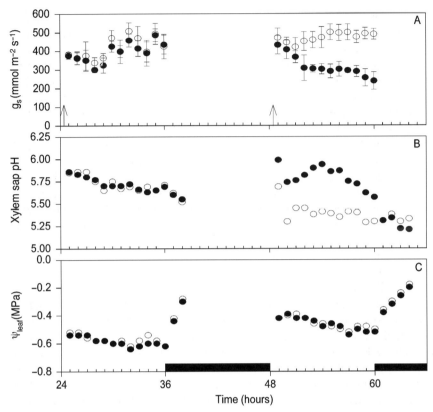

Figure 5.1 Effects of drying half the roots of tomato plants rooted in soil. Stomatal conductance (A), xylem sap pH (B) and leaf water potential (C) of plants watered daily on one (●) or both (○) sides of the split-pot are shown. In (A), points are means ± S.E. of 5 leaflets per leaf. Dark shading on the time axis indicates the night period while arrows indicate when the plants were watered. Two days in the soil drying cycle are shown (days 2 and 3). Effects of soil drying on stomatal behaviour are apparent on day 3 and this response is co-incident with an alkalinisation of the xylem sap. Note that shoot water status is regulated by these responses (W.Y. Sobeih, I.C. Dodd, M.A. Bacon, W.J. Davies, unpublished data).

water use. In addition to this, stomatal responses to drying only a few percent of the root mass are apparent when plants are growing at low VPD. High transpiration fluxes in the field can dilute out the signalling molecules and de-sensitise the response (Tardieu *et al.*, 1991).

Figure 5.1 shows how only very mild soil drying will generate a chemical signal (increased xylem sap pH) which apparently regulates stomatal behaviour such that plants watered with 50 per cent less water than control plants (on half of the root system) show comparable shoot water status to control plants. For tomato plants in drying soil, the sensitivity of production of chemical signals varies as follows: pH > ethylene > ABA accumulation, and all of this can occur before shoot water relations are perturbed. This is not to say that water uptake is not reduced by soil drying; this is almost certain to be the case. But one or more of these chemical and hydraulic signals has reduced stomatal conductance and transpiration so that shoot water status is regulated. This is a markedly different situation to that regularly suggested in general water relations texts, namely that stomatal behaviour is regulated by shoot water relations, rather than the reverse. We show in Figure 5.1 that in a situation where irrigation water is in short supply, we can water half of the root system of a tomato plant and we can retain control over shoot water balance. We will show below that this is important if we are to sustain high quality yield under deficit irrigation.

Figure 5.1 shows that the plant receiving water on only half of its roots exhibits almost perfect isohydric behaviour. The coincidence in time between the increase in xylem sap pH and the reduction in leaf conductance is highly suggestive of stomatal control via a pH signal, but it is necessary to perturb this relationship, for example via the application of pH buffers, before we can be confident of cause and effect.

Stomatal closure is a common response to root zone stresses including soil drying, soil flooding and soil compaction. Early work (reviewed in Davies *et al.*, 1994) implicated the plant hormone ABA in long-distance signalling in response to all of these stresses. If ABA is removed from xylem sap of droughted maize plants, then the antitranspirant activity of this sap is also removed. ABA concentration in droughted xylem sap and the degree of transpiration limitation that this causes is consistent with the degree of stomatal response to a xylem sap with artificial ABA added. Despite this powerful evidence, some studies fail to show an increase in sap ABA concentration that is coincident with observed limitations in conductance, and this has led some to propose the existence of additional, as yet unidentified antitranspirants in the sap. While this may be the case, there is an alternative explanation which also involves the action of ABA.

Recent studies have highlighted the importance of xylem sap pH in modulating stomatal response to xylem-borne signals generated by roots in drying soils (Wilkinson and Davies, 1997; Wilkinson *et al.*, 1998). Xylem sap becomes significantly more alkaline in response to soil drying and the proposal is that the mildly acidic hormone partitions into alkaline compartments in well-watered plants where the sap is comparatively acidic (pH 6) but as sap and apoplastic pH increases, then pH gradients are disrupted, ABA resides for longer in the apoplast and more penetrates to the sites of action on the guard cells. An increase in xylem sap pH can therefore promote closure of stomata through increased ABA penetration to guard cells without the necessity for extra ABA accumulation. Indeed, even well-watered plants contain enough ABA to limit stomatal opening if the compound can penetrate to the guard cells. Xylem/apoplastic pH changes, believed to be leaf-sourced, also occur in response to changes in atmospheric conditions such as PPFD/VPD and correlate with reductions in stomatal aperture and leaf growth rate (Wilkinson and Davies, 2002; Davies *et al.*, 2002).

How these pH changes arise is not known. Reductions in H^+-ATPase activity coupled with activation of mechano-sensitive Ca^{++} channels that alter ion transport systems such as K^+-H^+ symport would be expected to alter the pH of xylem sap in stressed plants (Netting, 2000). Perturbations in the ionic composition of xylem sap will probably influence its pH. Reduced nitrate concentration and associated changes in organic acid components in particular may cause a pH shift towards alkalinity (Wilkinson and Davies, 2002). Reduced nitrate supply rapidly induces nitrate reductase activity (NRA) to shift from the leaf to the root in many species. Products of NRA, such as malate, are then loaded in to the xylem and increase sap pH (Patonnier *et al.*, 1999). Wilkinson and Davies (2002) proposed that soil-drying induced xylem sap alkalinisation may result from the switch in NRA from shoots to roots and the subsequent loading of malate into the xylem.

The anti-transpirant activity of the pH increase is indirect and requires well-watered, wildtype ABA concentrations (Wilkinson *et al.*, 1998). Increases in apoplastic pH reduce the ability of leaf cells to remove xylem- and leaf-sourced ABA from the apoplast (Wilkinson and Davies, 1997). They also reduce the export of ABA out of the leaf *via* the phloem, and may decrease rates of ABA degradation (Jia and Zhang, 1997). The net result is an accumulation of ABA to physiologically active concentrations in the leaf apoplast, often in the absence of measurable increases in bulk tissue [ABA]. Recent work has shown that when root-sourced xylem [ABA] is low, a concomitant increase in pH can cause the sap to become enriched with ABA from stores in xylem parenchyma cells as it moves within the stem towards the leaves of maize (Sauter and Hartung, 2002). Direct measurements of apoplastic pH and [ABA] near to target tissues in shoots of whole plants are needed to test this hypothesis. Bacon *et al.* (1998) have shown convincingly that pH-induced redistribution of ABA can also limit leaf growth of maize plants experiencing drying soil. This is important in the context of the control of growth and development and water use efficiency of plants under deficit irrigation. Our enhanced understanding of the mechanistic basis of this and other potential chemical signalling mechanisms seems to provide a number of relatively simple means of manipulating growth of plants in the field. These may be via targeted application of water to roots (see below) or by application of non-pernicious compounds to leaves and roots with the aim of modifying xylem/apoplastic pH (see Wilkinson, this volume). Particular fertiliser treatments will be useful in this regard, as will the use of mulches to slow soil drying around particular parts of the root system.

The importance of root-sourced ABA in regulating stomatal apertures of plants growing in drying soil has recently been re-examined with conflicting conclusions drawn (Borel *et al.*, 2001; Holbrook *et al.*, 2002). Importantly, grafting studies with ABA-deficient mutants revealed that the stomatal response to drying soil can be strongly influenced by the capacity of the shoot to synthesise ABA (Holbrook *et al.*, 2002). The signal that prompted the enrichment of xylem sap with shoot-sourced ABA was not identified. An as yet unidentified 'source' of ABA ('ABA-adduct'; Netting, 2000) that may release ABA under certain conditions and the glucose ester of ABA (ABA-GE) have also been implicated in long-distance signalling (Sauter *et al.*, 2002). All of this information suggests a multi-component signalling system with contributions to the signalling process from the roots, the shoots and also from cells in the pathway between the two. Different combinations of climatic and edaphic conditions will generate different contributions to the signalling process from different chemical species and from different locations within the plant. Sauter *et al.* (2001) have shown how this might work for ABA (Figure 5.2) but one can imagine other networks for other hormones which will interact with the ABA network and be influenced by other factors such as the nutrient status of the soil and the evaporative demand of the atmosphere. While Assmann (1999) has shown that for *Arabidopsis*, ABA is not involved in the stomatal response of that plant to VPD, others (Davies *et al.*, 2002) have shown how VPD will influence the ABA chemical signalling network. We should not be surprised that atmospheric and edaphic stresses combine in their impact on shoot growth and functioning; indeed, it is vital that the plant is able to integrate the influence of these signals, if it is to optimise water use efficiency as the environment changes and the plant develops.

In the discussion so far we have highlighted the influence of chemical signalling on water use efficiency via a direct effect on stomatal behaviour and gas exchange and through an influence of this regulation on the water status of the shoot. We have shown (Figure 5.1) how shoot water balance can be maintained by the stomatal response and this regulation can then impact on both the growth of the plant with some roots in drying soil and on the morphological development of plant parts. Many plants will respond to deficit irrigation by showing a significant shoot water deficit (see below). If chemical signalling can be properly exploited through novel management or irrigation techniques (see below), then the development of this deficit can be avoided or at least minimised until the soil water deficit becomes very severe. We have chosen not to deal in detail with other factors that will influence stomatal behaviour and water use efficiency (e.g. CO_2 concentration and VPD), but it should be noted that there are effective models of stomatal behaviour based entirely on responses to these two variables (e.g. the Ball Berry model) and these models often make a good job of explaining many features of the plant's day-to-day stomatal response. Recently Dewar (2002) and others have shown how chemical control of stomatal responses can be linked to the Ball Berry type models to improve their performance under many conditions. The analysis suggests that stomata will respond to a combination of the edaphic and the climatic environments to protect the processes in the plant that are

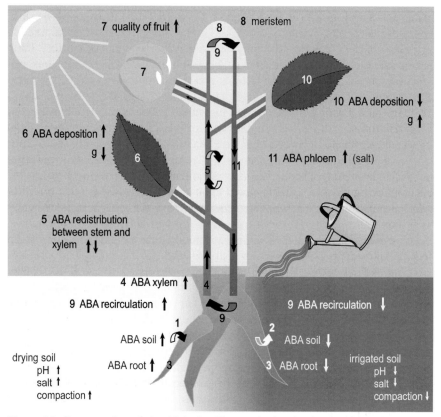

Figure 5.2 Representation of the ABA signalling mechanism in the droughted plant. Eleven component parts of the network are highlighted, each of which can affect the signalling process. See Sauter *et al.* (2001) for further details of the signalling process.

most sensitive to dehydration. We show below how we can exploit both stomatal and developmental responses to the environment to save water in irrigation and to protect the plant from the damaging effects of deficit such that yield can be protected, water use efficiency increased and the quality of yield enhanced.

5.3 Regulated Deficit Irrigation (RDI) and Partial Rootzone Drying (PRD) Irrigation

5.3.1 Regulated Deficit Irrigation

Water use for agriculture is subject to increasing scrutiny from government and environmentalists, and irrigators are under pressure to demonstrate that water use is both efficient and economically worthwhile. One consequence of this close scrutiny of resource usage has been an assessment of the relative impact of cropping system inputs on sustainability and profitability. Crop performance is, of course, subject to many influences such as cultivar geno-type, presence or absence of rootstocks, environmental variables like tempera-ture, humidity, precipitation and wind, and management variables such as pruning, nutrition and irrigation. It is really only irrigation that offers the potential for both the short- and long-term management of vegetative and reproductive development, and irrigators are increasingly exploring new ways to manipulate the water use efficiency of crops through irrigation manage-ment. Underpinning new irrigation techniques is an understanding of the accompanying physiological responses that allow prediction of crop response and assistance in exploring minimum water requirements for economically viable returns.

Regulated deficit irrigation (RDI) is an irrigation technique that relies on precise knowledge of the phenology of vegetative and reproductive develop-ment for effective implementation. RDI was developed for high density orchards of tree crops such as pear, apple and peach where the balance between vegetative and reproductive development is critical. Excessive vegetative vigour results in mutual shading and consequent effects on long-term fruitful-ness (Chalmers et al., 1981; Chalmers, 1986). Vigour reduction could be achieved through mechanical shoot and root pruning, branch manipulation, application of growth regulators, manipulating crop load, fertilizer management or use of dwarfing rootstocks (Goodwin and Boland, 2002). While some of these alternatives may be effective, there are penalties in terms of labour costs and inflexibility and longer establishment times in the case of the use of dwarfing rootstocks. On the other hand, manipulation of vigour through irrigation management offers more flexibility with minimal labour inputs.

The basic principle of RDI is that water is withheld or reduced during a period when vegetative growth is normally high and fruit growth is low. A normal irrigation regime is then resumed during the later period of rapid fruit growth (Figure 5.3). The major impact of the water deficit is to reduce vegetative growth with little effect on fruit development. Goodwin and Boland (2002) have discussed RDI effects on the efficiency of irrigation water use and provide many examples where WUE has doubled when compared with standard irrigation practice. For example, large commercial crops of canning peaches and WBC pears were grown with irrigation inputs of 6 and 4 ML/ha compared with a normal input of 10 and 7 ML/ha respectively (Mitchell and Charmers, 1982; Mitchell et al., 1989). These improvements are due not only to reductions in canopy size but also to reduced leaf stomatal conductance during the RDI period (Boland et al., 1993).

RDI also finds extensive application in the production of winegrapes (McCarthy, 1997; Coombe and McCarthy, 2000; McCarthy et al., 2002), but

Peach shoot and fruit growth and the timing of RDI

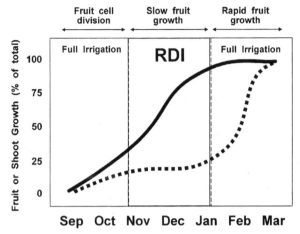

Figure 5.3 Typical shoot (solid line) and fruit (dotted line) growth patterns of peach. (Figure reproduced courtesy of Dr Ian Goodwin.)

here the aims are slightly different to those when applying RDI to pome and stone fruit. In grape, reproductive and vegetative growth occur more or less concurrently with early cell division phases of berry growth coinciding with the period of maximum canopy growth (Mullins *et al.*, 1992). During later stages of berry development when sugar accumulation is occurring, canopy growth has slowed or ceased. There is thus a competition for resources during the critical early phases of berry development and the deliberate withholding of water during early berry development, with a view to restraining vegetative vigour and improving the efficiency of water use may have undesirable effects on berry yield. However, berry growth tends to be less sensitive to water deficit than vegetative growth, and through careful attention to the regulation of minimum soil water tensions it is possible to achieve the desired balance between vegetative and berry growth (McCarthy *et al.*, 2002).

Goodwin *et al.* (2002) suggest minimum soil moisture tensions during the RDI period of 100, 200 and 400 kPa in sandy, loam and clay soils respectively. Winegrape production offers additional challenges to vineyard managers who may be using irrigation management to improve water use efficiency because of the growing recognition of the link between irrigation and fruit, and ultimately wine, quality. Part of this influence on grape quality derives from the effects of early season water deficit on berry size. Smaller berries have a higher surface to volume ratio which tends to intensify colour and flavour components present in the skin. Matthews *et al.* (1990) attempted to relate water deficits experienced at different times to wine quality characters and noted differences in appearance, flavour, taste and aroma that could be attributed to the time that the vines experienced water deficits. Water deficits applied later in the season, although they may offer the opportunity to save water, are likely to be difficult to manage because of the larger canopies developed during the early part of the season and the reduced ability to ripen the crop due to inhibition of photosynthesis (Wample and Smithyman, 2002).

5.3.2   Partial Rootzone Drying

Regulated deficit irrigation was originally conceived as a technique to reduce vegetative vigour and to change the resource allocation between vegetative and reproductive development in stone and pome fruit (Li *et al.*, 1989; Williamson and Coston, 1990; Irvin and Drost, 1987; Chalmers *et al.*, 1981). It also can result in significant improvements in WUE because of the reduced water allocation and plant water use during the RDI period. However, the possibility of having to refill the soil profile after the RDI period may mean that water saved during the whole season may not be great. Successful application of RDI requires careful attention to the timing of the water deficit period and to the degree of stress that is allowed to develop. Furthermore, in the case of RDI application to winegrapes, some yield loss may be experienced (McCarthy *et al.*, 2002; Wample and Smithyman, 2002), thereby reducing the possibility of improvements in WUE. In winegrapes the principal reason for using RDI is to reduce vigour and to manipulate fruit quality through effects on berry size. Changes in WUE are often a secondary consideration.

There is another strategic irrigation management tool now available to irrigators that is aimed more directly at manipulating WUE. This is partial rootzone drying (PRD). Work on developing PRD commenced in the 1980s when it was shown that in grapevine there were significant diurnal changes in leaf ABA and stomatal conductance and that transport of ABA from the roots was able to explain in large part these observations (Loveys, 1984a,b; Loveys and Düring, 1984). It was then suggested that if ways could be found to manipulate the delivery of root-sourced ABA then it might be possible to gain some control over development processes such as expression of shoot vigour (Loveys, 1991). This possibility was given added weight by observations showing that, if only part of the root system dried and the remaining roots were kept well watered, chemical signals produced in the drying roots reduced stomatal aperture and leaf growth (Gowing, 1990). At the same time the fully hydrated roots maintained a favourable water status throughout the aerial parts of the plant. In other words, it was possible to separate the biochemical responses to water stress from the hydraulic effects of reduced water availability.

In addition to reduced shoot extension Dry and Loveys (1999) also noted that stomatal conductance was inhibited when part of the grapevine root system was dried. A surprising finding was that in grapevines the effect was transient, and despite the fact that part of the root system remained dry, stomatal conductance, photosynthesis and growth of grapevine canes returned to pre-treatment levels within a few weeks despite the fact that there was no change in irrigation regime and that one part of the root system was watered and one part remained dry (Dry and Loveys, 2000). Subsequently it was shown that the production of abscisic acid (ABA) in the drying roots was also transient, and that the recovery in stomatal conductance and shoot growth commenced when ABA production in the drying roots slowed (Loveys *et al.*, 2000), suggesting that ABA produced in the roots was at least partly responsible for the observed canopy responses. Considerable additional evidence has since been accumulated to support this idea. The transient nature of the response seemed to preclude its use in any practical way until it was realized that by alternating the wetted and dried part of the roots on a regular basis it was possible to sustain the effect of the partial wetting over an indefinite period (Dry and Loveys, 1998; Loveys *et al.*, 2000). Kang *et al.* (2001) have similarly shown, in experiments with hot pepper plants (*Capsicum* sp.), that by alternately changing the wetted half of the root system at every irrigation, WUE was significantly enhanced when compared with a fully irrigated control and that root to shoot dry weight ratios were increased. These responses were not evident in plants where the same half of the root system was irrigated throughout the growing season and the remainder of the roots were dry, suggesting that the root-derived signals driving the stomatal responses were not sustained, in the same way as had been observed in grapevine.

Evidence for involvement of root-sourced ABA in the PRD response of grapevine has been provided by Stoll *et al.* (2000) who showed that the ABA content of xylem sap from field-grown Cabernet Sauvignon vines was significantly higher with PRD irrigation than in fully irrigated controls and that this corresponded with a decrease in stomatal conductance. Furthermore, we have also shown that it is possible to mimic the effects of partial root drying by applying low concentrations of ABA to part of a grapevine root system. After ABA was applied to fully irrigated plants, stomatal conductance fell in a very similar manner to that of plants which were subject to partial root drying (Figure 5.4). The changes in the ABA content of xylem sap and roots were similar for the exogenous ABA and PRD treatments compared with fully irrigated controls. Bulk leaf ABA changes were relatively small, however.

It had been noted previously that PRD did not result in large changes in leaf ABA when compared with plants that had been subject to an overall water deficit (Stoll *et al.*, 2000). It may be assumed that stomatal control can be achieved with relatively small changes in bulk leaf ABA when synthesis is initiated by changes in water availability to only part of the root system. ABA uptake from the xylem would be regulated to some extent by the reduced transpiration, and balanced by the active catabolism of ABA in this tissue.

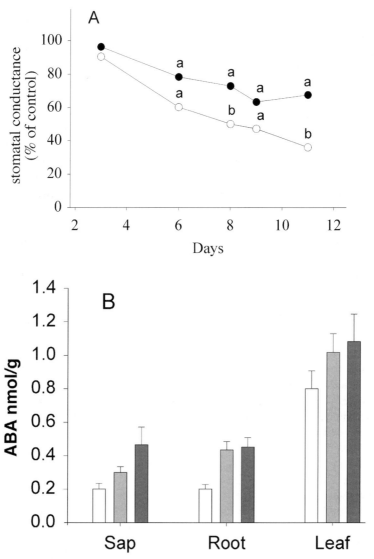

Figure 5.4   A. Changes in stomatal conductance of Cabernet Sauvignon potted vines resulting from drying part of the root system (open symbols) or application of ABA solution (3 × 10⁻⁶M) to part of the root system (solid symbols) B. Abscisic acid (ABA) concentration in sap expressed from petioles, roots and leaves of Cabernet Sauvignon vines. Unshaded bars = control (all roots watered); grey shading = part of root system irrigated with ABA solution (3 × 10⁻⁶M); black shading = part of root system allowed to dry. Bars are standard error of the mean (n=4)

Large changes in leaf ABA probably occur only as a result of a fall in leaf turgor, a situation which we are trying to prevent in PRD irrigation. Significantly, the pH of the sap from PRD vines was higher than that from control vines. The consequences of changes in xylem sap pH have been summarized by Davies *et al.* (2002) who showed that increases in xylem and therefore in leaf apoplastic pH reduce partitioning of ABA into the symplast away from sites of action on the stomatal guard cell. This also means that even the relatively low ABA concentrations present in the xylem of well-watered plants can close the stomata, if the pH relations of the leaf allow the ABA to access the guard cells.

Grapevines may be particularly sensitive to stomatal regulation through changes in the pH of the apoplast for a number of reasons. First, the concentration of ABA in the xylem of well-watered plants is quite high in comparison with many other species. Loveys (1984a) documented that the ABA concentration of xylem sap of several winegrape cultivars varied between 2×10^{-7} M and 5×10^{-7} M and that by supplying ABA to excised leaves via the xylem within this concentration range it was possible to bring about reductions in stomatal conductance. In further experiments with excised vine leaves Loveys (1991) showed that by supplying ABA (3.6×10^{-7} M) via the xylem, stomatal response to humidity change was enhanced when compared with leaves supplied with water containing no ABA. These experiments suggest that in grapevine the normal concentration of ABA in the apoplast needs to change by only a small amount to bring about significant stomatal response and that stomatal response to humidity change is dependent on the presence of ABA at this resting concentration. In experiments with soybean, Bunce (1998) similarly concluded that ABA was involved in stomatal response to leaf-to-air water vapour pressure difference. By contrast, Assmann *et al.* (2000) have shown that ABA-deficient and ABA-insensitive mutants of *Arabidopsis* have a wild-type stomatal response to humidity, appearing to rule out a role for ABA in this species.

Second, the pH of grapevine xylem sap is quite acidic when compared with reported values for other species. Stoll *et al.* (2000) reported that the pH of vine sap varied from 4.2 to 4.8, whereas Correia *et al.* (1999) reported pH values of 6.4 to 6.8 for lupin, Thomas and Eamus (2002) 5.0 to 6.5 for a range of Australian savannah trees, 5.2 to 6.0 for *Forsythia x intermedia* (Davies *et al.*, 2002), 5.6 for tomato (Liao *et al.*, 2000), and 5.8 to 6.6 for soybean (Liu *et al.*, 2003). The pH of grapevine sap is close to the pKa value of ABA (4.8), meaning that relatively small changes in sap pH will have quite large effects on the proportion of ABA present as its anion, whereas at the higher pH values typical of most other species the same shift will have a much smaller proportional effect on ABA ionization. Does this mean that plants that have xylem pH values close to 4.8 are particularly sensitive to stomatal regulation brought about by shifts in sap pH? Under these circumstances no net change in bulk leaf ABA may be necessary for stomatal regulation under a PRD irrigation regime. Although large changes in the ABA content of the drying grapevine roots have been demonstrated during PRD irrigation (Loveys *et al.*, 2000) the changes in leaves are relatively small (Stoll *et al.*, 2000).

More recent experiments with citrus may provide an even more extreme example. PRD irrigation was applied to field grown mature Valencia orange trees so that control (100% of soil surface wetted) and PRD (50% of soil surface wetted) trees received the same amount of water. Stomatal conductance was significantly lower in the PRD trees compared with control (Figure 5.5A). Furthermore, the intensity of the response was dependent on ambient atmospheric vapour pressure deficit (VPD) (Figure 5.5B). Although the ABA content of the roots on the drying side of the trees was significantly higher than in control roots there was no significant difference in leaf ABA between treatments or between sides of the tree (Figure 5.6). Another interesting observation from these citrus experiments was that the difference in stomatal conductance between control and PRD trees was evident most strongly in the morning, on the eastern sun-exposed face of the canopy. By afternoon, when VPD had risen significantly, stomatal conductance was low in both control and PRD trees on the now exposed western side of the trees and there was no significant difference between the treatments. These data have interesting implications for the efficiency of water use on either side of the trees. Under conditions of high VPD, stomatal conductance was always low on the western side of the trees, despite the abundant availability of water. Jifton and Syvertsen (2003) have shown that when citrus stomatal conductance is low, leaf temperatures of sun-exposed leaves were up to 6°C above ambient and that there were major non-stomatal limitations to photosynthesis. These conditions of high leaf temperatures, elevated leaf-to-air vapour pressure gradients and low assimilation would contribute to a lowering of WUE at the leaf level. The consequences of this for whole tree WUE and the interactions with physiological changes brought about by deficit irrigation practices such as PRD which accentuate stomatal closure needs to be assessed.

The linking of the intensity of the PRD stomatal response to VPD has also been noted in pears (Davies *et al.*, 2002) and this would be an cffcctive mechanism for increasing WUE since at low VPD differences in stomatal conductance, and therefore transpiration, between fully watered and PRD trees would be small but as VPD increased, the relative transpiration of PRD trees would progressively reduce because of the greater degree of stomatal closure. Differences in carbon assimilation would therefore occur only at higher VPD values. It is tempting to attribute this sensitizing of stomata to VPD to an increase in the ability of the xylem to supply ABA since we know that plants under PRD irrigation have a higher xylem ABA content (Stoll *et al.*, 2000), but as Assmann *et al.* (2000) point out a number of other mechanisms may drive stomatal response to humidity. These may involve a

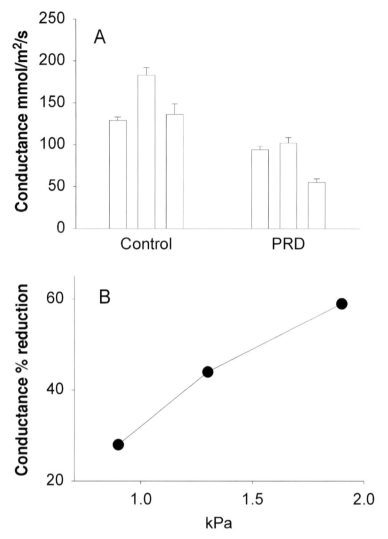

Figure 5.5   A. Stomatal conductance of Valencia orange trees measured on three separate days of control (irrigated both sides) and PRD treatments. Ten leaves were measured on each of four replicate trees. Bars are standard error of the means. B. Reduction in stomatal conductance resulting from the PRD treatment as a function of atmospheric VPD during the measurement period.

sensing of transpiration rates through the stomata or cuticle or perhaps the water potential of the leaf or the stomatal guard cells. PRD does not alter leaf water potential (Stoll *et al.*, 2000) and so a mechanism based on a root-derived chemical signal rather than a hydraulic one seems most likely. Munns and King (1988) have presented some evidence for the presence of a root-sourced

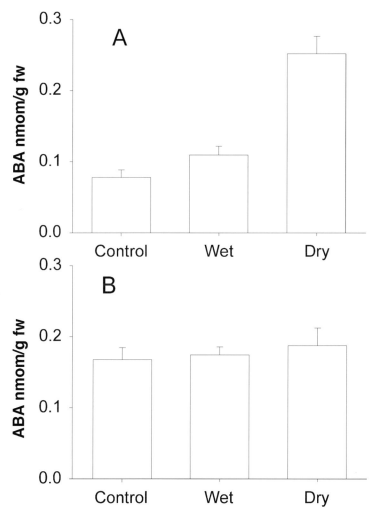

Figure 5.6   A. Abscisic acid (ABA) concentration in the roots of Valencia orange trees subject to PRD irrigation. B. Abscisic acid concentration in the leaves of Valencia orange trees subject to PRD irrigation. Wet side and dry side of the PRD trees and control (both sides irrigated) are shown. Bars are standard error of the mean (n=4)

chemical signal other than ABA which is responsive to soil water potential. This has never been further characterized, although it may be analogous to the ABA-adduct described by Netting *et al.* (1992). The possibility also exists for ABA being a secondary message, synthesized in leaf tissues in response to a primary chemical message originating in roots. Synthesis of the ethylene precursor 1-aminocyclopropane-1-carboxylic acid (ACC) increases in roots in response to stress (Gomex-Cadenas *et al.*, 1996; Jackson, 1997) and is transported to shoots where ethylene, released from ACC, can induce ABA synthesis (Grossmann and Hansen, 2001). Cytokinins have also been shown to influence stomatal behaviour and Stoll *et al.* (2000) have shown that the inhibitory effects of PRD on grapevine stomatal conductance can be overcome with exogenous application of the synthetic cytokinin benzyl adenine. Furthermore, PRD was shown to reduce endogenous concentrations of zeatin and zeatin riboside in roots and in shoots. The antagonistic effects of ABA and zeatin riboside on stomatal conductance have also been noted by Hansen and Dörffling (2003) who applied these substances to excised sunflower shoots at concentrations equivalent to those in xylems sap of stressed or unstressed plants.

PRD thus appears to be an effective technique for improving WUE by changing the relationship between stomatal conductance and ambient evaporative conditions. On average, conductance will be lower in PRD plants than in fully irrigated controls. However, the opposite condition where PRD plants have higher stomatal conductance than controls can also occur, but paradoxically, WUE is still improved. This can occur at very low water application rates and results from water stress developing in the conventionally irrigated vines but not in the PRD vines. In PRD water is applied to half the surface area compared with conventional irrigation and evaporation from the soil surface and lack of penetration of the irrigation water to the roots can limit water availability to the control vines.

In an extensive field trial De La Hera Orts *et al.* (2002) compared the effects of PRD and conventional drip irrigation with mature Mataro vines in south-eastern Spain. Both control and PRD vines received the same amount of irrigation water (1.4 ML/ha) during the growing season. Berry yield was significantly ($P < 0.05$) greater in the PRD vines (8.2 t/ha compared to 5.7 t/ha for the controls). Since all vines received the same amount of water, WUE was correspondingly increased. The more effective use of this water by the PRD vines was further demonstrated by the fact that canopy area per vine and pruning weight per vine were significantly ($P < 0.05$) enhanced. We have similarly shown with field-grown Cabernet Sauvignon vines that by applying the same amount of water either in a conventional way (two drippers per vine, whole of rootzone irrigated) or by PRD (single dripper per vine, half of rootzone irrigated at any one time), the effect of that water is dependent on its absolute amount. At high water application rates (4 to 5 L per vine per day) stomatal conductance of PRD vines was significantly ($P < 0.05$) less than the controls and water penetrated to a depth of 0.75 m in both cases. This may be analogous to many of the split-root pot experiments where one pot is kept at field capacity and the second pot is allowed to dry (Dry and Loveys, 2000a,b). As the total water application rates were reduced the stomatal conductance of the two treatments became similar at about 2.5 L/vine/day and at even lower rates of about 1.5 L/vine/day the PRD vines actually had significantly higher stomatal conductance than controls. At this low irrigation rate water to the deeper roots (>0.45 m) was not being replenished in the control vines whereas in the PRD vines water was still penetrating to 0.75 m, as judged by capacitive sensors installed at 100 mm intervals from 100 mm to 1 m.

In longer-term experiments with field grown Riesling vines we have also found that PRD influences root development, causing an increase in root development in deeper soil layers. At depths of 0.4 to 0.7 m, roots were more abundant in PRD-irrigated vines ($P < 0.05$), but significantly less so in the 0 to 0.4 m zone. This tendency to produce more roots in deeper soil layers may influence the efficiency of use of irrigation water by making more water available from greater soil depths. A similar effect on root distribution was also observed in pot experiments when one half of the root system was allowed to dry and the other remained well watered. Roots that had access to water, in both the wet container and in lower layers of the dry container, showed enhanced growth relative to roots in fully watered controls (Dry and Loveys, 2000b).

There seems little doubt that PRD can bring considerable benefits to grape growers in terms of improved WUE, but can other crops benefit from this new irrigation method and is it possible to predict how a crop plant may respond to any form of deficit irrigation in terms of changed WUE? The transmission of root-derived chemical signals such as ABA to leaves seems to be central to the PRD response and in grapevine there is ample evidence that this happens. The concentration of ABA in grapevine xylem contents, even in fully irrigated plants, is sufficient to modulate stomatal conductance and, when raised only slightly in response to PRD treatment, is able to bring about further reductions in stomatal aperture and small changes in bulk leaf ABA. These small changes may be accompanied by shifts in xylem sap pH and possibly reductions in cytokinins, all of which may contribute to the final expression of stomatal conductance and responsiveness to environmental change. An effective predictor of likely response to PRD may therefore be the presence of ABA in the xylem sap at a concentration close to that needed for stomatal modulation.

If this is the case then it might be predicted that apricot (*Prunus armeniaca*) would respond poorly to partial root drying. Loveys *et al.* (1987) showed that the ABA concentration in apricot xylem sap was only about 10% of that necessary to elicit a stomatal response and that there were only minor diurnal and seasonal changes in bulk leaf ABA when compared with grapevine. It was concluded that ABA did not play a major role in short term or seasonal stomatal responses. In contrast to grapevine, apricot leaves accumulated large amounts of osmotically-active solutes allowing maintenance of stomatal conductance, even at low leaf water potentials, through a decrease in osmotic potential and a corresponding increase in the ability to maintain turgor at low water potentials. These two species have adopted different strategies in dealing with the problem of maintaining turgor under a range of atmospheric and soil water potentials. Grapevine stomatal response is finely tuned to respond quickly to variations in atmospheric VPD and soil matric potential through root-derived chemical signals, and therefore maintains its leaf water potential within a relatively narrow range. Apricot, on the other hand, allocates a proportion of its fixed carbon to the accumulation of high concentrations of solute (sorbitol) and is able to maintain leaf and stomatal turgor, and therefore gas exchange, under conditions of falling leaf water potential. Short term responses to root-derived chemical signals may not be necessary under this strategy.

Augé and Moore (2002) similarly drew distinctions between these mechanisms in studies with a range of temperate deciduous tree species. They concluded that species showing considerable osmotic adjustment showed only small reductions in stomatal conductance in response to falling soil matric potential, whereas species that showed little foliar osmotic adjustment had a heightened stomatal sensitivity under similar conditions of soil drying. Since the improvements in WUE resulting from PRD come from the stimulation of partial stomatal closure it seems unlikely that species relying exclusively on osmotic adjustment will respond, at least in the short term. However, the stimulation of osmotic adjustment in leaf tissue by exposure of roots to reduced water availability may involve the same signalling mechanisms that are invoked under PRD.

Osmotic adjustment in corn seedlings can be stimulated by exogenous application of ABA (Zhao *et al.*, 1995; Wang *et al.*, 2003) and ABA and osmolytes such as proline and glycinebetaine have been shown to accumulate together in water stressed tissues (Abernethy and McManus, 1998). By contrast, Ismail and Davies (1997) showed that the degree of osmotic adjustment in *Capsicum* species was inversely related to the ability of the roots to supply ABA to the shoots. Even in species which rely on osmotic adjustment for turgor maintenance during severe water stress, root production and transport of ABA in the xylem may constitute the first line of defence as soil water deficits are encountered (Ali *et al.*, 1999). The effectiveness of rapid responses to prevailing ambient conditions in controlling grapevine leaf gas exchange is illustrated by experiments of Yunusa *et al.* (2000). These authors used heat pulse sap flow sensors to determine transpiration and showed that even when maximum atmospheric VPD increased from about 1.5 kPa to 4.5 kPa there was little change in transpiration due to a tight linking of stomatal conductance to the atmosphere.

In its present form, implementation of PRD requires that an irrigation system is installed so that close to 50% of the root system is irrigated and 50% is drying at any one time. In most instances above-ground drip or sub-surface drip lines will be used which naturally produce wetted and drying zones which are disposed in a horizontal plane. Plants growing under rain-fed conditions will also experience soil at a range of matric potentials but these will more frequently be disposed in a vertical plane. For example, significant rain events may bring the entire soil profile to field capacity and from then on soil drying will occur from the top down as a result of evaporation from the soil surface and water use by roots that are more abundant in soil layers near the surface. Conversely, lesser rain events falling on a dry soil will create the reverse situation where surface soil layers become wet and soil deeper in the profile will remain dry. Both of these situations have been described by Burgess *et al.* (1998) who used sap flow gauges to show that water is redistributed within root systems according to the prevailing water potential gradients. These authors also considered the possibility, and provided evidence, that water not only moved within roots but also from roots into soil.

This raises the possibility that ABA, generated in drying roots as a result of irrigation management protocols like RDI and PRD, may be transferred to the soil. ABA has been shown to be present in soil and that its breakdown in this medium was least under acid or saline conditions and that it may originate from both plant and microbiological sources (Hartung *et al.*, 1996). This rhizosphere ABA may be available to the plant when the soil is rewetted by irrigation or by rain events, since ABA is readily taken up through roots. This redistribution of water, and presumably water-soluble compounds like ABA, within roots may be important to our understanding of the effects of irrigation management techniques like PRD and RDI. In addition, water redistribution will maintain the viability of roots in dry soil, lessening the potential for root damage under PRD irrigation and maintaining them in a condition where they are best able to take advantage of minor rain events or the resumption of irrigation during the next PRD cycle.

In experiments with mature PRD-irrigated Valencia orange trees, we installed sap flow sensors in the roots on either side of a tree. Some were oriented to detect flow from root to shoot and some had a reverse orientation to detect flow from the shoot towards root tips. Maximum flow on the irrigated side of the trees occurred the day after an irrigation (Figure 5.7A) and over the next five days flow decreased. Maximum flow on any day occurred at about solar noon. Flow at a reduced amplitude was also detected in the drying roots, again with maxima on each day at about solar noon. Reverse flow on the non-irrigated side of the trees was also detected, with maxima occurring soon after midnight (Figure 5.7B), suggesting bi-directional flow in these roots. There was no significant reverse flow in the roots on the irrigated side of the trees. Analysis of ABA in the roots of the control (irrigated both sides) and PRD trees showed that on most days concentrations were significantly higher in the drying roots (Figures 5.7C and D). Assuming that the ABA present in these roots, or a proportion of it, was available for xylem transport it is likely that the supply of ABA to the leaves would have been increased by the PRD treatment.

While there is no doubt that RDI and PRD can bring improvements in WUE as a result of the demonstrated effects on canopy area and stomatal

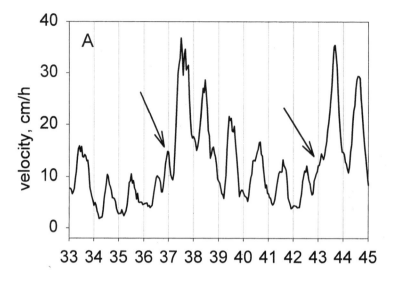

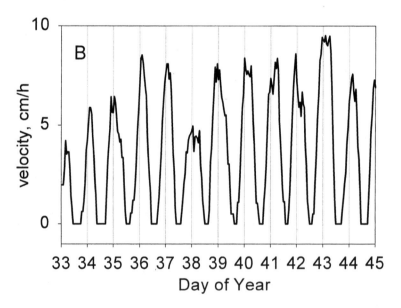

Figure 5.7 A. Sap flow velocity in roots of Valencia orange on the irrigated side of trees. Arrows show times when irrigation occurred. B. Sap flow velocity in roots of Valencia orange on the non-irrigated side of trees. C. Abscisic acid (ABA) concentration in the roots of control (both sides irrigated) Valencia orange trees. D. Abscisic acid concentration in roots from the non-irrigated side of PRD Valencia orange trees. Bar represents LSD (P=0.05) n=4.

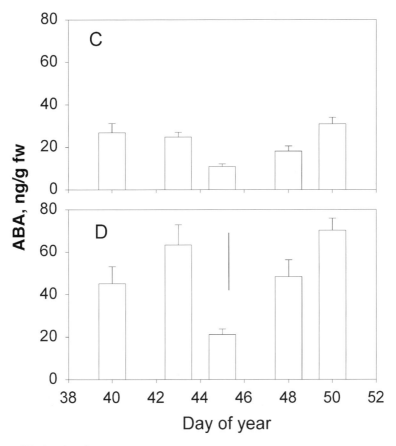

Figure 5.7   (continued)

conductance with minimal effects on crop yield, the magnitude of the improvements, as judged by comparison with current industry practice, may be deceptive. A plentiful supply of cheap water and a need to maximize yield and profit have, in the past, often resulted in inefficiencies in irrigation practice. The benchmarks, against which novel irrigation practice can be judged, may comprise non-plant components of orchard water use such as soil evaporation, runoff, drainage and storage which were poorly quantified or unknown. Modern irrigation practice seeks to understand and control these losses through closed water transport pathways, pressurized delivery systems and irrigation scheduling based on measurement of soil water or some aspect of plant performance. An integral part of understanding the water balance of an orchard is some measure of plant water use as opposed to these non-plant components. This information is vital to gaining an understanding of all components that contribute to the efficiency of water use in plant production systems. It may be that gains in efficiency will come as much from engineering solutions that address problems associated with irrigation water transmission and delivery as from introduction of techniques aimed at modifying the plant response to water.

Acknowledgements

We would like to acknowledge the financial support of the Grape and Wine Research and Development Corporation (Australia) and Land and Water Australia and Dr Ian Goodwin for permission to reproduce Figure 5.3.

References

Abernethy G.A. and McManus M.T. (1998) Biochemical responses to an imposed water deficit in mature leaf tissue of *Festuca arundinace*. *Environmental and Experimental Botany* **40**, 17–28.

Ali, M., Jensen, C.R., Mogensen, V.O., Andersen, M.N. and Henson, I.E. (1999) Root signalling and osmotic adjustment during intermittent soil drying sustain grain yield of field grown wheat *Field Crops Research* **62**, 35–52.

Anon., (1999) *Water and the Australian Economy*. Australian Academy of Technological Sciences and Engineering.

Assmann, S.M., Snyder, J.A. and Lee, Y.R.J. (2000) ABA-deficient (aba1) and ABA-insensitive (abi1-1, abi2-1) mutants of *Arabidopsis* have a wild-type stomatal response to humidity. *Plant, Cell and Environment* **23**, 387–395.

Augé R.M and Moore J.L. (2002) Stomatal response to non-hydraulic root-to-shoot communication of partial soil drying in relation to foliar dehydration tolerance. *Environmental and Experimental Botany* **47**, 217–229.

Bacon, M.A., Wilkinson, S. and Davies, W.J. (1998) pH-regulated leaf cell expansion in droughted plants is abscisic acid dependent. *Plant Physiology* **118**, 1507–1515.

Boland, A.M., Mitchell, P.D., Jerie, P.H. and Goodwin, I. (1993) The effect of regulated deficit irrigation on tree water use and growth of peach. *Journal of Horticultural Science* **68**, 261–274.

Borel, C., Frey, A., Marion-Poll, A., Simmoneau, T. and Tardieu, F. (2001) Does engineering abscisic acid biosynthesis in Nicotiana plumbaginifolia modify stomatal response to drought? *Plant, Cell and Environment* **24**, 477–489.

Bunce, J.A. (1998) Effects of humidity on short term responses of stomatal conductance to an increase in carbondioxide concentration. *Plant, Cell and Environment* **21**, 115–120.

Burgess, S.O., Adams, M.A., Turner, N.C. and Ong, C.K. (1998) The redistribution of soil water by tree root systems. *Oecologia* **115**, 306–311.

Chalmers, D.J. (1986) Research and progress in cultural systems and management in temperate fruit orchards. *Acta Horticulturae* **175**, 215–225.

Chalmers, D.J., Mitchell, P.D. and van Heek, L. (1981) Control of peach tree growth and productivity by regulated water supply, tree density and summer pruning. *Journal of the American Society of Horticultural Science* **106**, 307–312.

Coombe, B.G. and Mccarthy, M.G. (2000) Dynamics of grape berry growth and physiology of ripening. Australian *Journal of Grape and Wine Research* **6**, 131–135.

Correia, M.J., Rodrigues, M.L., Osório, M.L. and Chaves, M.M. (1999) Effects of growth temperature on the response of lupin stomata to drought and abscisic acid. Australian. *Journal of Plant Physiology* **26**, 549–559.

Davies, W.J., Wilkinson, S. and Loveys, B.R. (2001) Stomatal control by chemical signalling and the exploitation of this mechanism to increase water use efficiency in agriculture. *New Phytologist* **153**, 449–460.

Davies, W.J., Tardieu, F. and Trejo, C.L. (1994) How do chemical signals work in plants that grow in drying soil? *Plant Physiology* **104**, 309–314.

De La Hera Orts, M.L., Perez Prieto, L.J., Fernandez, J.I., Martinez Cutillas, A., Lopez Roca, J.M. and Gomez Plaz, E. (2002) Partial Rootzone Drying. Una Experiencia Espanola para la variedad Monastrell. *Fruticultura Profesional* **131**, 70–76.

Dewar, R.C. (2002) The Ball-Berry-Leuning and Tardieu-Davies stomatal models: synthesis and extension within a spatially aggregated picture of guard cell function. *Plant, Cell and Environment* **25**, 1383–1398.

Dry, P.R. and Loveys, B.R. (1999) Grapevine shoot growth and stomatal conductance are reduced when part of the root system is dried. *Vitis* **38**, 151–156.

Dry, P.R. and Loveys, B.R. (2000a) Partial drying of the rootzone of grape. I. Transient changes in shoot growth and gas exchange. *Vitis* **39**, 3–7.

Dry, P.R. and Loveys, B.R. (2000b) Partial drying of the rootzone of grape. II. Changes in the pattern of root Development. *Vitis* **39**, 9–12.

Gomez-Cadenas, A., Tadeo, F.R., Talon, M. and Primo-Millo, E. (1996) Hormones from roots as signals for the shoots of stressed plants. *Plant Physiology* **112**, 401–408.

Goodwin, I. and Boland A.M. (2002) Scheduling deficit irrigation of fruit trees for optimizing water use efficiency. Water Reports, FAO Publication number 22, Rome, pp. 67–79.

Gowing, D.J.G, Davies, W.J. and Jones, H.G. (1990) A positive root-sourced signal as an indicator of soil drying in apple, Malus x domestica-Borkh. *Journal of Experimental Botany* **41**, 1535–1540.

Granier, A. (1985) Une novelle méthode pour la mesure du flux de sève brute dans le tronc des arbres. *Annales des Sciences Forestières* **42**, 193–200.

Green, S.R. and Clothier, B.E. (1988) Water use by Kiwifruit vines and apple trees by the heat pulse techniques. *Journal of Experimental Botany* **39**, 115–121.

Grossmann, K. and Hansen, H. (2001) Leaf abscission induced by ethylene in water-stressed intact seedlings of Cleopatra mandarin requires previous abscisic acid accumulation in roots. *Physiologia Plantarum* **113**, 9–14.

Hansen, H. and Dorffling, K. (2003) Root-derived trans-zeatin riboside and abscisic acid in drought-stressed and rewatered sunflower plants: interaction in the control of leaf diffusive resistance? *Functional Plant Biology* **30**, 365–375.

Hartung, W., Sauter, A., Turner, N.C., Fillery, I. and Heilmeiser, H. (1996) Abscisic acid in soils: What is its function and which factors and mechanisms influence its concentration? *Plant and Soil*, **184**, 105–110.

Holbrook, N.M., Shashidhar, V.R., James, R.A. and Munns, R. (2002) Stomatal control in tomato with ABA-deficient roots: response of grafted plants to soil drying. *Journal of Experimental Botany* **53**, 1503–1514.

Irving, D.E. and Drost, J.H. (1987) Effects of water deficit on vegetative growth, fruit growth and fruit quality in Cox's Orange Pippin apple. *Journal of Horticultural Science* **62**, 427–432.

Ismail, M.R. and Davies, W.J. (1997) Water relations of Capsicum genotypes under water stress. *Biologia Plantarum* **39**, 293–297.

Jackson, M.B. (1997) Hormones from roots as signals for the shoots of stressed plants. *Trends in Plant Science* **2**, 22–28.

Jarvis, P.G. and McNaughton, K.G. (1986) Stomatal control of transpiration – scaling up from leaf to region. *Advances in Ecological Research* **15**, 1–49.

Jia, W.S. and Zhang, J.H. (1997) Comparison of exportation and metabolism of xylem-delivered ABA in maize leaves at different water status and xylem sap pH. *Plant Growth Regulation* **21**, 43–49.

Jifton. J.L. and Syvertsen, J.P. (2003) Moderate shade can increase net gas exchange and reduce photoinhibition in citrus leaves. *Tree Physiology* **23**, 119–127.

Jones, H.G. (1985) Partitioning stomatal and non-stomatal limitations to photosynthesis. *Plant, Cell and Environment* **8**, 95–104.

Kang, S., Zhang, L.Hu, X., Li, Z. and Jerie, P.H. (2001) An improved water use efficiency for hot pepper grown under controlled alternate drip on partial roots. *Scientia Horticulturae* **89**, 257–267.

Liao, M.T., Hedley, M.J., Woolley, D.J., Brooks, R.R. and Nichols, M.A. (2000) Copper uptake and translocation in chicory (*Cichorium intybus* L. cv Grasslands Puna) and tomato

(*Lycopersicon esculentum* Mill. cv Rondy) plants grown in NFT system. II. The role of nicotianamine and histidine in xylem sap copper transport. *Plant and Soil* **223**, 243–252.

Li, S-H., Huguet, J-G., Schoch, P.G. and Orlando. P. (1989) Response of peach tree growth and cropping to soil water deficit at various phenological stages of fruit development. *Journal of Horticultural Science* **64**, 541–552.

Liu, F.L., Jensen, C.R. and Andersen, M.N. (2003) Hydraulic and chemical signals in the control of leaf expansion and stomatal conductance in soybean exposed to drought stress. *Functional Plant Biology* **30**, 65–73.

Loveys, B.R. (1984a) Diurnal changes in water relations and abscisic acid in field-grown *Vitis vinifera* cultivars 3. The influence of xylem-derived abscisic acid on leaf gas exchange. *New Phytologist* **98**, 563–573.

Loveys, B.R. (1984b) Abscisic acid transport and metabolism in grapevine (*Vitis vinifera* L.). *New Phytologist* **98**, 575–582.

Loveys, B.R. (1991) How useful is a knowledge of ABA physiology for crop improvement? In: Davies, W.J. and Jones, H.G. (eds.). *Abscisic Acid: Physiology and Biochemistry*: 245–260. BIOS Scientific Publishers.

Loveys, B.R. and Düring, H. (1984) Diurnal changes in water relations and abscisic acid in field-grown *Vitis vinifera* cultivars 2. Abscisic acid changes under semi-arid conditions. *New Phytologist* **97**, 37–47.

Loveys, B.R., Robinson, S.P. and Downton, W.J.S. (1987) Seasonal and diurnal changes in abscisic acid and water relations of apricot leaves (*Prunus armeniaca* L.). *New Phytologist* **107**, 15–27.

Matthews, M.A., Ishii, R., Anderson, M.M. and O'Mahony, M. (1990) Dependence of wine sensory attributes on vine water status. *Journal of the Science of Food and Agriculture* **51**, 321–335.

Meyer, W.S. (1994) Optimising rice production within sustainable environmental constraints. RIRDC Project CS1–2A, Final Report. CSIRO Division of Water Resources.

Meyer, W.S., Dugas, W.A., Barrs, H.D., Smith, R.C.G. and Fleetwood, R.J. (1990) Effects of soil type on soybean crop water use in weighing lysimeters. 1 Evaporation. *Irrigation Science* **11**, 69–75.

Meyer, W.S., Dunin, F.X., Smith, R.C.G., Shell, G.S.G. and White, N.S. (1987) Characterising water use by irrigated wheat at Griffith, New South Wales. *Australian Journal of Soil Research* **25**, 499–515.

Mitchell, P.D. and Chalmers D.J. (1982) The effect of reduced water supply on peach tree growth and yields. *Journal of the American Society of Horticultural Science* **107**, 853–856.

Mitchell, P.D., van den Ende, B., Jerie, P.H. and Chalmers, D.J. (1989) response of 'Bartlett' pear to withholding irrigation, regulated deficit irrigation and tree spacing. *Journal of the American Society of Horticultural Science* **114**, 15–19.

McCarthy, M.G. (1997) The effect of transient water deficit on berry development of cv. Shiraz (Vitis vinifera L.). *Australian Journal of Grape and Wine Research* **3**, 102–108.

McCarthy, M.G., Loveys, B.R., Dry, P.R. and Stoll, M. (2002) Water Reports, FAO Publication number 22, Rome, pp. 79–87.

Mullins, M.G., Bouquet. A. and Williams, L.E. (1992) *Biology of the Grapevine*. Cambridge University Press.

Munns, R.E. and King, R.W. (1988) Abscisic acid is not the only stomatal inhibitor in the transpiration stream of wheat plants. *Plant Physiology* **88**, 703–708.

Netting, A.G. (2000) pH, abscisic acid and the integration of metabolism in plants under stressed and non-stressed conditions: cellular responses to stress and their implication for plant water relations. *Journal of Experimental Botany* **51**, 147–158.

Netting, A.G., Willows, R.D. and Milborrow, B.V. (1992) The isolation and identification of the prosthetic group released from a bound form of abscisic acid. *Plant Growth Regulation* **11**, 327–334.

Patonnier, M.P., Peltier, J.P. and Marigo, G. (1999) Drought-induced increase in xylem malate and mannitol concentrations and closure of Fraxinus excelsior L-stomata. *Journal of Experimental Botany* **50**, 1223–1229.

Quarrie, S.A. (1991) Implications of genetic differences in ABA accumulation for crop production. In: Davies, W.J. and Jones, H.G. (eds.). *Abscisic Acid: Physiology and Biochemistry*:

137–152. Bios Scientific Publishers.

Sauter, A. and Hartung, W. (2002) The contribution of internode and mesocotyl tissues to root-to-shoot signalling of abscisic acid. *Journal of Experimental Botany* **53**, 297–302.

Sauter, A., Davies, W.J. and Hartung, W (2001) The long-distance abscisic acid signal in the droughted plant: the fate of the hormone on its way from root to shoot. *Journal of Experimental Botany* **52**, 1991–1997.

Sauter, A., Dietz, K.J. and Hartung, W. (2002) A possible stress physiological role of abscisic acid conjugates in root-to-shoot signalling. *Plant, Cell and Environment* **25**, 223–228.

Skewes, M. and Meissner, A. (1997) Irrigation benchmarks and best management practices for winegrapes. Primary Industries and Resources SA Technical report No 259.

Stoll, M., Loveys, B.R. and Dry P.R. (2000) Hormonal changes induced by partial rootzone drying of irrigated grapevine. *Journal of Experimental Botany* **51**, 1627–1634.

Tardieu, F., Katerji, N., Bethenod, O., Zhang, J. and Davies W.J. (1991) Maize stomatal conductance in the field – its relationship with soil and plant water potentials, mechanical constraints and ABA concentration in the xylem sap. *Plant, Cell and Environment* **14**, 121–126.

Thomas, D.S. and Eamus, D. (2002) Seasonal patterns of xylem sap pH, xylem abscisic acid concentration, leaf water potential and stomatal conductance of six evergreen and deciduous Australian savanna tree species. *Australian Journal of Botany* **50**, 229–236.

Wample, R.L. and Smithyman, R. (2002) Regulated deficit irrigation as a water management strategy in *Vitis vinifera* production. Water Reports, FAO Publication number 22, Rome, pp. 89–101.

Wang, ZL, Huang, B.R., Xu, Q.Z. (2003) Effects of abscisic acid on drought responses of Kentucky bluegrass. *Journal of The American Society for Horticultural Science* **128**, 36–41.

Williamson, J.G. and Coston, D.C. (1990) Planting method and irrigation rate influences vegetative and reproductive of peach planted at high density. *Journal of the American Society of Horticultural Science* **115**, 207–212.

Wilkinson, S. and Davies, W.J. (1997) Xylem sap pH increase: A drought signal received at the apoplastic face of the guard cell which involves the suppression of saturable ABA uptake by the epidermal symplast. *Plant Physiology* **113**, 559–573.

Wilkinson, S. and Davies, W.J. (2002) ABA-based chemical signalling: the co-ordination of responses to stress in plants. *Plant, Cell and Environment* **25**, 195–210.

Wilkinson, S., Corlett, J.E, Oger, L. and Davies, W.J. (1998) Effects of xylem pH on transpiration from wild-type and flacca tomato leaves: a vital role for abscisic acid in preventing excessive water loss even from well-watered plants. *Plant Physiology* **117**, 703–770.

Wolters, W. (1992) Influences on the efficiency of irrigation use. International Institute for Land Reclamation and Improvement. Publication 51, Wageningen, The Netherlands.

Yunusa, I.A.M, Walker, R.R, Loveys, B.R. and Blackmore, D.H. (2000) Determination of transpiration in irrigated grapevines: comparison of the heat pulse technique with gravimetric and micrometeorological methods. *Irrigation Science* **20**, 1–8.

Zhao, K., Fan, H. and Harris, P.J.C. (1995) Effect of exogenous ABA on the salt tolerance of corn seedlings under salt stress. *Acta Botanica Sinica* **37**, 295–300.

6 Agronomic approaches to increasing water use efficiency

Peter J. Gregory

6.1 Summary

Until recently, agronomists have been concerned predominantly with yield. However, many practices which increase yield also contribute to increasing the efficiency with which water is used. Increasing scarcity of fresh water means that agriculture is increasingly one of several competing users for water. As a result, there is greater interest in using it more efficiently. Agronomic water use efficiency (WUE) is commonly defined in terms of the yield per unit area divided by the amount of water used to produce the yield. Improved WUE results from either crop improvement that increases yield per unit of water transpired (increased transpiration efficiency), or from crop management practices that maximize transpiration relative to other losses, or both. Variation within crop species and between cultivars exists for transpiration efficiency and there is a wide range of management practices that can reduce the loss of water by evaporation from the soil surface (including mulching, applying fertilizers, and sowing early) and/or increase the amount of water available to a crop (including supplementary irrigation, fallowing and cultivating, and deep rooting cultivars). The effectiveness of these interventions at specific locations depends upon soil properties (moisture characteristic curve and hydraulic conductivity), crop characteristics (growth of the canopy and distribution of roots) and climatic factors (rainfall amount and distribution, and potential rate of evaporation). Analysis of these interacting factors shows that the scope for reducing soil evaporation (and hence increasing WUE) was greatest on clayey soils with frequent rain showers and low evaporative demand, and least on sands with infrequent rain and high evaporative demand.

6.2 Introduction

Agriculture is the largest user of fresh water globally, accounting for some 80% consumed annually (Shiklomanov, 2003). However, water resources are finite and as the world's population continues to grow and demands for water for domestic and industrial purposes increase, then inter-sectoral competition is also likely to increase resulting in relatively less water being available to support systems of food production (Yudelman, 1998). In particular, the quantity and quality of water for irrigation will become more limited. National scale estimates of water consumption (e.g. Falkenmark, 1997; Wallace, 2000) demonstrate that at present about 7% of the world's population live in areas where there is water scarcity. The most acute areas of water scarcity are in North Africa where per capita water availability is already less than the basic requirement of $1000\,\text{m}^3$ per annum. There is also some current scarcity throughout southern Africa and the Middle East (water availability $<2000\,\text{m}^3$ per person per annum). By 2050, though, the prediction is that about 18% of the world's population will have insufficient water to meet their daily needs and that as much as 66% of global population may experience some water stress (Wallace, 2000). Acute water scarcity will be common over large regions of Africa (North, East and southern) and the Middle East, with some scarcity in West Africa, China, the Far East and parts of the Indian sub-continent. Such considerations have drawn attention to the importance of agricultural water requirements and to the need for improvements in the efficiency with which water is used to produce agricultural products (Wallace and Gregory, 2002).

Global water use efficiency is the result of both local water use efficiency and decisions regarding the manner in which water is allocated for different purposes. Agriculture is increasingly one of a number of competing users and, whether for rainfed or irrigated systems of production, must play its part in improving the efficiency of resource use. This chapter focuses on the largely local scale improvements that can be achieved on-farm through improved crop management. It considers both rainfed and irrigated systems of production and suggests those environments in which some improvements in yield per unit of water available might be achieved. Finally, the chapter discusses some of the wider issues involved in the sustainability of systems of crop production as water use efficiency is increased.

6.3   Agronomic definitions of water use efficiency

Until comparatively recently, agronomists have been concerned pre-dominantly with yield of economically important plant parts; efficiency of resource use has been a secondary or tertiary consideration. Much of the early impetus to considering efficient use of water came from the expansion of irrigated area in the USA and the need to estimate the water requirements of crops, but this has extended to rainfed systems of production and concepts of efficient use of water (see Tanner and Sinclair (1983) for a brief overview).

Early work by Lawes (1850) and Briggs and Shantz (1913) demonstrated that growth and transpiration were linked but that the amount of water transpired by plants to produce a unit of dry matter differed with seasons, sites and species. For example, Lawes (1850) found that wheat and peas produced 4.04 and 3.86 g dry matter per kg water transpired, respectively when unmanured, but 4.50 and 4.74 g kg^{-1}, respectively, when manured. The term 'water use efficiency' (WUE) is often used to describe this basic relation between growth and water use. Strictly it is not an 'efficiency' because true efficiency is a comparative term (i.e. dimensionless) requiring a theoretical maximum value, so that some workers prefer to use the term water use 'coefficient' or 'ratio'. Moreover, as Sinclair *et al.* (1984) point out, it 'has been used interchangeably to refer to observations ranging from gas exchange by individual leaves for a few minutes to grain yield response to irrigation treatments through an entire season'.

For the agronomist, WUE is usually a seasonal value defined in terms such as:

$$\text{WUE} = \frac{\text{yield per unit area}}{\text{water used to produce yield}} \qquad (6.1)$$

The numerator and denominator can both be expressed in several ways. Yield is often expressed as grain yield although total above-ground dry matter is also commonly used. In rainfed agriculture in many parts of the world, straw often has an economic value as great as that of grain because it is used to sustain livestock. In such contexts, therefore, total shoot biomass is an appropriate term to use. Similarly, the quantity of water used to produce the yield may also be expressed in several ways. Commonly it is measured as the residual term in the soil water balance equation and expressed as the sum of evaporation directly from the soil surface (E_s) and transpiration (T) during the growing season. It may also be expressed as T alone or, depending on the production system, may take account of total water input into the system (e.g. rainfall plus irrigation). If the denominator is written in terms of all of the components of loss in the soil water balance equation (i.e. including runoff (R) and drainage below the root zone (D)), then equation 6.1 can be expressed in terms of the biomass (M) produced per unit of water lost from the soil profile:

$$\text{WUE} = \frac{M}{(E_s + T + R + D)} \qquad (6.2)$$

Equation 6.2 can then be rearranged to give:

$$\text{WUE} = (M/T) \cdot (1/[1 + (E_s + R + D)/T]) \qquad (6.3)$$

As Gregory *et al.* (1997) have noted, this equation makes clear that improved WUE can come about by either crop improvement that increases (M/T; referred to as the transpiration efficiency) or agronomic management practices that maximize T by reducing the other losses. If the total water supply is increased, WUE will only be increased if T is increased proportionately more than (E_s + R + D). Equation 6.3 highlights the potentially important role that both improved crop genotypes and improved agronomic management can play in improving the WUE of crop production (see Passioura, this volume).

6.4   Crop and genotype influences on transpiration efficiency

As shown by other authors in this volume, carbon assimilation and transpiration by plants are intimately connected processes. If it is assumed that the production of dry matter by a crop in a field is similar to the net assimilation of individual leaves, then the amount of crop dry matter produced per unit of water transpired (the transpiration efficiency, M/T) will be determined by:

1. The gradient of CO_2 concentration between the ambient atmosphere and the intercellular surfaces within the leaf.
2. The gradient of water vapour concentration between the ambient atmosphere and the substomatal cavity within the leaf.
3. The nature of the carbon products constituting the plant biomass.

Large differences exist between crop species in the concentration of CO_2 within the leaf principally depending on the photosynthetic pathway. Typically, the concentration in C_4 species is less than half that of C_3 species so that, if grown under the same conditions, C_4 crops such as maize and sorghum will have a WUE that is some two to three times that of C_3 crops such as wheat and barley. However, there are also differences in the gradient of CO_2 concentration within species giving rise to genotypic variation. In C_3 species, the primary carboxylating enzyme, ribulose bisphosphate carboxylate discriminates against the naturally occurring ^{13}C in favour of the ^{12}C during photosynthesis. This can be detected by carbon isotope discrimination, Δ, and means that M/T and Δ should be negatively related (Farquhar and Richards, 1984). Studies in glasshouses with 14 wheat genotypes, selected on the basis of variation in Δ of dry matter, found substantial genotypic variation in M/T (where M was above-ground biomass) ranging from 4.06 to 5.27 g kg^{-1} in well-watered plants and 4.69 to 5.95 g kg^{-1} in water-stressed plants, and that as expected from theory M/T was negatively related to Δ in both cases (Condon *et al.*, 1990). Field studies with cereals, too, have demonstrated differences in transpiration efficiency although the relationship between M/T and Δ was positive (Condon *et al.*, 1987; López-Castañeda and Richards, 1994). Two factors may account for the positive relationship. First, because the gradient of water vapour between leaf and atmosphere changes during the season (the air generally becomes drier with time) any differences in the pattern of growth of genotypes will lead to different M/T without necessarily affecting Δ. Second, variations may occur in boundary layer conductance of leaves associated with small field plots. López-Castañeda and Richards (1994) found that transpiration efficiency was positively related to an index combining plant height and anthesis date suggesting that these two factors are important in the field. Early flowering cultivars that elongate first can maintain a cooler canopy with a greater boundary layer conductance than late flowering cultivars and thereby have a higher transpiration efficiency. Genotypic differences in transpiration efficiency and Δ have also been found in groundnut (Hubick et al., 1986), upland rice (Dingkuhn et al., 1991) and upland cotton (Leida et al., 1999), although several authors point out that the usefulness of the trait for breeding drought-tolerant cultivars will have to wait for a better understanding of the factors affecting Δ under field conditions.

Except in very water stressed conditions, the temperature of the crop canopy is frequently close to air temperature over extended periods. This means that the difference in water vapour concentration between inside the leaf and the ambient air can be approximated by the saturation deficit of the ambient air to give the expression:

$$M/T = k/(e^* - e) \tag{6.4}$$

where e^* is the saturation vapour pressure of the air, e is the actual vapour pressure of the air, and k is a constant specific to the genotype (Tanner and Sinclair, 1983). It is clear from this equation that transpiration efficiency will be higher in humid regions than in arid regions because the saturation deficit $(e^* - e)$ will be lower. Typically, the saturation deficit changes during the growing season with the air becoming drier as the season progresses. This means that growth early in the season will have a higher transpiration efficiency than later growth. Seasonal changes can have a marked effect on transpiration efficiency. For example, Monteith (1986) found that saturation deficit in India and West Africa decreases from 3–4 kPa in the dry season to 0.5–1.0 kPa in the rainy season and that this change accounted for about 0.5 to 0.33 of the response of crop dry matter production to the rains. Even in more temperate regions, seasonal differences in saturation deficit can affect M /T (Day et al., 1987). Table 6.1 summarizes values of M/T and k found for a range of crops. Clearly, k is greater in C_4 than C_3 crops.

There are several practical difficulties in calculating $(e^* - e)$ for crop canopies (Tanner and Sinclair, 1983). Firstly, it should be estimated for only the daylight hours when transpiration is occurring because inclusion of night time values when the stomata are closed will tend to underestimate the day time value. Secondly, the value of e^* is only likely to correspond to the internal water vapour concentration of the leaf if the canopy completely covers the ground (i.e. leaf area index for most cereals $\geq \approx 3$–4). In sparse canopies, different effects may result depending on whether the soil surface is dry or not. If the surface is wet, evaporation directly from the soil surface may appreciably reduce the saturation deficit around the lower leaves in comparison with the bulk atmosphere and thereby increase the transpiration

Table 6.1 Experimentally derived values of WUE (based on shoot biomass (M) and transpired water (T)) and k determined for a range of crops.

Crop	Location	Year	WUE (kg ha^{-1} mm^{-1})	k (Pa)	Author
Barley, Julia	Rothamsted, UK	1976 1979	33.3 43.5	2.9 2.9	Day et al. (1987)
Barley, Beecher	East Beverley, Western Australia	1988	48.4	3.0	Gregory et al. (1992)
Maize, various	Various states, USA	1912, 1974 and 1975	Range 20 to 55	Range 8.2 to 12.0	Tanner and Sinclair (1983)
Oat, Leanda	Goettingen, Germany	1976, 1977, 1982 and 1983	41	3.3	Ehlers (1989)
Potato, Russet Burbank	Wisconsin, USA	1972 1973 1976	55 36 42	6.5	Tanner (1981)
Sunflower, Beauty and Cannon	Tatura, Victoria, Australia	1988	2.41 pre-anthesis 1.88 post-anthesis	3.6	Sadras et al. (1991)
Wheat, Gutha	East Beverley, Western Australia	1988	38.5	2.4	Gregory et al. (1992)

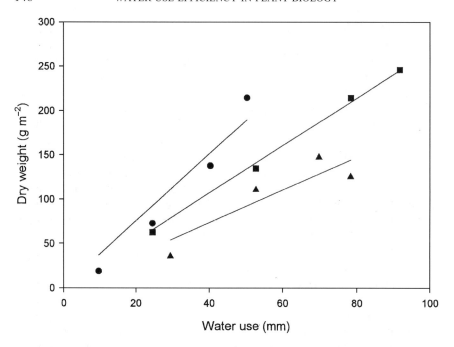

Figure 6.1 Effects of row spacing (● 150 cm, ■ 75 cm, and ▲ 37.5 cm) on the relation between accumulated dry weight and accumulated water use determined with a neutron probe for stands of pearl millet grown in Niamey, Niger (from Azam-Ali *et al.*, 1984).

of that part of the canopy and the overall transpiration efficiency. If on the other hand the surface is dry, saturation deficit is likely to be greater than in denser canopies leading to lower transpiration efficiency. Azam-Ali *et al.* (1984) grew pearl millet on a deep sand in West Africa and found, though, that these effects were small compared to the changed seasonal pattern of water use by the crops. In widely spaced plants, the stomata remained open for longer so that despite the dry surface the overall WUE was higher. This effect is shown in Figure 6.1; as row spacing increased from 0.375 m to 1.5 m, so the transpiration efficiency increased from 20 to 30 kg ha^{-1} mm^{-1} and WUE increased from 21 to 47 kg ha^{-1} mm^{-1}. To the extent, then, that saturation deficit is affected by row spacing, M/T may be affected by agronomic management as well as biological and meteorological factors.

The proportions and chemical composition of the individual components constituting the total crop biomass will also influence transpiration efficiency. Most calculations of transpiration efficiency are made on the basis of above ground biomass and ignore the roots. In large crops, roots may be only about 10% of the total crop biomass at maturity (Gregory, 1994), but in many regions they may be a substantially greater percentage. For example, Gregory and Eastham (1996) found that root weight of wheat and lupin crops grown on duplex soils in Western Australia comprised 30 and 50% of total plant weight, respectively, during vegetative growth and 20 and 35% during grain filling. So, root mass may be important on occasions and supposed differences between sites and seasons in transpiration efficiency may reflect differences in root growth. For example, López-Castañeda and Richards (1994) found that transpiration efficiency based on above ground biomass of two barley and two wheat cultivars was, on average 23% less at one site than another. However, the site with the apparently lower transpiration efficiency had greater root mass and when this was accounted for, the difference between sites was substantially reduced to 9.3%. Similarly, while Siddique et al. (1990b) noted a slight tendency for lower WUE on a shoot biomass basis in older cultivars of wheat, complementary studies showed that old cultivars partitioned a higher proportion of their total biomass into roots. When calculated on the basis of total dry matter, the WUE of old and modern cultivars was similar (Siddique et al., 1990a).

The chemical composition of the biomass will also affect the transpiration efficiency because 1 g of primary photosynthate is equivalent to about 0.83 g of carbohydrate, 0.4 g of protein and 0.33 g of lipid (Penning de Vries, 1975). This is a major reason why oilseed crops have much lower transpiration efficiencies than cereal crops and why the transpiration efficiency may apparently change during a season in such crops. For example, Sadras et al. (1991) found that transpiration efficiency based on above ground biomass of two sunflower cultivars was, on average 22% lower in the grain filling period than in the pre-anthesis period. However, when the transpiration efficiency was re-calculated to allow for the cost of oil synthesis and the seasonal change in saturation deficit, there was no significant difference between the pre-anthesis and post-anthesis periods.

6.5 Management of water in rainfed crop production

Many workers have commented on the small proportion of water potentially available to crops that is actually transpired in some environments. For example, Allen (1990) measured evaporation and transpiration of barley crops in northern Syria and, by difference, found that transpiration was a small component (<35%) of total water use. Similarly, in Niger where rainfall frequently occurs as intensive showers, transpiration is normally less than evaporation from the soil surface (Wallace et al., 1993) and on sloping land (2–3%) in farmers fields can be as low as 6% of rainfall (Rockström, 1997). In the same environment drainage is, in many cases, almost equal to the water lost as soil evaporation and transpiration (Gaze et al., 1997; Rockström, 1997). Such findings have led many to conclude that the efficiency with which water is used to produce crops could be significantly improved in many rainfed environments (Payne, 1997).

Because of the difficulty of measuring transpiration, some workers have attempted to determine relations between growth and rainfall in rainfed environments. For example, Stephens and Lyons (1998) found that wheat yields in Australia were strongly related to total rainfall and to the seasonal distribution of rainfall through the year. Such temporal relations varied spatially and the correlations increased as the climate became more water-limiting and as the water-holding capacity of the soils increased. Autumn rainfall that allowed early sowing and rain in spring were important for higher yields. French and Schultz (1984) analysed many experiments with wheat in South Australia and concluded that a 'potential yield' could be defined given as a line with a slope of 2 g grain mm^{-1} rain above a base of 110 mm rain. The rainfall used was that during the growing season and the base figure was believed to be the amount of evaporation directly from the soil surface (E_s). They pointed out, though, that E_s ranged from 30 to 170 mm depending on seasonal rainfall and soil type. Similarly, Cooper et al. (1987a) found a range of 124–172 mm for barley crops grown at two sites in northern Syria with different amounts of fertilizer. This marked inter-seasonal variability of E_s was confirmed in a simulation analysis of wheat yields at three sites in Western Australia using long-term daily historical weather records for an 82–87-year period (Asseng et al., 2001). On a sandy soil at a site with a mean growing season rainfall of 322 mm (range 112–535 mm), they found that for crops grown without N fertilizer, mean E_s was 168 mm (range 82–218 mm). Asseng et al. (2001) found that with no fertilizer applied, E_s was correlated with the amount of seasonal rainfall with about 25% of rainfall > 200 mm lost as E_s. Similarly, E_s was highest at the high rainfall site and lowest at the low rainfall site. With high N fertilizer inputs the correlation of E_s with rainfall disappeared because of increased growth and coverage of the soil surface.

Transpiration as a proportion of (E_s + T) also increases with rainfall (Figure 6.2) so that in low rainfall environments a larger proportion of rainfall is not transpired and crop production is, therefore, smaller (Gregory et al., 1997). There are several reasons for this low utilisation of water in crop production which are well illustrated by the Kenyan data in Figure 6.2. First, more extensive canopies in wetter seasons reduce E_s as a result of several factors including: (i) shading of the soil surface, (ii) impeded aerodynamic transfer of water vapour away from the soil surface, (iii) humidification of the air in the canopy immediately above the surface, and (iv) water extraction by roots, which reduces soil hydraulic conductivity and the upward flux of water through the soil matrix towards the soil surface. Second, wetter seasons tend to have more rain per rainfall event, resulting in a greater depth of wetting which increases the total storage and reduces the proportion of rainfall lost as E_s in the few days immediately following rainfall (when E_s is relatively fast). Finally, wetter seasons tend to have more frequent rain events which results in a smaller average E_s per rain event. Figure 6.2 also shows that crop management can

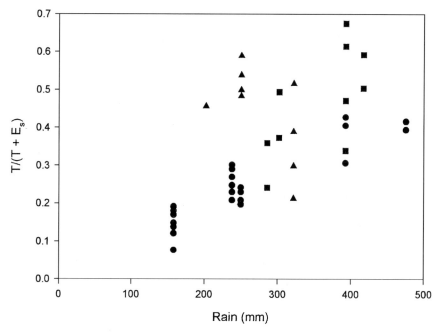

Figure 6.2   Relation between $T/(T + E_s)$ and rainfall for Kiboko, Kenya (● data from Pilbeam *et al.*, 1995), Breda and Tel Hadya, Syria (■ data from Gregory *et al.*, 1984; Brown *et al.*, 1989; Allen, 1990), and East Beverley, Merredin and Wongan Hills, Western Australia (▲ data from Gregory *et al.*, 1992; Yunusa *et al.*, 1993; Asseng *et al.*, 2001).

have a substantial effect on the amount of transpiration (and thus crop biomass) in a given growing season, and there is currently interest in optimising crop management to improve both WUE and yield.

6.5.1   Increasing T relative to other losses

The principal ways in which evaporation from the soil surface might be reduced or the total water supply of the soil increased are summarised in Table 6.2. Mulching with crop residues is an obvious way to reduce E_s and it may have other desirable effects such as reducing runoff, increasing infiltration, and decreasing surface temperatures. In terms of water conservation, the main effect of mulches is to reduce the rate of evaporation when the soil surface is damp (first-stage drying) and thereby to extend the duration of this stage (Papendick *et al.*, 1990). If drying persists, then the initial high rate of water loss from an unmulched soil is partially compensated for by the increased resistance to water flow provided by dry soil, so that over time the cumulative losses start to converge. For example, Bristow *et al.* (1986) in simulations of mulching in the Pacific Northwest USA showed that mulching significantly

Table 6.2   Agronomic practices available for improving water use efficiency in rainfed cropping systems.

Reducing evaporation from the soil surface	Increasing total water supply
Mulching	Rain harvesting
Early sowing	Irrigation
Cultivars with rapid early growth	Cultivation to increase infiltration and reduce runoff
Modifying plant population and spacing	Weed control
Application of fertilizer	Fallowing
	Application of fertilizer
	Cultivars with deep root systems

reduced E_s during the winter months when there were frequent rain showers but increased it in the summer months. Overall, mulching reduced E_s by 36%. Under field conditions, rain showers interrupt a drying cycle resulting in multiple wetting and drying cycles, and the mulch layer is degraded by living organisms. Jalota *et al.* (2001) working in Ludhiana, India found that the effects of straw mulching were complex and depended on the soil type, rainfall pattern, and evaporative demand. Over a short period, mulching increased water storage under all conditions but over a longer period straw mulching was best only if evaporative demand was low and rainfall frequent. Under conditions of high evaporative demand, tillage or straw mulching resulted in little long-term increase in water storage in sandy soils. Straw incorporation was advantageous for long-term water storage depending upon the combination of soil texture, rainfall and evaporative conditions. In some regions with silty soils (e.g. the Pacific Northwest, USA), shallow tillage is itself effective in producing a dry soil mulch which can conserve appreciable quantities of water by disrupting the micropores through which water moves to the surface to evaporate (Papendick *et al.*, 1990).

In many rainfed regions of the world, though, crop residues are in short supply as they are fed to animals or broken down by animals such as termites, and the labour required to collect and spread the quantities of material necessary to appreciably increase water storage is unavailable. For these reasons, much crop management research has focussed on using the crop canopy itself to cover the soil surface to reduce E_s, and to increase T, yield and WUE. Increasing the early growth of the canopy when the soil surface is usually damp and the saturation deficit is low has proved effective in increasing WUE especially on the finer textured soils common in northern Syria (Cooper *et al.*, 1987b). Early sowing of crops is a very important means of maximising crop yield and WUE. In South Australia, French and Schultz (1984) found that for each week that sowing was delayed after the optimum, grain yield was reduced by 200–250 kg ha^{-1}. Similarly in Western Australia, early sowing of wheat and lupin crops has been found to increase grain yield substantially (Gregory and Eastham, 1996; Anderson, 1992) accompanied by improvements in WUE at some sites depending on soil and rainfall (Eastham and Gregory, 2000; Tennant and Hall, 2001). In a detailed study using ventilated chambers and microlysimeters over two seasons, Eastham et $al.$ (1999) measured greater losses of water by E_s in both years in the early part of the season beneath late-compared with early-sown crops of wheat and lupin. The more advanced canopy development associated with early sowing reduced E_s during the energy-dependent, first stage of evaporation and the greater losses beneath late-sown crops were not sustained as surface soil water contents declined. Loss et $al.$ (1997) also found that sowing date of faba bean affected the partitioning of water use. $(E_s + T)$ from sowing to maturity decreased as sowing was delayed but when E_s before emergence of the later sowings was taken into account, water use did not differ between sowing times and all soil moisture was depleted by harvest. WUE decreased significantly with sowing date at all sites from values typically about 25–30 kg dry matter $ha^{-1} mm^{-1}$ in early May to values about 10–15 kg $ha^{-1} mm^{-1}$ in late June.

Although many studies in Mediterranean regions have shown little or no differences in seasonal $E_s + T$ between crops or agronomic treatments (e.g. Cooper et $al.$, 1987b; Siddique et $al.$, 1990b), small differences may be important for other reasons. For example, in both seasons of the study by Eastham et $al.$ (1999), $E_s + T$ for early-sown crops was greater than that from late-sown crops when soil water availability was high suggesting that early sowing may increase early season water use and thus reduce drainage losses compared with late-sown crops. In this environment, drainage resulting in secondary salinity is an important issue so that crop husbandry that maximizes T and minimizes D is essential (see section 6.8). A small but consistent difference between seasonal water use of chickpea and lentil crops was also found from three experiments over 12 seasons in northern Syria (Zhang et $al.$, 2000). Mean $E_s + T$ was 268 mm for chickpea and 259 mm for lentil with an average depth of extraction of 1.2 m for chickpea and 0.8 m for lentil.

While early sowing is generally beneficial to yield and WUE, Cooper et $al.$ (1987b) point out two practical limitations to its optimisation. First, in many parts of the world initial rains may be unreliable so that early sown crops may germinate and then suffer severe drought or die in subsequent weeks. Second, many farmers prefer to allow early rains to germinate weed seeds which can then be killed by tillage before the crop is sown. This is particularly important in areas of high weed infestation and in crops and regions where hand weeding rather than herbicides are used to control weeds after emergence.

In some regions there is scope to improve growth and WUE by altering row spacing and adjusting plant populations but this is often difficult to achieve in practice because the optimum population for grain yield is often related to the water supply available for the season. Many crop plants have sufficient plasticity of form that yields remain unaffected by a wide range of row spacing and seed density combinations. For example, Yunusa *et al.* (1993) found no significant effect of up to a four-fold increase in row spacing (0.09 to 0.36 m) on wheat yield at two sites in Western Australia and E_s was similarly unaffected. Grain yields in eight other experiments over three years were similarly unaffected probably because in this environment the period when the soil surface is wet is short, coupled with the need to intercept 90% of the incident radiation to significantly reduce E_s. In these crops the leaf area index rarely exceeded two which intercepted only about 50% of the incident radiation and most of the seasonal evaporation from the soil occurred when the soil was in the second phase of drying. In such circumstances, row spacing is unlikely to significantly change E_s.

Contrasting results were obtained by Payne (1997) in West Africa with pearl millet. He found that increasing the density of 'hills' (several seeds are planted in a slightly raised area) from 5000 to 20 000 ha^{-1} increased yield and WUE significantly under a range of fertility conditions. The popular notion that wide spacings were necessary to safeguard against crop failure in dry years was not supported by these experiments because even in a very dry year higher plant densities yielded more. However, there is a need for caution when advocating alterations to traditional planting patterns especially when moving from extensive agricultural practices to more intensive ones. For example, in a series of trials with sorghum over a four-year period, Rees (1986) found the highest overall average yields (about 1.5 t ha^{-1}) with plant densities in the range 40 000 to 80 000 plants ha^{-1}, substantially more than the 1.1 t ha^{-1} obtained with 10 000 plants ha^{-1}. The coefficient of variation of yield, though, increased from about 63% at 10 000 plants ha^{-1} to about 95% at 80 000 plants ha^{-1}. This meant that although a low plant density did not yield as much as a high density in favourable conditions, it did not fail in harsh conditions, and yield was stable although at a low level.

Where soil nutrients are deficient for maximum growth of crops, application of fertilizers and manures may not only result in increased growth but also in increased WUE. Fertiliser use may increase slightly the total amount of water used (e.g. Cooper *et al.*, 1987b for barley in Syria; Ogola *et al.*, 2002 for maize in the UK), but the principal effect is to increase early canopy growth so that it shades the surface and thereby reduces E_s as a proportion of the total water that is evaporated. This effect of modest applications of fertilizer has been well documented in several studies (Cooper *et al.*, 1987a,b; Latiri-Souki *et al.*, 1998; Ogola *et al.*, 2002) and is illustrated in Figure 6.3 and Table 6.3. Similarly in a study of pasture production by mixed swards, no significant difference in water use was observed between systems with either no or high applications of superphosphate and potassium fertilizers over three years despite large positive effects of fertilizer on biomass production (Bolger and Turner, 1999). However, the beneficial effect of fertilizer in increasing growth and reducing E_s is not

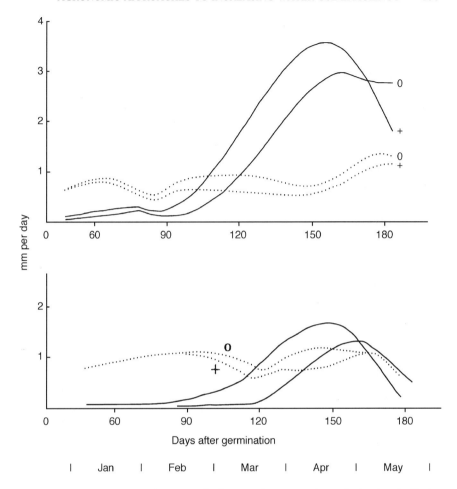

Figure 6.3   Seasonal pattern of daily rates of soil evaporation (– –) and transpiration (——) of barley grown with (+) and without (o) fertilizer at Jindiress (upper figure) and Breda (lower figure) in northern Syria (from Cooper *et al.*, 1987a).

universal (Gregory *et al.*, 1997). Figure 6.3 shows differences between the results obtained in Kenya and Syria. In Syria, a greater proportion of the total evaporation is transpired than in Kenya because the low evaporative demand results in a smaller proportional loss of rainfall even though the soil surface is effectively wet for much of the growing season. Moreover, in Kenya the majority of E_s occurs during periods when it is limited by the movement of water through the soil rather than by the amount of radiation incident on the soil surface (i.e. E_s is largely supply limited in Kenya rather than demand limited as in Syria).

In semi-arid production systems, the efficiency of N and P fertilizers depends on the amount of water available to the crop so that the response to N,

Table 6.3 Effects of modest applications of fertilizer on shoot dry matter, water use and water use efficiency (WUE) for crops of barley at Breda, Syria (from Cooper et al., 1987b) and pearl millet at Sadore and Dosso, Niger (ICRISAT) (from Gregory et al., 1997).

Crop	Season	Rainfall (mm)	Fertilizer	Dry matter (t ha^{-1})	Water use (mm)	WUE (kg ha^{-1} mm^{-1})
Barley	1981/82	324	+	6100	231	26.4
			-	4540	231	19.7
	1983/84	204	+	2880	176	16.3
			-	1340	171	7.8
Millet	1984	260	+	4750	165	28.8
			-	2417	163	14.8
	1985	380	+	5000	247	20.2
			-	3100	270	11.5
	1986	440	+	3850	268	14.4
			-	1140	211	5.4

in particular, is variable and limited in dry years. Typically, crop response to N increases with increasing rainfall while response to P decreases on P-deficient soils, (Jones and Wahbi, 1992). Studies in the Sahel have also concluded that soil fertility is often a more important factor in rangeland and crop productivity than rainfall, so that effective management of water cannot be achieved without also managing soil nutrient constraints (Payne, 2000). The limiting factors to crop growth at different times during any particular season could be either water or nutrient availability or both. A practical problem to be resolved in many semi-arid regions is how to afford and apply the optimum amount of fertilizer to produce an economically viable yield in a given season. For example, Sadras (2002) analysed financial returns over a 40-year period from fertilizer applications in an area with low and erratic rainfall in south-eastern Australia and found that a very low input of N fertilizer (5 kg N ha^{-1}) ensured the greatest economic stability at all sites examined. Most scientific analysis is based on hindsight but farmers must operate without this benefit so that conservative practices tend to dominate.

6.5.2   Increasing the total water supply

Table 6.2 shows that many management practices are used to increase the water available to crops and, provided that this water is transpired and not lost as E_s, drainage or runoff, increased growth and WUE will result. The collection of water by rain-harvesting in some areas and its diversion to crop-producing areas of land has been practised in Mediterranean and some African countries for many centuries. Supplementary irrigation can increase crop yields significantly though the benefits vary from year to year depending on rainfall. For example, in northern Syria, Zhang et al. (2000) found that on a clay soil one or two applications of irrigation at flowering and/or during pod-filling increased grain yield by 92% for chickpea and by 70% for lentil. The increase was smaller in wet years than dry years, and supplementary irrigation stabilised grain yield of both crops in a very dry season in which the crops might otherwise have failed. In contrast, Latiri-Souki *et al.* (1998) found no beneficial effect of irrigation to durum wheat on a clay soil but yield was increased on a sandy soil.

Weed control is an essential way of ensuring that the water stored in soil is used by crops. Weeds can considerably decrease crop growth and WUE particularly in food legume crops which have slower initial growth than many cereals. For example, Cooper *et al.* (1987b) showed that weed control in lentil almost doubled dry matter production (from 775 to 1429 kg ha^{-1} averaged over fertilizer and cultivation treatments) while seasonal water use was only slightly affected (234 mm without weeds and 268 mm with weeds). As a result, WUE was also almost doubled (from 2.9 to 5.9 kg ha^{-1} mm^{-1}) by controlling weeds.

Cultivation and fallowing are often used in combination to increase the supply of water to crops though the traditional role of fallowing is decreasing in many regions as production is intensified. Fallowing is used to increase the amount of water available for the next season although the amount available is highly dependent on soil depth, soil type, and whether weeds are controlled. Analysis by Papendick *et al.* (1990) showed the importance of the additional water stored in the profile for the production of wheat in the Pacific Northwest, USA. Wheat yields were linearly related to the percentage of precipitation stored in the soil at the end of the fallow season (the fallow efficiency) to give a WUE for grain of 18 kg ha^{-1} mm^{-1} at two sites with precipitation of 587 and 720 mm. Soil moisture was also conserved by bare fallows in Botswana (Jones and Sinclair, 1989). In each of ten trials, a sorghum crop following bare fallow out-yielded one following a cereal crop (maize or sorghum). Long-term means yields were 1730 and 860 kg ha^{-1} so that a fallow/sorghum rotation gave twice as much crop half as often; this pattern of production is unacceptable to subsistence farmers. Moreover, because the water storage capacity of the soil profile is limited in comparison with the rainfall, a short-season legume (cowpea) sorghum rotation had greater benefit. Similar considerations have led to the development of cereal/legume rotations in many regions (e.g. Australia).

In a long-term study of tillage systems on three-course rotations on a fine clay in northern Syria, Pala *et al.* (2000) found no advantage from deep tillage, with either disc or chisel plough, on profile water storage compared to either minimum or zero tillage (Table 6.4). However, they noted that there were no intense rain storms during their experiments and suggested that deep tillage with its associated surface roughness night increase infiltration and reduce runoff if intense rain fell between the primary and secondary tillage

Table 6.4 Effects of soil tillage with two crop rotations (1 wheat, chickpea, watermelon; 2 durum wheat, lentil, watermelon) on profile soil water content (mm) at sowing and harvest for three seasons at Tel Hadya, northern Syria (from Pala *et al.*, 2000).

Tillage	Rotation	1987/88		1989/90		1991/92	
		Sowing	Harvest	Sowing	Harvest	Sowing	Harvest
Zero tillage	1	429.5	410.6	452.2	426.9	422.9	399.0
Ducks-foot	1	409.7	409.6	443.0	427.9	456.7	410.9
Chisel	1	395.5	383.6	454.1	419.9	450.2	403.1
Deep disc	1	395.1	392.0	415.7	400.6	391.6	359.6
Zero tillage	2	417.2	440.7	510.9	448.0	424.8	425.6
Ducks-foot	2	421.6	425.8	512.3	425.0	442.3	397.1
Chisel	2	414.9	450.6	490.6	432.5	448.0	418.2
Deep disc	2	421.8	436.7	465.3	416.4	423.8	406.3

applications. Similarly, Jones (2000) at a drier site also in northern Syria showed that zero-till systems with retention of barley residues may marginally enhance stored soil water, but that the effect on barley yields, either monocropped or rotated with vetch, was not significant. In contrast, Kilewe and Ulsaker (1984) working on alfisols in Kenya where infiltration is reduced in the early part of the rainy seasons by surface sealing caused by intense rainfall found that cultivation and surface structuring could reduce runoff substantially. Cultivation in wide furrows decreased runoff and allowed infiltration to continue long after the rain had ceased. In turn, maize yields were increased by the formation of ridges and furrows (by almost 50% in one season and three-fold in another) and WUE similarly increased.

Deep root systems have also been suggested to be advantageous in maximising the uptake of water by crops. On deep sand soils in West Africa where drainage is common, Payne *et al.* (1990) determined that increased rooting depth and root length density were associated with reduced drainage below the root zone and more soil water extraction from upper and lower layers during dry periods. In such environments, improved crop management may have a greater influence on D than E_s by promoting root growth that will utilise the water before it can drain. In several studies conducted on pearl millet grown on farmer-managed fields in Niger, about 80% of the seasonal rainfall >240 mm was lost by drainage whereas for crops that were more intensively managed on experiments at research centres there was much less drainage (Gregory *et al.*, 1997). Similarly, Gaze *et al.* (1997) found that once there was sufficient infiltration to cause drainage, <30% of the additional rainfall was lost as either T or E_s.

In parts of northern Syria with rainfall <350 mm, there is limited scope for root systems of barley crops to affect the amount of water used because there is little or no drainage and rooting depth and the seasonal depth of wetting are closely related (Gregory et al., 1984; Cooper et al., 1987b). In some years water from the previous season is stored below the depth of wetting and becomes available for crop uptake when the wetting front reaches it. This can occur when bare fallowing is practised and in continuous cropping when a year of high rainfall allows water to penetrate below the rooting depth. In general, this occurs only rarely so that water below the depth of wetting must be regarded as extractable once, or only once between years of high rainfall. Zhang et al. (2000), though, found that the depth of extraction of lentil crops (average 0.8 m) was less than that of chickpea crops (average 1.2 m) so that some water storage may occur beneath more shallow rooted crops and be available to subsequent cereal crops.

6.6 Management of water in irrigated crop production

About 20% of arable land in developing countries is irrigated and this produces some 40–55% of basic staples and value in several developing countries (Yudelman, 1998). Irrigated area increased rapidly between 1960 and the mid-1980s coincident with the introduction of new varieties and agri-chemicals. Over 70% of the increase in grain production between 1962 and 1990 came from increased yields in irrigated areas and <10% from increased area of production. Average yields on irrigated land in the early 1990s were $3.7 \, t \, ha^{-1}$ for rice and $2.4 \, t \, ha^{-1}$ for wheat compared with 2.4 and $1.7 \, t \, ha^{-1}$, respectively, on rainfed land. Irrigation, then, results in agriculture that uses more water but produces more output per unit land area (provided that salinization is prevented).

The definition of water use efficiency given in equation 6.1 has to be broadened when considering irrigated production systems because the efficiency of water use includes not only the on-field use of water but also the losses of water from reservoirs (S) and the losses during conveyancing (C) to the field. Storage and conveyancing efficiency can be expressed as:

$$(S + C)_{WUE} = \frac{(E_s + T + R + D)}{(E_s + T + R + D + S + C)} \tag{6.5}$$

Globally storage and conveyancing efficiencies are about 70% (Bos, 1985), implying a 30% loss of water before delivery to the field, Table 6.5 (Wallace, 2000). Irrigation efficiencies are defined as:

$$IRR_{WUE} = \frac{(E_s + T)}{(E_s + T + R + D)} \tag{6.6}$$

and are typically about 37%, giving on-field losses averaging 63% of the water delivered to the field, or 44% of the total water resource at source (Table 6.5). Similar losses to runoff and drainage have been reported in semi-arid

Table 6.5 Estimates of the water losses in irrigated and rainfed agriculture in semi-arid areas (from Wallace and Gregory, 2002).

	Irrigated agriculture – Fraction of available water[1] %	Rainfed agriculture – Fraction of rainfall %
Storage and conveyancing	30	0
Runoff and drainage	44	40–50
Evaporation (from soil or water)	8–13	30–35
Transpiration	13–18	15–30

[1] Rainfall and stored surface or groundwater

rainfed agriculture (Table 6.5). Because weather forecasts are rarely taken account of in irrigation management, water may be scheduled just before or after rain leading to substantial losses by drainage or runoff. Schulze *et al.* (2002) evaluated such losses for 127 quaternary catchments constituting the sugarcane belt of South Africa and showed that in median climatic conditions drainage losses can vary from 40 to nearly 300 mm per annum and runoff from 40 to 200 mm per annum. Losses were greater in coastal areas than those in interior and hot, dry areas where full irrigation is the norm. Losses decreased as the irrigation cycle was increased from 7 to 14 to 21 d. While there was a small yield penalty incurred by increasing the length of cycle, IRR_{WUE} and WUE were increased and losses reduced. However, some of the water lost by R and D may return to aquifers and streams from which it can be abstracted again so that the efficiency of water use over large areas may be greater than the estimates of field efficiency suggest. For example, in a large-scale irrigation scheme such as the Nile, the overall IRR_{WUE} may approach 80% despite lower values at individual field scale (Seckler, 1999). The IRR_{WUE} of the system will increase as the amount of reuse increases but is very dependent on how much water is recyclable. Every reuse requires energy and the quality may deteriorate with use if agri-chemicals are leached with the water.

The above IRR_{WUE} is the ratio of the total water evaporated to that supplied to the field. However, as with rainfed agriculture, a significant amount (30–50%) of the evaporation is from the soil or water supporting the crop (Wallace, 2000). Table 6.5 summarizes how the various efficiencies translate to losses in terms of the original water resource (rainfall or stored surface and/or ground-water). Less of the available water tends to be evaporated in irrigated systems, however, they generally have higher ground cover, so that the transpiration component is larger, with the net effect that the amount of available water transpired is broadly similar in both types of agriculture (15–30%).

In addition to the agronomic means of increasing WUE listed in Table 6.2, there are other technical, managerial and institutional improvements available for increasing the efficiency of use of irrigation water at a field level. Managerial and institutional options tend to be most effective in reducing losses and agronomic and technical options to maximizing transpiration and carbon fixed per unit of water transpired. Hiler and Howell (1983) note that as water becomes scarcer and pumping costs increase, yield per unit of applied water becomes relatively more important than yield per unit area in the thinking of producers. Improved methods of irrigation scheduling based on concepts such as stress day indices and soil water balance modelling allow real-time management decisions to be made and, together with improved systems for applying water through sprinklers and drippers, will allow further improvements in efficient water use in irrigated production systems.

6.7   Interactions between soils, climate and crop management

As will be clear from the preceding descriptions, while there is a variety of possible agronomic means (including choice of genotype) to influence WUE, the effectiveness of the intervention differs depending on several edaphic and climatic factors. Only recently have means been found to account for the major factors, principally through careful experimentation backed up by good process based models of the soil water balance and crop growth. The physical factors that appear to be important in determining the success of a management option include the moisture characteristic curve and hydraulic conductivity of the soil, the amount of crop cover and the distribution of roots, the quantity and temporal distribution of rain, and the potential rate of evaporation (Gregory *et al.*, 2000). The effects of these properties on the potential to alter WUE were examined using a simple dynamic water balance model in combination with experimental results from Syria, Kenya and Niger (Gregory *et al.*, 2000). The model simulated a 20 d period during which the rainfall was set to be equal to the potential evaporation. The output was expressed in terms of an index (I), which expresses the potential for reducing E_s by increasing the crop cover from 40 to 80%. The reduction in soil evaporation is normalized with respect to the appropriate evaporative demand to give an index between 0 and 1. A value for I of 1 indicates that the saving in E_s brought about by increasing ground cover is that which would be expected if the saving was equal to the reduction in potential evaporation from the soil surface. Conversely, a value of <1 indicates that the saving in E_s resulting from increased ground cover is less than would be expected from the reduction in potential soil evaporation. Such values arise in situations where E_s is limited in part by the availability of water, rather than the supply of energy. In such circumstances, it might reasonably be expected that the effect of the crop on E_s would be influenced by the distribution of roots.

Results of the simulations for three soil types, rain applied at 2, 10 or 20 d intervals, two rainfall intensities and with either a surface-oriented or deep root system are shown in Figure 6.4. In all cases, rewetting the soil every 2 d

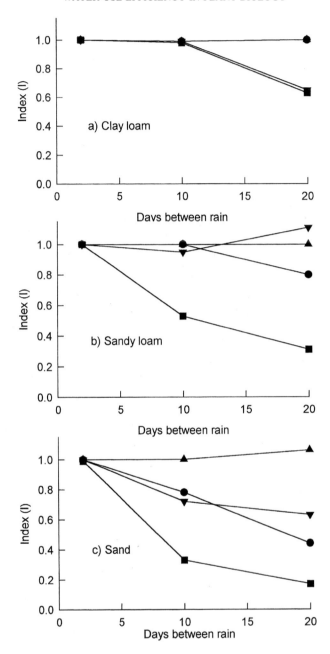

Figure 6.4 The effect of soil type, rainfall distribution and root distribution on the index (I). The simulations used shallow roots with rainfall of either 2 mm d^{-1} (•) or 6 mm d^{-1} (■) and deep roots with rainfall of either 2 mm d^{-1} (▲) or 6 mm d^{-1} (▼) (from Gregory *et al.*, 2000).

maintained the surface sufficiently wet for E_s to be unrestricted by the supply of water (i.e. $I = 1$). For the sandy soil, extending the interval between rains allowed the soil surface to dry out between rain events such that E_s was limited by the supply of water. In such cases, there is less scope for reducing E_s by increasing ground cover (i.e. I fell substantially below 1). This effect was even more marked for scenarios with the higher evaporative demand. The distribution of roots had a pronounced effect. In the 'shallow root' simulations, there was much more rapid uptake of water by roots from soil close to the surface, which had the effect of 'throttling' the water supply available for direct evaporation from the soil surface. One of the consequences of this was to reduce substantially the scope for reducing E_s by increasing the size of the canopy. In the most extreme case (i.e. with sandy soil, high evaporative demand and 20 d rainfall interval), the potential saving of E_s by manipulating ground cover was less than 20% of that when the soil surface was continuously wet. In contrast, the evaporation from the clay loam in conditions of low evaporative demand remained highly susceptible to changing ground cover, even when rainfall events were 20 d apart.

In a few cases, it was found that the saving in E_s brought about by increasing the ground cover was actually greater than that expected when the soil surface remained freely-evaporating throughout. This was most evident in the sandy soils when there were long dry periods and low potential evaporation. This is believed to occur because the rate of evaporation from the soil surface became water limited sooner with 0.8 ground cover than when the cover was sparser, because of more rapid soil drying by root water uptake.

Gregory et al. (2000) concluded that the scope for reducing E_s was greatest on clayey soils with frequent rain showers and low evaporative demand, and least on sands with infrequent rain and high evaporative demand. The analysis also demonstrated that drying of the soil surface by roots was another important means of reducing E_s, especially when the evaporative process was water limited. The distribution of roots had least effect on E_s in the clay loam because its higher unsaturated conductivity at modest suctions meant that there was much less time spent under conditions when E_s was water-limited, and therefore susceptible to the effects of soil drying by uptake of water by roots.

Asseng et al. (2001), too, used a model to investigate the degree of variation in yield, WUE and nitrogen use efficiency (NUE) to be expected under natural conditions in Western Australia. They used the Agricultural Production Systems Simulator (APSIM) to analyse the effects of rainfall on growth, WUE and NUE at three locations (average annual rainfall of 461, 386 and 310 mm) and two soil types that had contrasting available water holding capacities in the root zone (sand with 55 mm and clay with 109 mm). Historical weather records were used but genotypes and crop management were taken as current. Output from the model clearly showed the inherently high degree of seasonal variability in yield, WUE and NUE and their dependence on soil type, N fertilizer application rate, and the amount and distribution of rainfall. For example, yield on the sand in the medium rainfall zone with $150\,kg\,N\,ha^{-1}$ ranged from 0.3 to $4.5\,Mg\,ha^{-1}$ with a median of $2.1\,Mg\,ha^{-1}$, evapotranspiration ranged from 104 to 299 mm (mean 229 mm) and E_s ranged from 80 to 177 mm (mean 138 mm). The clay soil was more productive in terms of grain yield, WUE and NUE at the high and medium rainfall locations but generally less productive in the low rainfall location (Figure 6.5). The clay soil was productive at higher rainfall locations because there was less N leaching than on the sandy soil but less productive at the low rainfall location than the sandy soil because of higher E_s and relatively less water use in the post-anthesis period. Figure 6.5 shows that WUE on the clay soil was unaffected by the amount of N fertilizer applied at the low rainfall location and that it was greater on the sandy soil at this location. At the medium and high rainfall locations, WUE on the clay soil was increased by N applications up to $60\,kg\,N\,ha^{-1}$ but on the sandy soil WUE continued to increase as N rate increased. These results and those of Gregory et al. (2000) demonstrate the many factors that influence WUE and the role that process-based models may have in assessing the crop management practices most likely to optimize yields and WUE.

6.8   WUE and trade-off issues in a wider context

When considering how to gain best advantage from the limited quantity of water available, a number of 'trade-off' issues are important. For example, if E_s is managed to minimize it and to maximize T, then humidity within the canopy may be reduced with adverse consequences for M/T (see Wallace and Verhoef, 2000). Of greater practical importance, though, is the effect that crop management practices have on runoff and drainage. Although agricultural production may result in lowered groundwater in some regions (e.g. parts of central India), clearance of land for agriculture generally results in increased drainage. For example, in Australia the clearance of native, perennial vegetation and its replacement by annual crops and pastures has had a significant effect on the water cycle (Farrington et al., 1992). The consequence has been a rise of groundwater and the spread of secondary salinization resulting in the failure of vegetation to grow in large areas. Data show that agronomic management has some potential to increase water use of annual crops and pastures in some environments (e.g. Angus et al., 2001) particularly on deep sands and loamy sands (by 40–70 mm, Tennant and Hall, 2001). On the more widespread gradational and duplex soils of Australia, chemical and physical properties restrict water infiltration and root penetration and limit the potential to increase water use to the order of 5–15 mm in short season, low rainfall environments and about 40 mm in long season, high

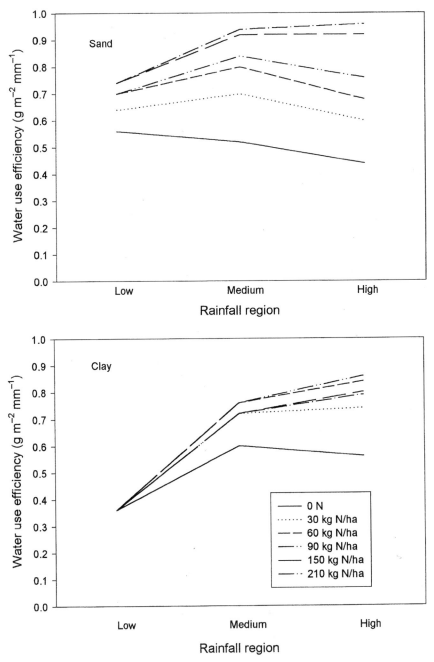

Figure 6.5 Effects of rainfall and rate of N fertilizer application on the simulated long-term average WUE of wheat crops grown on sandy and clayey soils (from Asseng *et al.*, 2001).

rainfall environments (Tennant and Hall, 2001). Such limited ability to reduce drainage on such common soils has led to the search for new farming systems, incorporating perennial species (Cocks, 2003). The deep-rooted perennial pasture plant, lucerne (*Medicago sativa* L.), has been shown in some studies to have some potential to reduce drainage (Ward *et al.*, 2001) especially when compared to annual pasture species such as subterranean clover. However, in a region where recharge can vary between 0 and 250 mm annually, the experimentally determined reduction of 50 mm may be insufficient to have long-term impact on the rise of groundwater. Ward *et al.* (2003) modelled the impact of changing the length of the lucerne phase in rotation with a phase of annual crops over 81 years for the 300 and 500 mm rainfall zones. They found that drainage leakage was more readily controlled in the drier environment, on a duplex soil, where a rotation involving 3 years of lucerne followed by 3 years of crop resulted in 95% less drainage. On an acid loamy sand in the wetter environment, though, the same rotation reduced leakage by only 40%.

Increasingly water is being considered as an economic good and concepts and tools are being developed that will allow it to be so treated (Hoekstra and Hung, 2003). At a global scale, the volume of water used in the production of a product constitutes the virtual water 'hidden' or 'embodied' in the product. For water-scarce countries, it might be possible to achieve some measure of water security by importing water-intensive products instead of producing them domestically (Hoekstra and Hung, 2003). This has some attractions in relation to crop products where the virtual water content of grains is high (1 kg of grain requires 1000 to 2000 kg water to produce in temperate regions and 3000 to 5000 kg water in arid regions). Trade in water-intensive products is an attractive means of increasing water security in water-poor regions of the world because it overcomes many of the difficulties of cost involved in trading real water. Hoekstra and Hung (2003) calculate that in the period 1995 to 1999, 13% of the water used for crop production was not used for domestic consumption but was exported. This global virtual water trade varied between countries with large net exports from Argentina, Australia, Canada, India, Thailand and the United States, and large imports into China, Indonesia, Japan, the Netherlands, the Republic of Korea and Sri Lanka. Developed countries generally had a more stable virtual water trade balance than developing countries.

Finally, the increasing desire to develop systems of production that are sustainable will require a change of emphasis by producers and consumers from systems where the level of production is paramount to systems where resources are used efficiently and on- and off-site effects minimized. Improving the efficiency with which water is used in this 'doubly-green revolution' poses a major challenge to agronomists.

References

Allen, S.J. (1990) Measurement and estimation of evaporation from soil under sparse barley crops in northern Syria. *Agricultural and Forest Meteorology*, **49**, 291–309.

Anderson, W.K. (1992) Increasing grain yield and water use of wheat in a rainfed Mediterranean type environment. *Australian Journal of Agricultural Research*, **43**, 1–17.

Angus, J.F., Gault, R.R., Peoples, M.B., Stapper, M. and van Herwaarden, A.F. (2001) Soil water extraction by dryland crops, annual pastures, and lucerne in south-eastern Australia. *Australian Journal of Agricultural Research*, **52**, 183–192.

Asseng, S., Turner, N.C. and Keating, B.A. (2001) Analysis of water- and nitrogen-use efficiency of wheat in a Mediterranean climate. *Plant and Soil*, **233**, 127–143.

Azam-Ali, S.N., Gregory, P.J. and Monteith, J.L. (1984) Effects of planting density on water use and productivity of pearl millet (*Pennisetum typhoides*). II. Water use, light interception and dry matter production. *Experimental Agriculture*, **20**, 203–214.

Bolger, T.P. and Turner, N.C. (1999) Water use efficiency and water use of Mediterranean annual pastures in southern Australia. *Australian Journal of Agricultural Research*, **50**, 1035–1046.

Bos, M.G. (1985) Summary of ICID definitions on irrigation efficiency. *ICID Bulletin*, **34**, 28–31.

Briggs, L.J. and Shantz, H.L. (1913) *The water requirements of plants. II. A review of literature.* Bulletin, USDA, Bureau of Plant Industry No. 285.

Bristow, K.L., Campbell, G.S., Papendick, R.I. and Elliott, L.F. (1986) Simulation of heat and moisture transfer through a surface residue-soil system. *Agricultural and Forest Meteorology*, **36**, 193–214.

Brown, S.C., Gregory, P.J., Copper, P.J.M. and Keatinge, J.D.H. (1989) Root and shoot growth and water use of chickpea (*Cicer arietinum*) grown in dryland conditions: effects of sowing date and genotype. *Journal of Agricultural Science, Cambridge*, **113**, 41–49.

Cocks, P.S. (2003) A revolution in agriculture is needed if we are to manage dryland salinity and related natural resource issues,in *Solutions for a better environment*. Proceedings of the 11th Australian Agronomy Conference, 2–6 February 2003, Geelong, Victoria. Australian Society of Agronomy, CDROM ISBN 0–9750313–0–9. Web site www.regional.org.au/au/asa.

Condon, A.G., Richards, R.A. and Farquhar, G.D. (1987) Carbon isotope discrimination is positively correlated with grain yield and dry matter production in field-grown wheat. *Crop Science*, **27**, 996–1001.

Condon, A.G., Farquhar, G.D. and Richards, R.A. (1990) Genotypic variation in carbon isotope discrimination and transpiration efficiency in wheat. Leaf gas exchange and whole plant studies. *Australian Journal of Plant Physiology*, **17**, 9–22.

Cooper, P.J.M., Gregory, P.J., Keatinge, J.D.H. and Brown, S.C. (1987a) Effects of fertilizer, variety and location on barley production under rainfed conditions in Northern Syria. 2. Soil water dynamics and crop water use. *Field Crops Research*, **16**, 67–84.

Cooper, P.J.M., Gregory, P.J., Tully, D. and Harris, H.C. (1987b) Improving water use efficiency of annual crops in the rainfed farming systems of West Asia and North Africa. *Experimental Agriculture*, **23**, 113–158.

Day, W., Lawlor, D.W. and Day, A.T. (1987) The effect of drought on barley yield and water use in two contrasting years. *Irrigation Science*, **8**, 115–130.

Dingkuhn, M., Farquhar, G.D., De Datta, S.K. and O'Toole, J.C. (1991) Discrimination of ^{13}C among upland rices having different water use efficiencies. *Australian Journal of Agricultural Research*, **42**, 1123–1131.

Eastham, J., Gregory, P.J., Williamson, D.R. and Watson, G.D. (1999) The influence of early sowing of wheat and lupin crops on evapotranspiration and evaporation from the soil surface in a Mediterranean climate. *Agricultural Water Management*, **42**, 205–218.

Eastham, J. and Gregory, P.J. (2000) The influence of crop management on the water balance of lupin and wheat crops on a layered soil in a Mediterranean climate. *Plant and Soil*, **221**, 239–251.

Ehlers, W. (1989) Transpiration efficiency of oat. *Agronomy Journal*, **81**, 810–817.

Falkenmark, M. (1997) Meeting water requirements of an expanding world population. *Philosophical Transactions of the Royal Society London*, **B352**, 929–936.

Farquhar, G.D. and Richards, R.A. (1984) Isotopic composition of plant carbon correlates with water-use efficiency of wheat genotypes. *Australian Journal of Plant Physiology*, **11**, 539–552.

Farrington, P., Salama, R.B., Watson, G.D. and Bartle, G.A. (1992) Water use of agricultural and native plants in a Western Australian wheatbelt catchment. *Agricultural Water Management*, **22**, 357–367.

French, R.J. and Schultz J.E. (1984) Water use efficiency of wheat in a Mediterranean type environment. 1. The relationship between yield, water use and climate. *Australian Journal of Agricultural Research*, **35**, 743–764.

Gaze, S.R., Simmonds, L.P., Brouwer, J. and Bouma, J. (1997) Measurement of surface redistribution of rainfall and modelling its effect on water balance calculations for a millet field on a sandy soil in Niger. *Journal of Hydrology*, 188–189, 267–284.

Gregory, P.J. (1994) Resource capture by root networks, in *Resource Capture by Crops*. Proceedings of the 52nd Easter School, University of Nottingham (eds J.L. Monteith, R.K. Scott and M.H. Unsworth), Nottingham University Press, pp. 77–97.

Gregory, P.J. and Eastham, J. (1996) Growth of shoots and roots, and interception of radiation by wheat and lupin crops on a shallow, duplex soil in response to time of sowing. *Australian Journal of Agricultural Research*, **47**, 427–447.

Gregory, P.J., Shepherd, K.D. and Cooper, P.J. (1984) Effects of fertilizer on root growth and water use of barley in northern Syria. *Journal of Agricultural Science, Cambridge*, **103**, 429–438.

Gregory, P.J., Simmonds, L.P. and Pilbeam, C.J. (2000) Soil type, climatic regime, and the response of water use efficiency to crop management. *Agronomy Journal*, **92**, 814–820.

Gregory, P.J., Simmonds, L.P. and Warren, G.P. (1997) Interactions between plant nutrients, water and carbon dioxide as factors limiting crop yields. *Philosophical Transactions of the Royal Society London*, **B352**, 987–996.

Gregory, P.J., Tennant, D. and Belford, R.K. (1992) Root and shoot growth and water and light use efficiency of barley and wheat crops grown on a shallow duplex soil in a Mediterranean-type environment. *Australian Journal of Agricultural Research*, **43**, 555–573.

Hiler, E.A. and Howell, T.A. (1983) Irrigation options to avoid critical stress: An overview. In *Limitations to efficient water use in crop production* (eds H.M. Taylor, W.R. Jordan and T.R. Sinclair), American Society of Agronomy, Madison, US, pp. 479–497.

Hoekstra, A.Y. and Hung, P.Q. (2003) Virtual water trade: a quantification of virtual water flows between nations in relation to international crop trade. In *Virtual Water Trade, Proceedings of the International Expert Meeting on Virtual Water Trade* (ed A.Y. Hoekstra), Value of Water Research Report Series No. 12, IHE, Delft, The Netherlands, pp. 25–41.

Hubick, K.T., Farquhar, G.D. and Shorter, R. (1986) Correlation between water-use efficiency and carbon isotope discrimination in diverse peanut (*Arachis*) germplasm. *Australian Journal of Plant Physiology*, **13**, 803–816.

Jalota, S.K., Khera, R. and Chahal, S.S. (2001) Straw management and tillage effects on soil water storage under field conditions. *Soil Use and Management*, **17**, 282–287.

Jones, M.J. (2000) Comparison of conservation tillage systems in barley-based cropping systems in northern Syria. *Experimental Agriculture*, **36**, 15–26.

Jones, M.J. and Sinclair, J. (1989) Effects of bare fallowing, previous crop and time of ploughing on soil moisture conservation in Botswana. *Tropical Agriculture*, **66**, 54–60.

Jones, M.J. and Wahbi, A. (1992) Site-factor influence on barley response to fertilizer in on-farm trials in northern Syria: descriptive and predictive models. *Experimental Agriculture*, **28**, 63–87.

Kilewe, A.M. and Ulsaker, L.G. (1984) Topographical modification of land to concentrate and redistribute runoff for crop production. *East African Agricultural and Forestry Journal*, **44**, 254–265.

Latiri-Souki, K., Nortcliff, S. and Lawlor, D.W. (1998) Nitrogen fertilizer can increase dry matter, grain production and radiation and water use efficiencies for durum wheat under semi-arid conditions. *European Journal of Agronomy*, **9**, 21–34.

Lawes, J.B. (1850) Experimental investigation into the amount of water given off by plants during their growth; especially in relation to the fixation and source of their various constituents. *Journal of the Horticultural Society of London*, **5**, 38–63.

Leidi, E.O., López, M., Gorham, J. and Gutiérrez, J.C. (1999) Variation in carbon isotope discrimination and other traits related to drought tolerance in upland cotton cultivars under dryland conditions. *Field Crops Research*, **61**, 109–123.

López-Castañeda, C. and Richards, R.A. (1994) Variation in temperate cereals in rainfed environments III. Water use and water-use efficiency. *Field Crops Research*, **39**, 85–98.

Loss, S.P., Siddique, K.H.M. and Tennant, D. (1997) Adaptation of faba bean (*Vicia faba* L.) to dryland Mediterranean-type environments. III. Water use and water-use efficiency. *Field Crops Research*, **54**, 153–162.

Monteith, J.L. (1986) Significance of the coupling between saturation vapour pressure deficit and rainfall in monsoon climates. *Experimental Agriculture*, **22**, 329–338.

Ogola, J.B.O., Wheeler, T.R. and Harris, P.M. (2002) Effects of nitrogen and irrigation on water use of maize crops. *Field Crops Research*, **78**, 105–117.

Pala, M., Harris, H.C., Ryan, J., Makboul, R. and Dozom, S. (2000) Tillage systems and stubble management in a Mediterranean-type environment in relation to crop yield and soil moisture. *Experimental Agriculture*, **36**, 223–242.

Papendick, R.I., Parr, J.F. and Meyer, R.E. (1990) Managing crop residues to optimize crop/livestock production systems for dryland agriculture. *Advances in Soil Science*, **13**, 253–272.

Payne, W.A. (1997) Managing yield and water use of pearl millet in the Sahel. *Agronomy Journal*, **89**, 481–490.

Payne, W.A. (2000) Optimizing crop water use in sparse stands of pearl millet. *Agronomy Journal*, **92**, 808–814.

Payne, W.A., Wendt, C.W. and Lascano, R.J. (1990) Root zone water balances of three low-input millet fields in Niger, West Africa. *Agronomy Journal*, **82**, 813–819.

Penning de Vries, F.W.T. (1975) The cost of maintenance processes in plant cells. *Annals of Botany*, **39**, 77–92.

Pilbeam, C. J., Simmonds, L.P. and Kavilu, A.W. (1995) Transpiration efficiencies of maize and beans in semi-arid Kenya. *Field Crops Research*, **41**, 179–188.

Rees, D.J. (1986) Crop growth, development and yield in semi-arid conditions in Botswana. 1. The effects of population density and row spacing on *Sorghum bicolor*. *Experimental Agriculture*, **22**, 153–167.

Rockström, J., (1997) On-farm agrohydrological analysis of the Sahelian yield crisis: Rainfall partitioning, soil nutrients and water use efficiency of pearl millet. PhD Thesis, University of Stockholm, Sweden.

Sadras, V. (2002) Interaction between rainfall and nitrogen fertilization of wheat in environments prone to terminal drought: economic and environmental risk analysis. *Field Crops Research*, **77**, 201–215.

Sadras, V.O., Whitfield, D.M. and Connor, D.J. (1991) Transpiration efficiency in crops of semi-dwarf and standard-height sunflower. *Irrigation Science*, **12**, 87–91.

Schulze, R.E., Lumsden, T.G. and Horan, M.J.C. (2002) Irrigation water use by sugarcane in South Africa: A case study on water use requirements, water use efficiencies, yield gains and water losses, in *Modelling as a Tool in Integrated Water Resources Management: Conceptual Issues and Case Study Applications* (ed. R.E. Schulze), Water Research Commission Report 749/1/02, Pretoria, South Africa, pp. 178–197.

Seckler, D. (1999) Water for food in 2025: The major Issues. Newsflow 2/99, 1–3. Global Water Partnership, Stockholm, Sweden.

Shiklomanov, I.A. (2003) *World Water Resources at the Beginning of the 21st Century*. Cambridge University Press (in press).

Siddique, K.H.M., Belford, R.K. and Tennant, D. (1990a) Root:shoot ratios of old and modern and semi-dwarf wheats in a Mediterranean environment. *Plant and Soil*, **121**, 89–98.

Siddique, K.H.M., Tennant, D., Perry, M.W. and Belford, R.K. (1990b) Water use and water use efficiency of old and modern wheat cultivars in a Mediterranean-type environment, *Australian Journal of Agricultural Research*, **41**, 431–447.

Sinclair, T.R., Tanner, C.B. and Bennett, J.M. (1984) Water-use efficiency in crop production. *Bioscience*, **34**, 36–40.

Stephens, D.J. and Lyons, T.J. (1998) Rainfall-yield relationships across the Australian wheatbelt. *Australian Journal of Agricultural Research*, **49**, 211–223.

Tanner, C.B. (1981) Transpiration efficiency of potato. *Agronomy Journal*, **73**, 59–64.

Tanner, C.B. and Sinclair, T.R. (1983) Efficient water use in crop production: research or research? in *Limitations to Efficient Water Use in Crop Production* (eds H.M. Taylor, W.R. Jordan and T.R. Sinclair), American Society of Agronomy, Madison, USA, pp. 1–27.

Tennant, D. and Hall, D. (2001) Improving water use of annual crops and pastures limitations and opportunities in Western Australia. *Australian Journal of Agricultural Research*, **52**, 171–182.

Wallace, J. S. 2000 Increasing agricultural water use efficiency to meet future food production. *Agriculture, Ecosystems and Environment*, **82**, 105–119.

Wallace, J.S. and Gregory, P.J. (2002) Water resources and their use in food production systems. *Aquatic Science*, **64**, 363–375.

Wallace, J.S., Lloyd, C.R. and Sivakumar, M.V.K. (1993) Measurement of soil, plant and total evaporation from millet in Niger. *Agricultural and Forest Meteorology*, **63**, 149–169.

Wallace, J.S. and Verhoef, A. (2000) Modelling interactions in mixed-plant communities: light, water and carbon dioxide, in *Leaf Development and Canopy Growth Marshall* (eds. B. Marshall and J.A. Roberts), Sheffield Academic Press, Sheffield, pp. 204–250.

Ward, P., Dolling, P. and Dunin, F.X. (2003) The impact of a lucerne phase in a crop rotation on groundwater recharge in south-west Australia, in *Solutions for a better environment*. Proceedings of the 11th Australian Agronomy Conference, 2–6 February 2003, Geelong, Victoria. Australian Society of Agronomy, CDROM ISBN 0-9750313-0-9. Web site www.regional.org.au/au/as

Ward, P.R., Dunin, F.X. and Micin, S.F. (2001) Water balance of annual and perennial pastures on a duplex soil in a Mediterranean environment. *Australian Journal of Agricultural Research*, **52**, 203–209.

Yudelman, M. (1998) Water and food in developing countries in the next century, in *Feeding a World Population of more than Eight Billion People* (eds. J.C. Waterlow, D.G. Armstrong, L. Fowden and R. Riley), Oxford University Press, Oxford, pp. 56–68.

Yunusa, I.A.M., Belford, R.K., Tennant, D. and Sedgley, R.H. (1993) Row spacing fails to modify soil evaporation and grain yield in spring wheat in a dry Mediterranean environment. *Australian Journal of Agricultural Research*, **44**, 661–676.

Zhang, H., Pala, M., Oweis, T. and Harris, H. (2000) Water use and water-use efficiency of chickpea and lentil in a Mediterranean environment. *Australian Journal of Agricultural Research*, **51**, 295–304.

7 Plant nutrition and water use efficiency

John A. Raven, Linda L. Handley and Bernd Wollenweber

7.1 Introduction

Measurements of the water lost during plant growth, given different nutrient availabilities, can be traced back to Woodward (1699; see Stanhill, 1985, 1986a,b). Woodward (1699) grew sprigs of *Mentha spicata* in water culture – and so he was also a pioneer in hydroponics.

Some of Woodward's data are presented in Table 7.1, with weights in grains converted to grams. Although Woodward was unable to present data on the nutrient status of the growth media or of the plants, it is likely that nutrient content increased in the order spring water = rain water < Thames water < Hyde Park Conduit water. Certainly the relative growth rate increases in parallel with the probable nutrient availability, as did water use efficiency, with all plant measurements on a fresh weight basis (Table 7.1). Woodward (1699) modified the Hyde Park Conduit water by adding soil, which yielded the highest relative growth rate among the highest water use efficiencies of any of the treatments. Distillate from the Hyde Park Conduit give low specific growth rates and water use efficiency, while the liquid residue from the distillation yielded a high relative growth rate and water use efficiency (Table 7.1). In retrospect, Woodward's work is seen to not discriminate transpiration from evaporation, and only replicated two of the experiments. We can also see that his data are consistent with the current view that water use efficiency increases when plant growth rate increases as a function of greater nutrient availability (Table 7.1; Montgomery and Kiesselbach, 1912; Briggs and Schantz, 1914; Kiesselbach, 1916; Tanner and Sinclair, 1983; Walker and Richards, 1985).

This chapter considers our current understanding of how water use efficiency interacts with plant nutrition at the single plant, symbiotic, population and community levels.

7.2 Defining the status of nutrient elements used by plants

7.2.1 The nutrient elements

There are three categories of elemental requirements for photosynthetic organisms (Marschner, 1995; Epstein, 2001). The first category comprises

Table 7.1 The experiments reported by Woodward (1699) rendered in SI units.

Nutrient medium	Plant[1] fresh weight at start/g	Plant fresh weight[2] at end/g	Increase in fresh weight (g)	Water lost during growth (g)	Water use efficiency:g fresh weight gain per kg water lost	Relative growth rate:g fresh weight gain per kg fresh weight per day
Spring water	1.7	2.7	1.0	165	6.1	6.0
Rain water	1.8	3.0	1.2	195	6.1	6.6
Thames water	1.8	3.5	1.7	162	10.5	8.6
Hyde Park (1)	8.2	16.5	8.3	920	9.0	12.5
Conduit water (2)	7.1	16.1	9.0	851	10.5	14.6
Hyde Park (1)	4.9	15.8	10.9	695	15.6	20.9
Conduit water plus soil (2)	6.0	24.4	18.4	813	22.7	25.1
Distillate from Hyde Park Conduit water	7.4	10.0	2.6	570	4.6	5.4
Residue from distilled Hyde Park water[3]	5.2	11.3	6.1	281	21.7	13.8

Notes
[1] Plants described by Woodward (1699) as 'common spear mint', i.e. *Mentha spicata*
[2] Length of growth 77 days for the Spring/Rain/Thames water experiments and 56 days for the Hyde Park Conduit water
[3] Very turbid, and as high-coloured (reddish) as ordinary beer'

elements essential for all O_2-evolvers. In approximate order of abundance (by atoms) in higher plants these are H, O, C, N, K, Ca, Mg, P, S, Cl, Fe, B, Mn, Zn, Cu, Mo and Ni. The next category comprises elements which are essential for some photosynthetic O_2-evolvers, and which may be beneficial for the growth of others, e.g. Na, Si, Co. Finally, there is an element for which even a significant beneficial role is in doubt, i.e. Se (Terry et al., 2000).

The roles of these elements are discussed by Marschner (1995) and Stewart and Schmidt (1999), and are listed in Table 7.2. These roles will be considered again in the context of the observed effects of nutrients on water use efficiency, and the mechanisms underlying these effects.

7.2.2 Sources on nutrient elements: spatial aspects

Plants which transpire, and for which water use efficiency is an important consideration, are generally rhizophytes. This means that the plants have roots or analogous structures in soil, which is the source of H_2O for transpiration, for hydration of plant tissues, and as substrate for photosynthetic reductant generation and O_2 evolution. Roots are also the main means of acquisition, from soil, of N, K, Ca, Mg, P, S, Cl, Fe, B, Mn, Zn, Cu, Mo, Ni, Na, Si, Co and Se (Table 7.2; cf. Raven, 1986, 1988; Raven, Wollenweber and Handley, 1992a, 1993; Raven and Yin, 1998). The shoots of rhizophytes harvest photons and CO_2 from the atmospheric environment in photosynthesis, necessitating H_2O loss in transpiration, and supplying almost all of the C and some of the O in the plant. The atmosphere can, under some circumstances, supply a significant fraction of some other elements such as N (Table 7.2). Typically, those interested in 'mineral nutrition' of land plants do not explicitly include CO_2 as a nutrient.

7.2.3 Sources of nutrient elements: speciation

Table 7.2 shows the chemical form of the essential nutrient elements, which are available to plants. Nitrogen is the element supplied via the largest range of chemical species (Table 7.2). The form of nitrogen available to the plant has significant effects on the content of other elements (for example, Mo, a component of the commonest form of nitrogenase and of nitrate reductase) and organic components (for example, more organic acid anions when NO_3^- is the N source than when N_2 (via diazotrophic symbioses), $CO(NH_2)_2$ or, especially, NH_4^+ is the N source) (Raven and Smith, 1976; Raven and Wollenweber, 1992).

The proton and the hydroxyl ion are a frequently overlooked pair of 'nutrients' which are conditional on the chemical speciation of the source of N, S and P (Raven and Wollenweber, 1992, cf. Raven, 1990). Quantitatively the most important of these is N. When the N source is NH_4^+, about 1.22 H^+

Table 7.2 Roles and sources of nutrient elements in plants (Marschner, 1995; Raven, 1988).

Element	Role(s)	Source(s)
^{1}H	Solvent H_2O; all organic compounds	H_2O in soil (atmospheric H_2O vapour for some algae, lichens)
^{1}O	Solvent H_2O; all organic compounds	H_2O (as for H); some from CO_2 (as for C)
^{1}C	Organic compounds	CO_2 in atmosphere for land plants (soil for amphibious *Lobelia, Isoetes*)
^{1}N	Nucleic acids, proteins, cofactors, alkaloids, etc.	N_2 (symbiotic diazotrophs in roots, rarely stem nodules), NH_4^+, NO_2^-, NO_3^-, organic N in soil; NH_3 from atmosphere
^{1}K	Cofactor for many enzymes, osmotic component	K^+ from soil
^{1}Ca	Signalling, osmotic component	Ca^{2+} from soil
^{1}Mg	Cofactor for many enzymes, osmotic component	Mg^{2+} from soil
^{1}P	Nucleic acids, cofactors, regulation	$H_2PO_4^-/HPO_4^-$, organic P in soil
^{1}S	Amino acids, lipids, osmotic component	SO_4^{2-} in soil (rarely HS^-, organic S)
^{1}Cl	Cofactor for photosystem II, osmotic component	Cl^- in soil
^{1}Fe	Cofactor for many redox reactions	Fe^{2+}, Fe^{3+} in soil
^{1}B	Cofactor in walls, membranes?	$B(OH)_3$ in soil
^{1}Mn	Cofactor for photosystem II	Mn^{2+} in soil
^{1}Zn	Zinc finger, enzyme cofactor e.g. Cu-Zn superoxide dismutase, carbonic anhydrase	Zn^{2+} in soil
^{1}Cu	Enzyme cofactor, e.g. cytochrome oxidase, CuZn superoxide dismutase	Cu^{2+} in soil
^{1}Mo	Enzyme cofactor, e.g. nitrate reductase, nitrogenase	MoO_4^{2-} in soil
^{1}Ni	Enzyme cofactor, e.g. urease	Ni^{2+} in soil
^{2}Na	Transport cofactor in C_4 plants	Na^+ in soil
^{2}Si	Mechanical, including anti-biophage, leaf posture	$Si(OH)_4$ in soil
^{2}Co	Vitamin B_{12}	Co^{2+} in soil
^{2}Se	Some peroxidases	SeO_4^{2-}, SeO_3^{2-}

[1] Element required by all photosynthetic organisms
[2] Element required by some photosynthetic organisms, with beneficial effect on growth in nature of some others.

are excreted from the roots per NH_4^+ taken up and assimilated (Raven and Wollenweber, 1992), so 1.22 OH^- are effectively consumed from buffers in the rooting medium. With N_2 (via diazotrophic symbioses) and to a lesser extent $CO(NH)_2$ as N sources, up to 1.0 H^+ are excreted per N assimilated, so that up to 1.0 OH^- are consumed from the medium per N assimilated. For

NO_3^- as N source, up to 0.78 OH^- are excreted from the roots per N assimilated, with up to 0.78 H^+ are consumed from buffers in the medium per N assimilated (Raven and Wollenweber, 1992). These impacts on soil pH as a function of nitrogen source are superimposed on any additional H^+ efflux related to the acquisition of iron and of phosphate. Some of this H^+ is accompanied by the organic anions generated in parallel with the H^+; the rest is exchanged for an extracellular cation, which is then stored in cells with the organic anion generated in the production of the H^+. An additional set of reactions adding H^+ to the soil, or removing H^+ from the soil, is the circulation of H^+ around growing root tips, with H^+ entry at the apex and H^+ efflux in more basipetal regions (Raven and Wollenweber, 1992).

These overall stoichiometries of nitrogen assimilation with proton efflux or influx during growth should not be confused with the mechanistic stoichiometries of proton transport in the operation of the transporters moving ammonium and nitrate into the cells (Williams and Miller, 2001; Kaiser et al., 2002).

7.2.4   Sources of nutrient elements: deficiency and excess

Restriction of the rate or extent of plant growth by limited nutrient availability has been recognized since the work of Liebig in 1840 (see Raven, 2001a). The imposition of restricted nutrient supply on higher plants is contentious, even in so simplified a culture system as hydroponics (Raven, 2001a). Growth rate can also be restricted by an excess of certain nutrients, acting osmotically if present in high enough concentrations (NaCl for glycophytes).

7.3   The observed effects on water use efficiency of nutrient element deficiency and excess and variation in nutrient element source

7.3.1   Methodological considerations

The methodology used for measuring water use efficiency are considered elsewhere in this volume, but some points are particularly important for consideration of interactions with nutrition. The 'standard' methods for long-term (growth) estimation of water use efficiency are measurements of water lost in transpiration per unit dry matter gain, and measurements of the $\delta^{13}C$ of plant organic C relative to source (atmospheric CO_2) $\delta^{13}C$ (Farquhar, Ehleringer and Hubick, 1989; Raven and Farquhar, 1990). A further method is based on the content of minerals ('ash'), or of specific mineral components (Si, K) in the dry matter (Hutton and Norrish, 1974; Masle, Farquhar and Wong, 1992; Mayland et al., 1993; Araus et al., 1998; Voltas et al., 1998; Merah, Deleers and Monneveux, 1999; Merah, 2001; Tsialtas, Kassioumi and

Veresoglou, 2002). The requirement for the use of this is based on uptake of minerals, which increases with the concentration of the minerals in the soil solution and also to the transpiration rate. This method can give useful rankings, which agree with those from the direct measurements of water use efficiency and measurements of $\delta^{13}C$, but is less useful in examining phenotypic variations in water use efficiency with changed growth conditions, e.g. water availability and atmospheric relative humidity.

7.3.2   Observations

The typical response to nutrient supplies giving less than the maximum growth rate is to reduce the water use efficiency (Table 7.3). This effect is best documented for nitrogen supply, with significant amounts of information for phosphorus and potassium. Fewer data are available for other nutrients (Table 7.3).

The general decrease in water use efficiency with decreasing nutrient availability means that there is also a parallel between water use efficiency and specific (relative) growth rate, with increased water use efficiency at higher growth rates. The significance of this correlation will be considered when we examine the effect of genetic, and non-nutrient environmental, factors on relative growth rate and on water use efficiency. An attempt will be made to distinguish specific effects of deficiency in a specific nutrient from the more general effect of reduced growth rate, in linking nutrient deficiency to decreased water use efficiency.

Fewer data are available as to the effect of nutrient excess on water use efficiency. However, Atkinson (1991) examined the effect of high and low Ca^{2+} concentrations in the rhizosphere on stomatal conductance and photosynthetic rate over a range of leaf-atmosphere differences in water vapour in the calcifuge *Lupinus lutens*, and showed that excess Ca^{2+} perturbed the regulation of water use efficiency.

Table 7.4 focuses on the impact of different nitrogen sources in the root medium and in the shoot environment (see Raven, Wollenweber and Handley, 1992a,b, 1993; Fernandes and Rossiello, 1995). The data on soil-derived nitrogen sources shows that the lowest water use efficiency is for N_2 in diazotrophic symbioses relative to combined (nitrate, ammonium) nitrogen sources. For plants without diazotrophic symbionts the general finding is that ammonium gives lower water use efficiencies than does nitrate in dicotyledons but similar values for grasses. As with changes in nitrogen availability with a single nitrogen source (usually nitrate), there is the complicating factor of lower growth rates with ammonium than with nitrate, and with N_2 than with either combined nitrogen source, for the dicotyledons, with water use efficiency paralleling relative growth rate. For the grasses these small differences in water use efficiency parallel small differences in growth rate among the grasses tested.

Table 7.3   Change in water use efficiency of plants as a function of nutrient availability.

Nutrient Element	Organism	Effect of relief of nutrient deficiency on water use efficiency	Reference
N	*Phleum pratense*	Increase	Bender and Berge (1979)
N	*Agropyron desertorum*	Increase	Power (1985)
N	*Alopecurus aundinaceus*	Increase	Power (1985)
N	*Bromus inermis*	Increase	Power (1985)
N	*Elytrigia intermedia*	Increase	Power (1985)
N	*Pascopyrum smithii*	Increase	Power (1985)
N	*Psathyrostachys juncea*	Increase	Power (1985)
N	*Stipa viridula*	Increase	Power (1985)
N	*Pinus pinaster*	Increase	Guehl *et al.* (1995)
N	*Quercus robur*	No effect	Guehl *et al.* (1995)
N	*Panicum maximum*	Increase	Pieterse, Rethman and Van Bosche (1997)
N	*Lolium perenne*	Increase	Stout and Schnabel (1997)
N	*Citrus parodisi*	No effect	Syvertson *et al.* (1997)
N	*Quercus robur*	No effect	Thomas and Gehler (1997)
N	*Populus trichocarpa**	Increase	Harvey and van den Driessche (1999)
N	*Quercus robur*	Increase	Welander and Ottoson (2000)
N	*Panicum virgatum*	Increase	Byrd and May (2000)
N	*Pennisetum glaucum*	Increase	Ashraf *et al.* (2001)
N	*Triticum aestivum*	Increase	Caviglia and Sodras (2001)
N	*Coffea canephora*	Increase	Da Matta *et al.* (2002)
N	*Populus tremuloides*	Increase	Des Rochers *et al.* (2003)
K	*Phleum pratense*	Increase	Bender and Berge (1979)
K	*Medicago sativa*	Increase	Walker and Richards (1885)
K	*Hordeum vulgare*	Increase	Anderson *et al.* (1992); Losch *et al.* (1992)
K	*Populus trichocarpa**	Increase	Harvey and van den Driessche (1999)
K	*Zea mays*	Increase[§]	Lips *et al.* (1990)
P	*Fragilaria anassana*	Increase	Chen and Lena (1997)
P	*Pennisetum glaucum*	Increase[†]	Payne *et al.* (1992, 1995); Brück *et al.*(2000)
P	*Triticum astivum*	Increase	Li *et al.* (2001)
Mo	*Trifolium alexandrium*	Increase	El-Bably (2002)
Zn	*Cicer arietinum*	Increase	Khan *et al.* (2003)

* Also hybrids of *Populus trichocarpa* with *Populus deltoides* and *Populus euamericana*
[§] Especially with NH_4^+ as nitrogen source
[†] Increased phosphate supply sometimes reduced water use efficiency when water was in ample supply, but always increases water use efficiency when water supply was limited.

Table 7.4 Water use efficiency as a function of the nitrogen source available to the plant. Growth at low salinity unless otherwise stated.

Organism	Order of decreasing water-use efficiency as a function of nitrogen source	Reference
Ricinus communis	$NO_3^- > NH_4^+$	Raven, Allen and Griffiths (1984); Allen and Raven (1987)
Phaseolus vulgaris	$NO_3^- > NH_4^+ > N_2$	Allen, Raven and Sprent (1998)
Arachis hypogea	$NO_3^- > N_2$	Hubick (1990)
Zea mays	$NO_3^- > NH_4^+$ (value with NH_4^+ increased more with added K^+ than that for NO_3^-)	Lips *et al.* (1990)
Helianthus annuus	$NO_3^- > NH_4^+$	Kaiser and Lewis (1991)
Triticum aestivum	$NH_4^+ = NO_3^-$	Leidi *et al.* (1991a,b)
Triticum aestivum	$NH_4^+ = NO_3^-$	Hawkins and Lewis (1993)
Lolium perenne	$NH_4^+ = NH_4^+ + NH_3^*) = (NH_4^+ NO_3^- + NH_3^*) = NH_4^+ NO_3^- = (NO_3^- + NH_3^*) > NO_3^- = NO_3^-$	Wollenweber and Raven (1993)
	$NO_3^- > (NH_4^+ NO_3^-) > NH_4^+$	
Cyanura scolymus	$NO_3^- > (NH_4^+ NO_3^-) > NH_4^+$	Elia, Santamaria and Serio (1996)
Quercus robur	$NH_4^+ = (NH_4^+ NO_3^-) = NO_3^-$	Thomas and Gehler (1997)
Casuarina equisetifolia	$NH_4^+ > NO_3^- > N_2$	Martinez-Carrasco *et al.* (1998)
Triticum aestivum	$NH_4^+ = (NH_4^+ + NH_3^*) = (NO_3^- NH_3^*) > NO_3^- = NH_4^+ NO_3^- = (NH_4^+ NO_3^- + NH_3^*)$	Yin and Raven (1998)
Zea mays	$NH_4^+ = (NH_4^+ + NH_3^*) = NO_3^- = (NO_3^- + NH_3^*) = NH_4^+ NO_3^- = (NH_4^+ NO_3^- + NH_3^*)$	Yin and Raven (1998)
Helianthus annuus	$NO_3^- > NH_4^+$ (low salinity); $NH_4^+ > NO_3^-$ (high salinity)	Ashraf (1999)
Beta vulgaris	$NO_3^- > NH_4^+$	Santamaria *et al.* (1999)
Apium graveolens	$NH_4^+ > NO_3^-$	Santamaria *et al.* (1999)
Foeniculura vulgare	$NH_4^+ > NH_4^+$	Santamaria *et al.* (1999)
Lycopersicon esculentum	$NO_3^- > NH_4^+$	Claussen (2002)
Populus tremuloides	$NH_4^+ = (NH_4^+ NO_3^-) = NO_3^- = CO(NH_2)_2$	Des Rochers, van den Driessche and Thomas (2003)

* NH_3 supplied in atmosphere around shoots

Soil nutrient supplies also influence the water lost per unit dry matter gain in the plants by influencing the extent of loss of organic matter to soils related to decreasing availability of, especially, P and Fe (Dinkelaker, Romheld and Marschner, 1989; Marschner, 1995; Ryan, Delhaize and Jones, 2001; Dakora and Phillips, 2002; Mahmood *et al.*, 2002; Neumann and Martinua, 2002; Skene, 2003). These losses can amount to 25% of net primary productivity for citrate efflux from cluster roots of *Lupinus albus* (Dinkelaker *et al.*, 1989). Less clearly related to nutrition are losses of organic carbon from above-ground parts as nectar and as volatile organic compounds (Baldoch *et al.*, 1999; Monson and Holland, 2001; Rosenstiehl *et al.*, 2003). These soluble or volatile organic carbon losses, like respiratory losses and (by many means of estimating water use efficiency) either fall (e.g. nutrient-related changes in longevity of leaves and fine roots) or alter the relationship of water use efficiency on a gross photosynthesis to that on a net primary productivity basis.

CO_2 is not considered in agricultural terms as a 'nutrient', which is accessible to supplementation, at least in economic terms, although it is manipulated in some horticultural contexts. However, atmospheric CO_2 is increasing and the extent to which such increments in CO_2 lead to increments of C_3 plant productivity is a function of, *inter alia*, soil nutrient availability (Beerling and Woodward, 2000).

There is potentially an economy of nitrogen use in C_3 plants as CO_2 increases, as the same net photosynthetic rate can be maintained with less nitrogen in RUBISCO. Another potential trade-off in resource use and photosynthetic rate as a function of increased CO_2 involves the water use efficiency, and there is evidence consistent with increased water use efficiency in higher CO_2, especially if nutrients are limiting (Beerling and Woodward, 2001). As we shall see, the trade-offs of water use efficiency and nutrient use efficiency can be modelled (Wright, Reich and Westoby, 2003).

7.4   Water use efficiency and nutrient availability to (and content in) the plant: mechanistic considerations

7.4.1   Does nutrient content relate to water use efficiency via effects of water influx on solute influx from the soil solution?

One of the methods of estimating water use efficiency mentioned above uses the 'ash' content or the content of some 'ash' component such as K or Si. This method depends on there being a relationship between water uptake by roots and the uptake of the solutes in the soil solution. Clearly there is not generally an uptake of solutes from the soil solution with water in the concentrations found in the soil solution, and ash content (= elements other than C, H, O, N, although ash contains some O and H as oxides and hydroxides of metals). Ashing of plant samples at temperatures at or above 520°C volatilizes organic components, leaving K^+, Na^+, Ca^{2+}, Mg^{2+}, Cl^-, $H_2PO_4^-/HPO_4^{2-}$ and SO_4^{2-}, with electroneutrality maintained by $OH^- + HCO_3^- + CO_3^{2-}$ (replacing the net organic negative charge). The ashing treatment also removes NH_4^+, NH_3 from organic N and NO_3^- via formation of NO_x (Wollenweber and Kinzel, 1988). In addition, dry-ash analysis might be used as a screening method for the investigation of the constraints of environmental stress as indicated by the relationships between ash content and carbon isotope discrimination (Δ) as shown in *Triticum durum* genotypes, where leaf ash correlated positively with grain Δ (Araus *et al.*, 1998). Prediction of grain yield through multiple linear regression analysis indicated further that grain ash might be used as complementary criterion to either grain Δ or leaf ash (Araus *et al.*, 1998). In addition, differences in leaf ash content between environments seem to be brought about by variations in accumulated transpiration during grain formation (Araus *et al.*, 2000) as grain Δ and ash content, and leaf dry-matter and flag leaf Δ were negatively correlated under non-droughted conditions (Merah *et al.*, 2001). These findings indicate that ash content might provide an alternative screening method in the improvement of drought tolerance and yield stability (Merah *et al.*, 2001).

There is considerable selectivity among solutes in their uptake, and nutrient elements occur in the biomass of a given genotype as a function of water content (K, N) or dry matter (C) rather than the amount of water transpired in generating the biomass (Barraclough and Leigh, 1993; Roderick *et al.*, 1999). We have seen that the relationship of ash, or ash component, content of dry matter to the water lost in production of that dry matter can be quite complex, and genotypic correlations are generally more robust than are those which relate to environmental effects on a given genotype (Masle *et al.*, 1992; Mayland *et al.*, 1993; Brown and Byrd, 1997; Araus *et al.*, 1998; Voltas *et al.*, 1998; Merah *et al.*, 1999; Merah, 2001; Tsialtas *et al.*, 2002).

The overall conclusions about mechanistic interactions between net nutrient influx and transpiratory water influx is that the relationships are generally complex, and a linear relationship between solute entry and water entry is very much the exception. However, silicon in some cereals (*Triticum, Hordeum*) shows changes in content with transpiration, which suggest uptake of water and silicic acid from the soil solution in proportion to their concentrations in the soil solution (Epstein, 2001; Raven, 2001b). In other experiments, barley cultivars were examined for silicon accumulation in dry matter as a possible indicator of water use efficiency, and for their organic ^{13}C composition (Walker and Lance, 1991). The results indicated that under mild conditions plants actively accumulate silicon, whereas under higher evaporative demand, silicon accumulates passively on the transpiration stream confirming that silicon could be used as a corroborative indicator of water use efficiency.

A related possibility in soil is the effect of transpiratory water flux on the supply of mobile solutes (especially nitrate) to the root surface (Marschner, 1995; cf. Raven and Handley, 1987). This is clearly not a major possibility in aerated hydroponic culture. In soils the question has recently been addressed in the context of nitrogen nutrition of C_3 plants growing in the present atmosphere (relatively high transpiration) and in a doubled CO_2 atmosphere (lower transpiration) (McDonald, Erickson and Knugger, 2002). Here the prediction is that nitrate supply to the root surface would be decreased on a plant dry matter gain basis in the high CO_2 plants because of the lower transpiration per unit dry matter gain. However, there is also the likelihood that decreased nitrogen content per unit dry matter in high CO_2 can result from a lower RUBISCO content or activity. By contrast, RuBP carboxylase is not always affected under these conditions (Fredeen et al., 1991).

7.4.2 Stomatal conductance

Nutrients can influence the development of stomata, altering stomatal density and hence potential maximum leaf conductance, and the achieved conductance under a given set of environmental conditions.

Losch et al. (1992) showed that essentially all of the effect of potassium deficiency in reducing transpiration rates in Hordeum vulgare could be attributed to lower stomatal density in the potassium-deficient plants. Greater reductions in photosynthetic capacity at the mesophyll level than in stomatal conductance means that the overall water use efficiency is increased while growth rate is decreased. The role of potassium in stomatal opening is involved in mechanisms by which the halophyte Aster tripolium can control the rate of transpiration by restricted stomatal opening, and hence increase water use efficiency at the expense of reduced growth rate, via the effect of increased Na^+ in the epidermal apoplasm reducing K^+ uptake into guard cells (Perera, de Silva and Mansfield, 1997; Robinson et al., 1997; Kerstiens et al., 2002). Here the decreased rate of transpiration per unit dry matter when there is high Na^+ in the leaf apoplast reduces any H_2O-linked Na^+ entry into the roots and, via the increased water use efficiency, decreases the increment of Na^+ per unit dry matter increase (Kerstiens et al., 2002).

Ca^{2+} is another ion which functions in the regulation of stomatal aperture via its concentration in the apoplasm of the leaf epidermis apoplasm (de Silva, Honour and Mansfield, 1996; de Silva, Hetherington and Mansfield, 1998). Work which bears on stomatal aperture, and hence potentially water use efficiency from the point of view of Ca^{2+} nutrition, has involved both calcifuge Lupinus luteus (Atkinson, 1991) and the calcicoles Centaurea scabiosa and Leontodon hispidus (de Silva and Mansfield, 1994; de Silva, Hetherington and Mansfield, 1996; Lionel et al., 2001). For the calcifuge, Atkinson (1991) found evidence that high levels of Ca^{2+} in the rhizosphere could reduce the coupling of stomatal conductance to assimilation and this might perturb water use efficiency, possibly by increasing apoplasmic Ca^{2+} in the leaf. For the calcifuges, evidence of sequestration of Ca^{2+} in the mesophyll cells and leaf trichomes indicates that the epidermal apoplasm might have 'normal' Ca^{2+} concentrations, and hence 'normal' stomatal function, despite high whole-plant Ca^{2+} levels. In some other cases where nutrient deficiency decreases water use efficiency measured as water loss per unit dry matter gain, it has been shown, in short-term gas exchange measurements on leaves, that photosynthesis is decreased more than transpiration. In such cases any influences of nutrient deficiency on increasing organic C loss by excretion, secretion or organ loss, or respiration, are not needed to (qualitatively) explain the decreased water use efficiency, but contribute to it quantitatively.

Mechanisms by which the nitrogen source (Table 7.4) could alter stomatal conductance include the involvement of active nitrate entry into guard cells in stomatal opening in *Arabidopsis* (Guo, Young and Crawford, 2003). However, as Guo *et al.* (2003) point out, under normal nutrient supply conditions the predominant inorganic anion involved in stomatal opening is chloride and the nitrate transporter is not essential. Accordingly, the nitrate transporter may not be involved in the differences between nitrate and ammonium nutrition with respect to stomatal function. Hawkins and Lewis (1993) showed that the stomatal conductance of NH_4^+-grown *Triticum aestivum* was 20 to 40% lower than that of NO_3^--grown plants, with increased difference between the two nitrogen sources at higher salinities.

7.4.3   Biochemical capacity for photosynthesis

Reduction in the biochemical capacity for photosynthesis as a result of nutrient-deficiency, results in photosynthesis taking place with a higher ratio of stomatal to biochemical conductance than in the case for nutrient-sufficient plants. This results in a lower water use efficiency, on a net photosynthesis basis, for the nutrient-deficient plants. It may be tempting to relate this response to the large fraction of leaf nutrients involved in producing the biochemical capacity of leaves. While it is not always the case that the quantitatively smallest consumers of a resource in plant functioning are most affected when availability of that resource is decreased, nitrogen deficiency does not restrict stomatal conductance (with low nitrogen requirements for production of the stomatal production) as much as biochemical capacity (with a much greater nitrogen requirement for chloroplast production).

The allocation of nitrogen between stomatal and mesophyll (and, in C_4, bundle sheath) cells must, however, be considered in the context of evolutionary trade-offs between water use efficiency and nitrogen use efficiency. This subject has received both experimental attention (Power, 1985; Patterson, Guy and Dang, 1997) and theoretical attention (Buckley,

Miller and Farquhar, 2002; Farquhar, Buckley and Miller, 2002; Wright, Reich and Westoby, 2003).

For C_4 plants, one mechanism whereby water use efficiency is decreased in nitrogen-deficient plants is, in some cases, increased leakiness of bundle sheaths to CO_2 (Fravolini, Williams and Thompson, 2002).

7.4.4 Xylem conductance

A reduced xylem conductance can influence water use efficiency via effects on leaf water potential, at a given soil water potential and atmospheric environment. Lower leaf water potential in plants with lower xylem conductance could impact on stomatal conductance and on the biochemical capacity for photosynthesis. Harvey and van den Driessche (1999) found that increased nitrogen supply decreased conductance via an increased tendency for cavitation as well as an intrinsic increase in conductance of uncavitated xylem as a result of larger vessel diameter.

7.4.5 Organic carbon loss

Organic carbon efflux as soluble low relative molecular mass compounds in some plants is accompanied by a deficiency in certain elements, with mechanistic linkages best seen for phosphorus and iron. For phosphorus the secretion of organic acid anions (with a variable ratio of anions to protons) releases phosphate from apatite and ferric complexes (Dinkelaker *et al.*, 1989; Marschner, 1995; Ryan *et al.*, 2001; Dakora and Phillips, 2002; Neumann and Martinoia, 2002). The extent to which these mechanisms are significant in arbuscular mycorrhizal organisms as well as in those with cluster roots, and the Brassicaceae, with infrequent mycorrhizal infection, needs further work. Iron acquisition in the majority of vascular plants uses mechanism 1, which uses organic acid and proton secretion, and surface reductases and soluble reductants, to release ferric iron from ferric oxides and hydroxides with uptake of ferrous iron (Marschner, 1995). Grasses secrete organic acid (including hydroxamic acids) siderophores, which chelate ferric iron from insoluble material, with subsequent uptake of ferric iron (Marschner, 1995). This increased loss of organic carbon from plants, which are low in phosphorus and/or iron, could contribute to the decreased dry matter gain per unit water lost in transpiration.

There does not seem to be a mechanistic basis for the increased loss of sugars from roots of ammonium-grown over those of nitrate-grown *Leptochloa fusca* (Mahmood *et al.*, 2002). This unexplained loss of sugar could contribute to the lower water use efficiency of ammonium-grown plants (Table 7.4). The nitrogen source may interact with phosphorus- or iron-deficiency related efflux of organic anions and protons since there is usually a greater net H^+ efflux with ammonium than with N_2 as nitrogen source, while there is generally a net H^+ influx with nitrate as nitrogen source. This is discussed in detail in section 7.2.3.

An increased turnover of leaves and roots is related to lower nutrient supply, so once more a loss of organic matter from the plants can decrease water use efficiency (Robinson, 1990; Eissenstat and Yardi, 1997; Franklin and Ågren, 2002). On the other hand, loss of soluble organic (and inorganic) matter from leaves can be restricted by scleromorphic/xeromorphic leaf structure, although the adoptive significance of such anatomy is still not clear (Beadle, 1966; Wright and Westoby, 2003; cf. Lamont, Cowling and Groom, 2002).

7.4.6   Inorganic carbon loss

Loss of a large fraction of net photosynthesis in 'dark' respiration in nutrient-deficient plants could decrease water use efficiency (higher water loss per unit dry matter gain) (Raven and Farquhar, 1990). The extent to which the alternate oxidase, rather than the cytochrome oxidase, pathway is engaged could be a factor in determining the effect of nutrient deficiency on respiration.

A more direct influence of respiration on water use efficiency as a function of nutrition could occur for NH_4^+ relative to NO_3^- as nitrogen source. Table 7.4 shows that growth with NH_4^+ as nitrogen source frequently involves a lower water use efficiency than growth with NO_3^-. Generally the plants with lower water use efficiency when growing with NH_4^+ also show a lower relative growth rate when growing with NH_4^+ rather than NO_3^-. One possible contributor to the lower water use efficiency of these plants when growing on NH_4^+, contrary to theoretical predictions (Raven, 1985), is that NH_4^+ is acting as a toxin as well as a nutrient. Among the costs of NH_4^+ as nitrogen source is recycling of NH_4^+/NH_3 across the plasmalemma, perhaps as a means of offsetting toxicity; this cycling has an energy cost, and thus increases respiratory losses of inorganic carbon (Raven, 1980; Kleiner, 1981; Britto et al., 2001; Kronzucker et al., 2001; Britto and Kronzucker, 2002).

7.5   The effect of interactions among organisms on nutrition and on water use efficiency

7.5.1   Symbioses

Dealing first with symbioses, which are generally held to be mutualistic, there are diazotrophic symbioses between vascular land plants and nitrogen-fixing bacteria such as rhizobia, cyanobacteria and *Frankia*. Diazotrophically growing plants have a lower water use efficiency than those grown on ammonium or nitrate (Table 7.4), in general agreement with prediction (Raven, 1985). Mycorrhizas can increase acquisition of phosphorus (especially arbuscular mycorrhizas). The increased plant nutrient content might be expected to increase water use efficiency (Table 7.3), although the dry matter allocated by the plant to mycorrhizal fungi is not readily accounted for in such computations.

The contribution of hyphae from two VAM fungi to water uptake and transport by the host plant was examined in *Lacuta sativa* cv. Romana plants, where roots were colonized by the VAM fungi *Glomus deserticola* or *Glomus fasciculatum*, or were left uninoculated but P-supplemented (Ruiz-Lozano and Azcon, 1995). Results indicated that much of the water supplied was taken up by hyphae in VAM plants. *G. deserticola* functioned efficiently under water limitation and mycelium from *G. fasciculatum*-colonized plants was very sensitive to water in the medium. *G. fasciculatum* caused a significant increase in net photosynthesis and rate of water use efficiency compared with *G. deserticola* and P-fertilized plants. In contrast, the *G. deserticola* treatment had the greatest effect on N, P and K nutrition, leaf conductance and transpiration. Since no differences in the intra- and extra-radical hyphal extension of the two endophytes were found, the results demonstrated that mycorrhizal hyphae can take up water and that there are considerable variations in both the behaviour of these two VAM fungi and in the mechanisms involved in their effects on plant water relations.

A well-documented case of ectomycorrhizal fungi or phosphorus content and water use efficiency in *Pseudotsuga menziesii* grown under water-sufficient conditions and with high and low phosphorus availability is provided in Guehl and Garbaye (1990). *Laccaria laccata* had a greater effect in increasing water use efficiency than did *Thelephora terrestris*, and comparison of the different phosphorus treatments with and without ectomycorrhizas shows that at least part of the increase in water use efficiency in mycorrhizal seedlings was a result of increased phosphorus content of the seedlings. Other data on comparing interactions of nutrient supply and mycorrhizality which relate to water use efficiency involve measurements of plant $\delta^{13}C$ rather than direct measurements of water use efficiency. Handley *et al.* (1993) examined arbuscular mycorrhizal (*Glomus clarum*) effects on *Ricinus communis*. Infected plants grew more rapidly and had increased phosphorus content and carbon isotope discrimination, suggestive of a *lower* water use efficiency in the infected plants (Handley *et al.*, 1993). Handley *et al.* (1993) also examined *Eucalyptus globus* with and without the ectomycorrhizal *Hydnaginm carneum* and given only inorganic nitrogen, but found no influence on the variables measured, including $\delta^{13}C$. Al-Karaki (1998) and Al-Karaki and Clark (1998) showed that arbuscular mycorrhizal *Triticum* had a higher water use efficiency than non-mycorrhizal plants under conditions of low phosphorus supply. Fonseca, Berbara and Daft (2001) found no effect of arbuscular mycorrhizal (*Glomus etunicatum*) infection on the $\delta^{13}C$ of the host *Sorghum bicolor*, but did influence the $\delta^{13}C$ of the non-host *Brassica napa*. More work is needed before general conclusions can be drawn about the nutritional and other influences of mycorrhizality on water use efficiency. In other experiments it was concluded that mycorrhizal fungi can modify leaf growth response to the root-to-shoot signal of soil drying, and that this mycorrhizal effect can occur independently of mycorrhizal effects on plant size or P nutrition (Auge *et al.*, 1995).

In addition, infection with endophytes can affect nutrient acquisition and WUE. A range of adaptations of endophyte-infected (i.e. *Neotyphodium* spp.) grasses to biotic and abiotic stresses has been identified but mechanisms of these adaptations are not clearly understood. In general, endophytes induce mechanisms of drought avoidance (morphological adaptations), drought tolerance (physiological and biochemical adaptations), and drought recovery in infected gasses. Mineral nutrition (nitrogen, phosphorus, calcium) affects production of alkaloids in these species. Recently, two endophyte-related mechanisms in tall fescue operating in response to phosphorus (P) deficiency have been identified (Malinowski and Belesky, 2000 and refs. therein). The first of these mechanisms, benefiting endophyte-infected plants grown under P deficiency, are altered root morphology (reduced root diameters and longer root hairs) and chemical modification of the rhizosphere resulting from exudation of phenolic-like compounds. The second mechanism consists of aluminium sequestration on root surfaces in endophyte-infected tall fescue, which also appears to be related to exudation of phenolic-like compounds with Al-chelating activity (Malinowski and Belesky, 2000).

Parasites of higher plants include viruses, fungi and other higher plants. Most information on the effect of parasites in altering the nutrient status of the host, and on growth and transpiration of the host, is available for plants, which are hosts to flowering plant parasites. Parasites, which have no, or very little, photosynthetic capacity (e.g. *Cassytha*, *Cuscuta*), mainly remove nutrients and water from the host phloem (Raven, 1983; Stewart and Press, 1990). It is not clear how such parasites impact on the water use efficiency of the host via (soil-derived) nutrient abstraction *per se* although dry matter gain by the host per unit water lost is decreased by diversion of organic carbon and soil-derived nutrient elements, but relatively little water, from host to parasite (Raven, 1983).

For photosynthetically competent parasites, the host supplies the water needed for parasite transpiration, the soil-derived nutrient elements used by the parasite, and some organic carbon (Raven, 1983). The simplest model for water and nutrient supply to photosynthetically competent higher plant parasites has a direct xylem element connection between host and parasite. A higher transpiration rate per unit dry matter increase in the parasite than in the host means that the parasite acquires more soil-derived nutrient elements per unit dry matter gain than does the host (Raven, 1983; Raven and Sprent, 1993). The supply of organic matter from host to parasite results from the organic matter content of the host xylem stream. The organic matter of xylem sap is, to a significant extent, related to organic nitrogen transfer from roots to shoots in the xylem following symbiotic diazotrophy in roots or assimilation of combined nitrogen in the roots. The plants with the lowest contribution of organic nitrogen to the organic solutes in the xylem sap are those using nitrate and with most of their nitrogen assimilation located in the shoot; these plants are mainly herbaceous (Raven and Smith, 1976). For such plants the diversion of organic matter from host to parasite must diminish the dry matter increment per unit water lost from the host in transpiration by the host.

This simple model of host-parasite connections is not universally applicable, since in some cases the transfer of water and solutes from host to parasite is not by direct connection of xylem conduits, but is via living parenchyma cells. Here the assumption of the parasite, and the host shoot, obtaining water and solutes in the same ratio as they occur in the xylem sap in the root. The flux of solutes and water from the host into parasitic flowering plants is considered by Hibberd and Jeschke (2001).

7.5.2   Populations

Most of the experiments, and much of the modelling, of water use efficiency as a function of nutrient availability has involved isolated plants. While this approximates to conditions for trees in savannas, and for large cacti (or their equivalents) in arid environments, the norm is for single-species stands to have significant interactions among individuals as far as the atmospheric environment is concerned in forming a canopy. The boundary layer associated with a canopy of limited roughness is, under given weather conditions, less than that for an individual of the same species. Accordingly, any influence of nutrition on water use efficiency at the isolated single plant level would be attenuated at the population level, provided that nutrient deficiency did not alter the surface roughness of the canopy.

7.5.3   Communities

Surface aerodynamic roughness of communities is generally greater than that of populations so that, at least for emergent species, there is a tendency toward applicability of an 'isolated plant' rather than a 'smooth canopy' model for the influence of stomatal behaviour. As for the impact of plant nutrition on community aerodynamic roughness, ombrotrophic mires are significantly nutrient-deficient communities. Paster et al. (2002) point out that there are alternative stable equilibria, with differing surface aerodynamic roughness, in peatlands, which are related to (atmospheric) nutrient supply and cycling.

These different equilibrium situations presumably have different water use efficiencies. While this is not, in general, of great significance to the vascular plants, mires can be subject to surface drying when water use efficiency can be a significant factor in performance. Interactions among plant species, which influence water use efficiency via nutrient availability, could, of course, also occur when canopies are aerodynamically 'smooth'. A prototype data set for such interactions is provided by Robinson et al. (2001).

7.5.4 Agriculture

Here the species interactions involve not just the crop plants, their symbiotic associates, their pests and competing weeds, but also *Homo sapiens*. Here increasing concerns about the availability of water resources for agriculture, and the environmental problems related to increased fertilizer usage, mean that both increased water use efficiency and increased nutrient use efficiency are aims of agricultural research.

Water use efficiency can be increased by the adoption of more intensive cropping systems in semiarid environments (bearing in mind the potential for salinization: Beresford et al., 2001) and increased plant populations in more temperate and humid environments (Falkenmark, 1997; Gregory et al., 1997; Wallace and Batchelor, 1997; Hatfield et al., 2001; Howell, 2001). Soil chemical and physical properties, for example water retention capacities, the available water in the soil profile, and the exchange rate between the soil and the atmosphere, in addition to the microbial biomass and the dynamics of the mineralization and formation of soil organic matter, are all important determinants of water and nutrient use by crops. Examples of recent advances in our understanding of these matters include the finding that soil matric potential rather than water content determines drought responses in *Lupinus angustifolius* (Jensen et al., 1998), and the possibility of increasing crop water use efficiency by 25 to 40% by modifying tillage (Hatfield et al., 2001).

Nutrient management practices to increase nutrient use by plants frequently involve combinations of the modification of the form and quantity of added nitrogen, and the timing of fertilizer application (Bulman et al., 1994). As modern crop genotypes, for example of cereals, were generally selected under conditions of high nutrient supply, there has been inadvertent selection for high fertilizer input. However, the expression of genetic variability for grain yield is largely dependent on the level of nitrogen fertilization. With high nitrogen input the variation in yield is largely explained by variations in the potential for nitrogen uptake, while with low nitrogen input the yield response is mainly a function of difference in nitrogen use efficiency. Spatial and temporal modification of fertilizer application regimes may increase nitrogen use in the grain, at unaltered levels of fertilization, through favouring uptake by roots over consumption by microbes, and through increased transport of nitrogen directly from the roots to the grain.

Thus, early applications of nitrogen at seeding or stem elongation can increase yield if they increase the rate at which the canopy closure occurs, thereby increasing the supply of photosynthate and hence, potentially, two components of yield, namely the number of fertile florets per spike and the number of grains per head (Rhodes and Mathers, 1974). After the boot stage, however, applied nitrogen has little, if any, effect on grain number, although it may increase the total dry matter of the crop and, in some circumstances, grain weight. The application of early-season nitrogen may, however, not always be sufficient to maintain protein accumulation throughout the grain-filling period. Late applications of nitrogen, for example as liquid ammonia, in early grain growth cannot influence canopy size and photosynthesis; furthermore, sink number is already determined and yield in terms of weight is largely sink limited. Applications of nitrogen fertilizer at the beginning of kernel growth have little or no effect on yield in terms of weight of grain, but have the greatest effect on protein content of the grain.

The practical optimization of fertilizer applications is complex. The use of decision-support systems, including precision farming tools such as GPS-derived digital maps of both soils (e.g. water retention capacity) and crops (yield mapping), as well as the use of optical sensors together with growth models, have been applied as tools to maximize the use of applied fertilizer. Much effort has also been put into varying the types of fertilizer, such as the use of liquid and stable manure, 'biocompost' and material from deep litter houses. Of importance also is the optimization of crop rotation systems, which permit improved nitrogen supply to crop plants at unchanged levels of fertilizer input. Modification of nutrient management practices can increase water use efficiency by 15 to 25%, with increases in crop yield (Gregory et al., 1997; Hatfield et al., 2001).

There has been apparently little experimentation on the water use efficiency of crops alone and in mixtures as a function of nutrient availability. However, in experiments with *Digitaria eriantha* ssp. *eriantha* and *Medicago sativa* grown separately and in mixtures under varying temperature, water and nitrogen regimes, Tow (1993) showed that the water use efficiency of the mixture was always similar to that of the most water-efficient monoculture for the particular treatment.

7.6 Conclusions

The available evidence shows that the restricted availability of soil-derived nutrients decreases, or less commonly has no effect on, plant water use efficiency. A decreased growth rate in nutrient-deficient plants is generally paralleled by decreased water use efficiency. It is difficult to compare effects of restricted growth rate resulting from nutrient deficiency with restrictions imposed by sub-optimal temperature, CO_2 and photosynthetically active radiation with respect to effects on water use efficiency since the non-nutrient constraints on growth rate have direct effects on the conductances and the driving forces for transpiratory water loss and for photosynthetic carbon gain.

More work is needed to extend the range on nutrients for which data are available for the effects of their deficiency on water use efficiency, and to address specifically the mechanistic aspects of the nutrition–WUE relationship, in experiments on plants under controlled conditions. Further information is also needed on nutrition–WUE effects in the field in 'natural' and agricultural situations.

Acknowledgements

John Raven gratefully acknowledges funding from AFRC, BBSRC and NERC for the study of carbon-nitrogen interactions in a range of photosynthetic organisms. Scottish Crop Research Institute is grant-aided by the Scottish Environment and Rural Affairs Department.

References

Al-Karaki, G.N. (1998) Benefit, cost and water-use efficiency of arbuscular mycorrhizal durum wheat grown under drought stress. *Mycorrhiza*, **8**, 41–45.

Al-Karaki G.N. and Clark R.B. (1998) Growth, mineral acquisition, and water use by mycorrhizal wheat grown under water stress. *Journal of Plant Nutrition*, **21**, 263–276.

Allen, S., and Raven, J.A. (1987) Intracellular pH regulation in *Ricinus communus* grown with ammonium or nitrate as N source: the role of long distance transport. *Journal of Experimental Botany*, **38**, 580–596.

Allen, S., Raven, J.A. and Sprent, J.I. (1988) The role of long-distance transport in intracellular pH regulation in *Phaseolus vulgaris* grown with ammonium or nitrate as nitrogen source, or nodulated. *Journal of Experimental Botany*, **39**, 513–528.

Anderson, M.N., Jensen, C.R. and Losch, R. (1992) The interaction effect of potassium and drought in field-grown barley. 1. Yield, water-use efficiency and growth. *Acta Agriculturae Scandinavica Section B Soil and Plant Science*, **42**, 34–44.

Araus, J.L, Amaro, T., Casadesus, J., Asbati, A. and Nachit, M.M. (1998) Relationships between ash content, carbon isotope ratio and yield in durum wheat. *Australian Journal of Plant Physiology*, **25**, 835–842.

Araus. J.L., Casadesus, J., Asbati. A. and Nachit, M.M. (2001) Basis of the relationship between ash content in the flag leaf and carbon isotope discrimination in kernels of durum wheat. *Photosynthetica*, **39**, 591–596.

Ashraf, M. (1999) Interactive effect of salt (NaCl) and nitrogen form on growth, water relations and photosynthetic capacity of sunflower (*Helianthus annuus* L.). *Annals of Applied Biology*, **135**, 509–513.

Ashraf, M., Ahmad, A. and McNeilly, T. (2001) Growth and photosynthetic characteristics in pearl millet under water stress and different potassium supply. *Photosynthetica*, **39**, 389–394.

Atkinson, C.J. (1991) The influence of increasing rhizospheric calcium on the ability of *Lupinus luteus* to control water-use efficiency. *New Phytologist*, **119**, 207–215.

Auge, R.M., Stodola, A.J.W., Ebel, R.C. and Duan, X. (1995) Leaf elongation and water relations of mycorrhizal sorghum in response to partial soil drying: two *Glomus* species at varying phosphorus fertilization. *Journal of Experimental Botany*, **46**, 297–307.

Baldoch, D.S., Fuentes, J.D., Bowling, D.R., Turnipseed, A.A. and Monson, R.K. (1999) Scaling isoprene fluxes from leaves to canopies: text cases over a boreal aspen and a mixed species temperate forest. *Journal of Applied Microbiology*, **38**, 885–898.

Barraclough, P.B. and Leigh, R.A. (1993) Critical plant K-concentration for growth and problems in the diagnosis of nutrient deficiencies by plant analysis. *Plant and Soil*, **156**, 219–222.

Beadle, N.C.W. (1966) Soil phosphate and its role in molding segments of the Australian flora and vegetation, with special reference to scleromorphy and xeromorphy. *Ecology*, **47**, 992–1007.

Beerling D.J. and Woodward F.I. (2001) *Vegetation and the Terrestrial Carbon Cycle. Modelling of the first 400 million years.* Cambridge University Press, Cambridge.

Bender, M.N. and Berge J.A. (1979) Influence of N and K fertilization and growth temperature on $^{13}C/^{12}C$ ratios of timothy (*Phleum pratense* L.). *Oecologia*, **44**, 117–118.

Beresford, Q., Beckle, H., Phillips, H. and Mulcock, J. (2001) *The Salinity Crisis: Landscapes, Communities and Politics.* University of Western Australia Press, Perth

Briggs, L.J. and Schantz, H.L. (1914) Relative water requirements of plants. *Journal of Agricultural Research*, **3**, 1–64.

Britto, D.T. and Kronzucker, H.J. (2002) NH_4^+ toxicity in higher plants: a critical review. *Journal of Plant Physiology*, **159**, 567–584.

Britto, D.T., Siddiqui, M.Y., Glass, A.D.M. and Kronzucker, H.J. (2001) Futile transmembrane NH_4^+ cycling: A cellular hypothesis to explain ammonium toxicity in plants. *Proceedings of the National Academy of Sciences of the United States of America*, **98**, 4255–4258.

Brown, R.H. and Byrd, G.T. (1997) Transpiration efficiency, specific leaf weight, and mineral concentration in peanut and pearl millet. *Crop Science*, **37**, 475–480.

Brück, H., Payne, W.A. and Sattelmacher, B. (2000) Effects of phosphorus and water supply on yield, transpirational water use efficiency, and carbon isotope discrimination of pearl millet. *Crop Science*, **40**, 120–125.

Buckley, T.N., Miller, J.M. and Farquhar, G.D. (2002) The mathematics of linked optimization for water and nitrogen use in a canopy. *Silva Fennica*, **36**, 639–669.

Bulman, P., Zarkadas, C.G. and Smith, D.L. (1994) Nitrogen fertilizer affects amino acid composition and quality of spring barley grain. *Crop Science*, **34**, 1341–1346.

Byrd, G.T. and May, P.A. (2000) Physiological comparisons of switchgrass cultivars differing in transpiration efficiency. *Crop Science*, **40**, 1271–1277.

Caviglia, O.P. and Sodras, V.O. (2001) Effect of nitrogen supply on crop conductance, water- and radiation-use efficiency of wheat. *Field Crops Research*, **69**, 259–266.

Chen, K. and Lena, R. (1997) Responses of strawberry to doubled CO_2 concentration and phosphorus deficiency. 2. Gas exchange and water consumption. *Gartenbauwissenschaft*, **62**, 90–96.

Claussen, W. (2002) Growth, water use efficiency, and proline content of hydroponically grown tomato plants as affected by nitrogen source and nutrient concentration. *Plant and Soil*, **247**, 199–209.

Dakora, F.D. and Phillips, D.A. (2002) Root exudates as mediators of mineral acquisition in low-nutrient environments. *Plant and Soil*, **245**, 35–47.

Da Matta, F.M., Loos, R.A., Silva, E.M., Loureiro, M.E. and Ducatti, C. (2002) Effect of soil water deficit and nitrogen nutrition on water relations and photosynthesis of pot-grown *Coffea canephora* Pierre. *Trees – Structure and Function*, **16**, 555–558.

de Silva, D.L.R. and Mansfield, T.A. (1994) The stomatal physiology of calcicoles in relation to calcium delivered in the xylem sap. *Proceedings of the Royal Society of London Series B – Biological Sciences*, **257**, 81–85.

de Silva, D.L.R., Hetherington, A.M. and Mansfield, T.A. (1996) Where does all the calcium go? Evidence of an important regulatory role for trichomes in two calcicoles. *Plant Cell and Environment*, **19**, 880–886.

de Silva, D.L.R., Honour, S.J. and Mansfield, T.A. (1996) Estimation of apoplastic concentrations of K$^+$ and Ca^{2+} in the vicinity of stomatal guard cells. *New Phytologist*, **134**, 463–469.

de Silva, D.L.R., Hetherington, A.M. and Mansfield, T.A. (1998) The regulation of apoplastic calcium in relation to intracellular signalling in stomatal guard cells. *Zeitschrift für Pflanzenernährung und Bodenkunde*, **161**, 533–539.

Des Rochers, A., van den Driessche, R. and Thomas, B.R. (2003) Nitrogen fertilization of trembling aspen seedlings grown on soils of different pH. *Canadian Journal of Forest Research*, **33**, 552–560.

Dinkelaker, B., Romheld, V. and Marschner, H. (1989) Citric acid excretion and precipitation of calcium citrate in the rhizosphere of white lupin (*Lupinus albus* L.). *Plant Cell and Environment*, **12**, 285–292.

Eissenstat, D.M. and Yanai, R.D. (1997) The ecology of root lifespan. *Advances in Ecological Research*, **27**, 1–60.

El-Bably, A.Z. (2002) Effect of irrigation and nutrition of copper and molybdenum on Egyptian clover (*Trifolium alexandrinum*). *Agronomy Journal*, **94**, 1066–1070.

Elia, A., Santamaria, P. and Serio, F. (1996) Ammonium and nitrate influence on artichoke growth rate and uptake of inorganic ions. *Journal of Plant Nutrition*, **19**, 1029–1044.

Epstein, E. (2001) Silicon in plants: Facts *vs.* concepts, in *Silicon in Agriculture* (eds. L.E. Datnoff, G.H. Snyder and G.H. Korndorfer), Elsevier, Amsterdam, pp. 291–317.

Falkenmark, M. (1997) Meeting water requirements of an expanding world population. *Philosophical Transactions of the Royal Society of London, Series B*, **352**, 929–936.

Farquhar, G.D., Ehleringer, J.R. and Hubick, K.T. (1989) Carbon isotope discrimination and photosynthesis. *Annual Review of Plant and Plant Molecular Biology*, **40**, 503–537.

Farquhar, G.D., Buckley, T.N. and Miller, J.M. (2002) Optimal stomatal control in relation to leaf area and nitrogen content. *Silva Fennica*, **36**, 625–637.

Fernandes, M.S. and Rossiello, R.O.P. (1995) Mineral nutrition in plant physiology and nutrition. *Critical Reviews in Plant Sciences*, **14**, 111–148.

Fonseca, H.M.A.C., Berbara, R.L.L. and Daft, M.J. (2001) Shoot δ^{15}N and δ^{13}C values of non-host *Brassica rapa* change when exposed to +/- *Glomus etunicatum* innoculum and three levels of phosphorus and nitrogen. *Mycorrhiza*, **11**, 151–158.

Franklin, O. and Ågren, G.I.. (2002) Leaf senescence and resorption as mechanisms of maximizing photosynthetic production during canopy development at N limitation. *Functional Ecology*, **16**, 727–733.

Fravolini, A., Williams, D.G. and Thompson, T.L. (2002) Carbon isotope discrimination and bundle sheath leakiness in three C$_4$ subtypes grown under variable nitrogen, water and atmospheric CO$_2$ supply. *Journal of Experimental Botany*, **53**, 2261–2269.

Fredeen, A.L., Gamon, J.A. and Field, C.B. (1991) Responses of photosynthesis and carbohydrate-partitioning to limitations in nitrogen and water availability in field-grown sunflower. *Plant, Cell and Environment*, **14**, 963–970.

Gregory, P.J., Simmonds, L.P. and Warren, G.P. (1997) Interactions between plant nutrients, water and carbon dioxide as factors limiting crop yields. *Philosophical Transactions of the Royal Society of London, Series B*, **352**, 987–996.

Guehl, J.M. and Garbaye, J. (1990) The effects of ectomycorrhizal status on carbon dioxide assimilation capacity, water-use efficiency and response to transplanting in seedlings of *Pseudotsuga menziesii* (Mirb) Franco. *Annales des Sciences Forestières*, **47**, 551–563.

Guehl, J.M., Fort, C. and Fehri. A. (1995) Differential response of leaf conductance, carbon-isotope discrimination and water-use efficiency to nitrogen deficiency in maritime pine and pedunculate oak plants. *New Phytologist*, **131**, 149–157.

Guo, F-Q., Young, J. and Crawford, N.M. (2003) The nitrate transporter AtNRT1.1(CHL1) functions in stomatal opening and contributes to drought susceptibility in *Arabidopsis*. *The Plant Cell*, **15**, 107–117.

Handley, L.L., Daft, M.J., Wilson, J., Scrimgeour, C.M., Ingleby, K. and Saltar, M.A. (1993) Effect of ectomycorrhizal and VA-mycorrhizal fungi *Hydnagium carneum* and *Glomus clarum* on the δ^{15}N and δ^{13}C of *Eucalyptus globus* and *Ricinus communis*. *Plant, Cell and Environment*, **16**, 375–382.

Harvey, H.P. and van den Driessche, R. (1999) Nitrogen and potassium effects on xylem cavitation and water-use efficiency in poplars. *Tree Physiology*, **19**, 943–950.

Hatfield, J.L., Sauer, T.J. and Prueger, J.H. (2001) Managing soils to achieve greater water use efficiency: A review. *Agronomy Journal*, **93**, 271–280.

Hawkins, H-J. and Lewis, O.A.M. (1993) Combination of NaCl salinity, nitrogen form and calcium concentration on the growth, ionic content and gaseous exchange properties of *Triticum aestivum* Lcv. Gamboos. *New Phytologist*, **124**, 161–170.

Hibberd, J.M. and Jeschke, W.D. (2001) Solute flux into parasitic plants. *Journal of Experimental Botany*, **52**, 2043–2049.

Howell, T.A. (2001) Enhancing water use efficiency in irrigated agriculture. *Agronomy Journal*, **93**, 281–289.

Hubick, K.T. (1990) Effects of nitrogen and water availability on growth, transpiration efficiency and carbon isotope discrimination in peanut cultivars. *Australian Journal of Plant Physiology*, **17**, 413–430.

Hutton, J.T. and Norrish, K. (1974) Silicon content of wheat husks in relation to water transpired. *Australian Journal of Agricultural Research*, **25**, 203–212.

Jensen, C.R., Mogensen, V.O., Poulsen, H.H., Henson, I.E., Aagot, S., Hansen, E., Ali, M. and Wollenweber, B. (1998) Soil water matric potential rather than water content determines drought responses in field-grown lupin *(Lupinus angustifolius* L.). *Australian Journal of Plant Physiology*, **25**, 353–363.

Kaiser, J.S. and Lewis, O.A.M. (1991) The influence of nutrient nitrogen on the growth and productivity of sunflower (*Helianthus annuus* var. Dwarf Sungold). *South African Journal of Botany*, **57**, 6–9.

Kaiser, B.N., Rawat, S.R., Siddiqi, M.Y., Masle, J. and Glass, A.D.M. (2002) Functional analysis of an *Arabidopsis* T-DNA 'Knockout' of the high-affinity NH_4^+ transporter AtAMTy1. *Plant Physiology*, **130**, 1263–1275.

Kerstiens, G., Tych, W., Robinson, M.F. and Mansfield, T.A. (2002) Sodium-related partial stomatal closure and salt tolerance of *Aster tripolium*. *New Phytologist*, **153**, 509–515.

Khan, H.R., McDonald, G.K. and Rengel, Z. (2003) Zn fertilization improves water use efficiency, grain yield and seed Zn content in chickpea. *Plant and Soil*, **249**, 389–400.

Kiesselbach, T.A. (1916) Transpiration as a factor in crop production. *Nebraska Agricultural Experimental Station Research Bulletin*, **6**.

Kleiner, D. (1981) The transport of NH_3 and NH_4^+ across biological membranes. *Biochimica et Biophysica Acta*, **639**, 41–52.

Kronzucker, H.J., Britto, D.T., Davenport, R.J. and Tester, M. (2001) Ammonium toxicity and the real cost of transport. *Trends in Plant Science*, **6**, 335–337.

Lamont, B.B., Cowling, R.M. and Groom, P.K. (2002) High leaf mass per area of related species assemblages may reflect low rainfall and carbon isotope discrimination rather than low phosphorus and nitrogen concentrations. *Functional Ecology*, **16**, 403–412.

Leidi, E.O., Silberbush, M. and Lips, S.H. (1991a) Wheat growth as affected by nitrogen type, pH and salinity. I. Biomass production and mineral composition. *Journal of Plant Nutrition*, **14**, 235–246.

Leidi, E.O., Silberbush, M. and Lips, S.H. (1991b) Wheat growth as affected by nitrogen type, pH and salinity. II. Photosynthesis and transpiration. *Journal of Plant Nutrition*, **14**, 247–256.

Li, F.M., Song, Q.H., Lie, H.S., Li, Fr. and Liu, X.L. (2001) Effects of pre-sowing irrigation and phosphorus application on watcr usc and yicld of spring whcat undcr scmi-arid conditions. *Agricultural Water Management*, **49**, 173–183.

Lionel, D., de Silva, R., Mansfield, T.A. and McAinsh, M.R. (2001) Changes in stomatal behaviour in the calcicole *Leontoden hispidus* due to the disruption by ozone of the regulation of apoplastic Ca^{2+} by trichomes. *Planta*, **214**, 158–162.

Lips, S.H., Leidi, E.O., Silberbush, M., Soares, M.I.M. and Lewis, O.A.M. (1990) Physiological aspects of ammonium and nitrate fertilization. *Journal of Plant Nutrition*, **13**, 1271–1289.

Losch, R., Jensen, C.R. and Andersen, M.N. (1992) Diurnal courses and factorial dependencies of leaf conductance and transpiration of differently potassium fertilized and watered field-grown barley plants. *Plant and Soil*, **140**, 205–224.

Mahmood, T., Woitke, M., Gimmler, H. and Kaiser W.M. (2002) Sugar exudation by roots of Kollar grass (*Leptochloa fusca* (L) Kunth) is strongly affected by the nitrogen source. *Planta*, **214**, 887–894.

Malinowski, D.P. and Belesky, D.P. (2000) Adaptations of endophyte-infected cool-season grasses to environmental stresses: Mechanisms of drought and mineral stress tolerance. *Crop Science*, **40**, 923–940.

Marschner, H. (1995) *Mineral Nutrition of Higher Plants*. Academic Press, London.

Martinez-Carrasco, R., Perez, P., Handley, L.L., Scrimgeour, C.M., Igual, M., del Molino, I.M., Sanchez, L. and De la Puente. S. (1998) Regulation of growth, water use efficiency and delta C-13 by the nitrogen source in *Casuarina equisetifolia* Forst. & Forst. *Plant Cell and Environment*, **21**, 531–534.

Masle, J., Farquhar, G.D. and Wong, S.C. (1992) Transpiration ratio and plant mineral content are related among genotypes of a range of species. *Australian Journal of Plant Physiology*, **19**, 709–721.

Mayland, H.F., Johnson, D.A., Asay, K.H. and Read, J.J. (1993) Ash, carbon-isotope discrimination, and silicon as estimators of transpiration efficiency in crested wheat grass. *Australian Journal of Plant Physiology*, **20**, 361–369.

McDonald, E.P., Erickson, J.E. and Kruger, E.L. (2002) Can decreased transpiration limit plant nitrogen acquisition in elevated CO_2? *Functional Plant Biology*, **29**, 1115–1120.

Merah, O. (2001) Carbon isotope discrimination and mineral composition of three organs in durum wheat genotypes grown under Mediterranean conditions. *Comptes Rendus de l'Academe des Sciences Serie III – Sciences de la Vie – Life Sciences*, **3224**, 355–363.

Merah, O., Deleens, E. and Monneveaux, P. (1999) Grain yield, carbon isotope discrimination, mineral and silicon content in durum wheat under different precipitation regimes. *Physiologia Plantarum*, **107**, 387–394.

Merah, O., Deleens, E., Souyris, I. and Monneveux, D. (2001) Ash content might predict carbon isotope discrimination and grain yield in durum wheat. *New Phytologist*, **149**, 275–282.

Monson, R.K. and Holland, E.A. (2001) Biospheric trace gas fluxes and their control over tropospheric chemistry. *Annual Review of Ecology and Systematics*, **32**, 547–576.

Montgomery, E.G. and Kiesselbach, T.A. (1912) Studies in the water requirements of corn. *Nebraska Agricultural Experimental Station Bulletin*, 128.

Neumann, G. and Martinoia, E. (2002) Cluster roots: an underground adaptation for survival in extreme environments. *Trends in Plant Science*, **7**, 162–167.

Paster, J., Peckham, B., Bridgham, S., Weltzin, J. and Chen, J. (2002) Plant community dynamics, nutrient cycling, and alternative stable equilibria in peatlands. *The American Naturalist*, **160**, 553–568.

Patterson, T.B., Guy, R.D. and Dang, Q.L. (1997) Whole-plant nitrogen- and water-relations traits, and their associated trade-offs, in adjacent muskeg and upland boreal spruce species. *Oecologia*, **110**, 160–168.

Payne, W.A., Drew, M.C., Hossner, L.R., Lascano, R.J., Onken, A.B. and Wendt, C.W. (1992) Soil phosphorus availability and pearl millet water-use efficiency. *Crop Science*, **32**, 1010–1015.

Payne, W.A., Hossner, L.R., Onken, A.B. and Wendt, C.W. (1995) Nitrogen and phosphorus uptake in pearl millet and its relationship to nutrient and transpiration efficiency. *Agronomy Journal*, **87**, 425–431.

Perera, L.K.R.R., de Silva, D.L.R. and Mansfield, T.A. (1997) Avoidance of sodium accumulation by the stomatal guard cells of the halophyte *Aster tripolium*. *Journal of Experimental Botany*, **46**, 707–711.

Pieterse, P.A., Rethman, N.N.G. and Van Bosch, J. (1997) Production, water use efficiency and quality of four cultivars of *Panicum maximum* at different levels of nitrogen fertilization. *Tropical Grasslands*, **31**, 117–123.

Power, J.F. (1985) Nitrogen- and water-use efficiency of several cool-season grasses receiving ammonium nitrate for 9 years. *Agronomy Journal*, **77**, 189–192.

Raven, J.A. (1980) Nutrient transport in micro-algae. *Advances in Microbial Physiology*, **21**, 47–226.

Raven, J.A. (1983) Phytophages of xylem and phloem: a comparison of animal and plant sap-feeders. *Advances in Ecological Research*, **13**, 135–234.

Raven, J.A. (1985) Regulation of pH and generation of osmolarity in vascular land plants: costs and benefits in relation to efficiency of use of water, energy and nitrogen. *New Phytologist*, **101**, 25–77.

Raven, J.A. (1986) Biochemical disposal of excess H^+ in growing plants? *New Phytologist*, **104**, 278–287.

Raven, J.A. (1988) Acquisition of nitrogen by the shoots of land plants: its occurrence and implications for acid-base regulation. *New Phytologist*, **109**, 1–20.

Raven, J.A. (1990) Sensing pH? *Plant Cell Environment*, **13**, 721–729.

Raven, J.A. (2001a) An aquatic perspective on the concepts of Ingestad relating plant nutrition to plant growth. *Physiologia Plantarum*, **113**, 301–307.

Raven, J.A. (2001b) Silicon transport at the cell and tissue level, in *Silicon in Agriculture* (eds. L.E. Datnoff, G.H. Korndorfer and G.H. Syder), Elsevier, Amsterdam, pp. 41–53.

Raven, J.A. and Farquhar, G.D. (1990) The influence of N metabolism and organic acid synthesis on the natural abundance of isotopes of carbon in plants. *New Phytologist*, **116**, 505–529.

Raven, J.A. and Handley, L.L. (1987) Transport processes and water relations, in *Frontiers of Comparative Plant Ecology* (eds. I.H. Rorison, J.P. Grime, R. Hunt, G.A.F. Hendry and D.H. Lewis), Academic Press, New York and London, pp. 217–233.

Raven, J.A. and Smith, F.A. (1976) Nitrogen assimilation and transport in vascular plants in relation to intracellular acid-base regulation. *New Phytologist*, **76**, 415–431.

Raven, J.A. and Sprent, J.I. (1993) Nitrogen assimilation and its role in plant water relations, in *Water Deficits. Plant Responses from Cell to Community* (eds. J.A.C. Smith and H. Griffiths), Bios Scientific Publishers, Oxford, pp. 205–219.

Raven, J.A. and Wollenweber, B. (1992) Temporal and spatial aspects of acid-base regulation. *Current Topics in Plant Biochemistry and Physiology*, **11**, 270–294.

Raven, J.A. and Yin, Z-H. (1998) The past, present and future of nitrogenous compounds in the atmosphere and their interactions with plants. *New Phytologist*, **139**, 205–219.

Raven, J.A., Allen, S. and Griffiths, H. (1984) N source, transpiration rate and stomatal aperture in *Ricinus*, in *Membrane Transport in Plants* (eds W.J. Cram, K. Janacek, R. Rybova and K. Sigler), Academia Praha, pp. 161–162.

Raven, J.A., Wollenweber, B. and Handley, L,L. (1992a) Ammonia and ammonium fluxes between photolithotrophs and the environment in relation to the global nitrogen cycle. *New Phytologist*, **121**, 5–18.

Raven, J.A., Wollenweber, B. and Handley, L.L. (1992b) A comparison of ammonium and nitrate as nitrogen sources for photolithotrophs. *New Phytologist*, **121**, 19–32.

Raven, J.A., Wollenweber, B. and Handley, L.L. (1993) The quantitative role of ammonia/ ammonium transport and metabolism by plants in the global nitrogen cycle. *Physiologia Plantarum*, **89**, 512–518.

Rhodes, A.P. and Mathers, J.C. (1974) Varietal differences in the amino acid composition of barley grain during development and under varying nitrogen supply. *Journal of the Science of Food and Agriculture*, **25**, 963–972.

Robinson. D. (1990) Phosphorus availability and cortical senescence in cereal roots. *Journal of Theoretical Biology*, **145**, 257–265.

Robinson, D.E., Wagner, R.G., Bell, F.W. and Swarton, C.J. (2001) Photosynthesis, nitrogen-use efficiency and water-use efficiency of jack pine seedlings in competition with four boreal forest plant species. *Canadian Journal of Forest Research*, **31**, 2014–2025.

Robinson, M.F., Very, A.A., Sanders, D. and Mansfield, T.A. (1997) How can stomata contribute to salt tolerance. *Annals of Botany*, **80**, 387–393.

Roderick, M.L., Berry, S.L., Sanders, A.R. and Noble, I.R. (1999) On the relationship between the composition, morphology and function of leaves. *Functional Ecology*, **13**, 696–710.

Rosenstiehl, T.N., Potosnak, M.J., Griffin, K.L., Fall, R. and Monson, R.K. (2003) Increasing CO_2 uncouples growth from isoprene emission in an agriforest ecosystem. *Nature*, **421**, 256–259.

Ruiz-Lozano, J.M. and Azcon, R. (1995) Hyphal contribution to water uptake in mycorrhizal plants as affected by the fungal species and water status. *Physiologia Plantarum*, **95**, 472–478.

Ryan, D.R., Delhaize, E. and Jones, D.L. (2001) Function and mechanisms of organic anion exudation from plant roots. *Annual Review of Plant Physiology and Plant Molecular Biology*, **52**, 527–560.

Santamaria, P., Elia, A., Serio, F., Gonnella, M. and Parente, A. (1999) Comparison between nitrate and ammonium nutrition in fennel, celery and swiss chard. *Journal of Plant Nutrition*, **22**, 1091–1106.

Skene, K.R. (2003) The evolution of physiology and development in the cluster root: teaching an old dog new tricks? *Plant and Soil*, **248**, 21–30.

Stanhill, G. (1985) The water-resource for agriculture. *Philosophical Transactions of the Royal Society of London, Series B*, **310**, 161–173.

Stanhill. G. (1986a) Water-use efficiency. *Advances in Agronomy* **39**, 53–85.

Stanhill, G. (1986b) John Woodward – A neglected 17th-century pioneer of experimental botany. *Israel Journal of Botany*, **35**, 225–231.

Stewart, G.R. and Press, M.C. (1990) The physiology and biochemistry of parasitic angiosperms. *Annual Review of Plant Physiology and Plant Molecular Biology*, **41**, 127–151.

Stewart, G.R. and Schmidt, S. (1999) Evolution and ecology of plant mineral nutrition, in *Physiological Plant Ecology* (ed. by M.C. Press, J.D. Scholes and M.G. Barker), Blackwell Science, Oxford, pp. 91–114.

Stout. W.L. and Schnabel. R.R. (1997) Water use efficiency of perennial ryegrass as affected by soild drainage and nitrogen fertilization on two floodplain soils. *Journal of Soil and Water Conservation*, **52**, 207–211.

Syvertson, J.P., Smith, M.L., Lloyd, J. and Farquhar, G.D. (1997) Net carbon dioxide assimilation, carbon isotope discrimination, growth and water-use efficiency of citrus trees in response to nitrogen status. *Journal of the American Society for Horticultural Science*, **122**, 226–232.

Tanner, C.B. and Sinclair, T.R. (1983) Efficient water use in crop production: research or re-search, in *Limitations to Efficient Water Use in Crop Production* (eds H.M. Taylor, W.R. Jordan and T.R. Sinclair), Publication of the American Society of Agronomy, American Soil Science Society and the American Crop Science Society, pp. 1–27.

Terry, N., Zayed, A.M., de Souza, M.P. and Tarun, A.S. (2000) Selenium in higher plants. *Annual Review of Plant Physiology and Plant Molecular Biology*, **51**, 401–432.

Thomas, F.M. and Gehler, R. (1997) Effects of different forms of nitrogen supply on the gas exchange of young pedunculate oaks (*Quercus robur* L.). *Zeitschrift für Pfanzenernährung und Bodenkunde*, **160**, 209–215.

Tow, P.G. (1993) Persistence and water-use efficiency of a tropical grass and lucerne on a solodic soil on the far north-west slopes of New South Wales. *Australian Journal of Experimental Agriculture*, **33**, 245–252.

Tsialtas, J.T., Kassioumi, M. and Veresoglou, D.S. (2002) Evaluating leaf ash content and potassium concentration as surrogates of carbon isotope discrimination in grassland species. *Journal of Agronomy and Crop Science*, **186**, 168–175.

Voltas, J., Ramagosa, I., Munoz, P. and Araus, J.L. (1998) Mineral accumulation, carbon isotope discrimination and indirect selection for grain yield in two-rowed barley grown under semi-arid conditions. *European Journal of Agronomy*, **9**, 147–155.

Walker, C.D. and Lance, R.C.M. (1991) Silicon accumulation and 13C composition as indices of water-use efficiency in barley cultivars. *Australian Journal of Plant Physiology*, **18**, 427–434.

Walker, G.K. and Richards, J.E. (1985) Transpiration efficiency in relation to nutrient status. *Agronomy Journal*, **77**, 263–269.

Wallace, J.S. and Batchelor, C.H. (1997) Managing water resources for crop production. *Philosophical Transactions of the Royal Society of London, Series B*, **352**, 937–947.

Welander, N.T. amd Ottoson, B. (2000) The influence of low light, drought and fertilization on transpiration and growth in young seedlings of *Quercus robur*. L. *Forest Ecology and Management*, **127**, 139–151.

Williams, L.E. and Miller, A.J. (2001) Transporters responsible for the uptake and partitioning of nitrogenous solutes. *Annual Review of Plant Physiology and Plant Molecular Biology*, **52**, 659–688.

Wollenweber, B. and Kinzel, H. (1988) Role of carboxylate in the nitrogen metabolism of plants from different natural habitats. *Physiologia Plantarum*, **72**, 321–328.

Wollenweber, B. and Raven, J.A. (1993) Nitrogen acquisition from atmospheric NH_3 by *Lolium perenne*: utilization of NH_3 and implications for acid-base balance. *Botanica Acta*, **106**, 42–51.

Woodward, J. (1699) Some thoughts and experiments concerning vegetation. *Philosophical Transactions of the Royal Society of London*, **21**, 193–227.

Wright, I.J. and Westoby, M. (2003) Nutrient concentration, resorption and lifespan: leaf traits of Australian sclerophyll species. *Functional Ecology*, **17**, 10–19.

Wright, I.J., Reich, P.B. and Westoby, M. (2003) Least-cost inputs of water and nitrogen for photosynthesis. *The American Naturalist*, **161**, 98–111.

Yin, Z-H. and Raven, J.A. (1998) Influences of different nitrogen sources in nitrogen- and water-use efficiency, and carbon isotope discrimination, in C_3 *Triticum aestivum* L. and C_4 *Zea mays* L. plants. *Planta*, **205**, 574–580.

8 Crop yield and water use efficiency: a case study in rice

Jianhua Zhang and Jianchang Yang

8.1 Water shortage in crop production: a growing crisis

The twentieth century witnessed great human effort to produce more food. In the 1950s and 1960s, the breeding of semi-dwarf rice and wheat cultivars that were lodging-resistant and required more nutrients, led to an unprecedented increase in grain production known as the First Green Revolution. As a result, in the second half of the twentieth century, many more agricultural practices – such as extensive irrigation, use of agricultural chemicals and hybrid cultivars – were adopted to exploit the yield potentials of such semi-dwarf cultivars.

At the beginning of the twenty-first century, another revolution – the struggle in agriculture for better use of water resources – has just begun. We may call it the Blue Revolution, if we are convinced that its importance is no less than the First Green Revolution. Norman E. Borlaug, awarded the Nobel Peace Prize in 1970 in recognition of his leading role in the First Green Revolution, advocated the 'Blue Revolution – more crop for every drop'. Why do we need a revolution to tackle this problem? The major problem is that the water resources for agricultural use are decreasing rapidly, especially in developing countries, while the world population continues to increase. In many parts of the world, economic development, increases in population and urbanization will place increasing pressure on water resources. The greenhouse effect and global weather changes may result in unpredicted and changing rainfall distribution patterns, and evaporation stress in some major agricultural areas may change significantly. We have seen many reports showing the retreat of glaciers in high mountainous areas and the disappearance of lakes. Ecological water requirements, such as the maintenance of lakes, rivers and underground water levels, will place even further pressures to develop sustainable agricultural water use.

The competition for available water resources has never been so crucial and fierce as it is today. Although 70% of the global surface is covered by water, only 2.5% of it is pure fresh water, of which most is in the Greenland and Antarctic ice caps. It is estimated that only 1% of global fresh water is actually available to us as soil moisture, lakes, rivers, reservoirs and shallow underground aquifers. This represents only about 0.007% of all the water on Earth! Agricultural water use (including precipitation and irrigation) consumes well over 70% of the available water resources in many countries. Rain-fed arable land constitutes about 83% of the world total and produces about 60% of total yield. In comparison, irrigated land accounts for about 17% of the world's cropped land but produces 40% of the food (Borlaug and Dowswell, 2000), illustrating the fact that irrigation is already an indispensable part of global agriculture.

Lack of water resources for irrigation is the major limiting factor for rain-fed agriculture. Even in areas where irrigation has been used extensively, the amount and acreage of irrigation is shrinking rather than expanding. As early as 1972, the UN-sponsored 'Human and Environment' conference warned that water shortage would become a social crisis similar to the oil crisis of the 1970s. The UN's 1997 Comprehensive Assessment of the Freshwater Resources of the World estimated that about one-third of the world's population lives in countries that are experiencing moderate-to-high water stress. By the year 2025, as much as 'two-thirds of the world's population could be under stress conditions' (WMO, 1997). It is foreseeable that such a water shortage might increasingly limit agricultural development and may put global food supplies in jeopardy. The optimistic view after the great success of the First Green Revolution and recent biotechnological advances may soon vanish with the developing economic stagnation in many areas of the world.

A good example illustrating the development of such a crisis is evident in China. China's total water resources are estimated as 2.8 trillion m^3, the sixth largest reserve in the world, but per capita this is only about one-third of the world average. China uses about 8% of world water resources and needs to feed 22% of the world population. To make things worse, China's water resources are far from evenly distributed. About 70% of the available water is concentrated in Southern China and the Yangtse Basin. Northern China has 64% of the cropped lands but possesses only 30% of the available water resources. The rainfalls are not evenly distributed either. About 70–90% of annual precipitation falls during June to September in most parts of China. Concentrated rainfalls in the summer make the river flows fluctuate significantly. Flooding and drying-up of rivers are a recurring problem in Northern China. From the late 1990s, an average of 30 million hectares of cropped land are under water shortage in Northern China resulting in a 10 million ton reduction in grain production each year.

8.2   Water-saving cultivation of rice: problems and possibilities

Rice is a luxury water user among all the crops, and consumes about half the total available water resources. In major rice-producing countries such as China, land cropped by rice accounts for about 28% of the total cropped area, but uses 70% of total water resources available to agriculture. Normally this would not be a problem because rice is mainly grown in areas where typhoons or monsoons in the summer bring abundant water to the paddy field. Extensive irrigation systems may also help maintain the paddy condition. As a result, rice production is relatively stable when compared to other crops.

In recent years, however, water shortage has begun to threaten stable rice production. One problem is the increasingly fluctuating water supply, partly due to the global weather changes and partly due to the uncontrolled deforestation in the upper-reaches of rivers, flooding and drought are becoming much more frequent in some areas. Another problem is linked to the increasing demand for water for other crops and purposes. Rice acreage is actually shrinking in some countries.

We have to use less water now for the rice crop before the problems get worse. There are several ways to tackle this problem. One is to breed cultivars that do not need so much irrigation. Although there have been many efforts to find drought-tolerant genes at the molecular level, the breeding of suitable cultivars that do not need the traditional paddy conditions remains a long way off. Rice, after all, is still a crop that needs abundant water supply and paddy conditions during most of its growing season. We know relatively so much about drought stress on many dry-land plants, but so little about rice, the so-called 'wetland' plant. Indeed, people are usually interested in knowing how rice survives and grows in the hypoxic or anoxic soil conditions and much less want to know how it will perform if watering is reduced.

As a realistic approach, we first of all need to investigate whether we can save some water using the current available cultivars. Much work in irrigation research on other land crops has shown that we can cut down water use substantially without much reduction in economic yield (e.g. Kang et al., 1998, 2000a; Zhang et al., 2001). This should be especially true for those 'luxury water users'. Rice should be tested for this hypothesis.

Another suggestion is to grow more upland rice that requires much less irrigation. Upland rice does not necessarily need paddy conditions and uses substantially less water (and less fertilizers). At present, however, upland rice cannot replace the lowland rice in most of the rice-producing areas. Two major problems limit its spreading at the moment. One is its low yield potential. Upland rice cultivars are much taller (up to 2 m) and will lodge easily if the head is too heavy. The second problem is the taste and eating quality of its cooked rice. It usually tastes harder and has less fragrance. In addition, upland rice requires more herbicides but less soil fertility. In the future, better upland rice cultivars may be bred to overcome its shortcomings (see, for example, Hirasawa et al., 1998) so that more upland rice will be grown.

However, upland rice should not replace lowland rice completely. Rice paddy fields have some important ecological functions in typhoon and monsoon-affected areas. Large areas of paddy fields serve as a huge reservoir during early summer when rainfall is abundant and prevent floods in these regions. The evapotranspiration from paddy rice also contributes to the moist atmosphere and local climate that are favorable to other vegetations in these regions.

What shall we do about the current lowland rice if a water shortage is possibly imminent? First, we need to know to what degree rice plants can withstand a water shortage and how they show it. We need to know to what extent water deficit causes irreversible damage, and which growth stage is most vulnerable to water deficit. Although some preliminary experiments have been performed in this respect (e.g. Jearakongman *et al.*, 1995; Boonjung and Fukai, 1996; Kobata *et al.*, 1996; Lilley and Ludlow, 1996; PerezMolphe-Balch *et al.*, 1996; Wopereis *et al.*, 1996; Baker *et al.*, 1997; Yeo *et al.*, 1997; Mitchell *et al.*, 1998; Scartazza *et al.*, 1998; Cabangon *et al.*, 2000; Fujii and Horie, 2001), more detailed work is still needed. Progress in plant stress physiology has enabled us to investigate more possible inter-regulations and interactions among different physiological processes. Detailed information on this aspect should help plant molecular biologists to understand possible roles of the genes they want to work on. Such information will also tell us whether it is possible theoretically to reduce irrigation without much reduction of economic yield.

Second, we need to screen the available rice cultivars in this region extensively for those which can perform better under reduced irrigation. Ideally we should breed some cultivars that can use less water. Perhaps due to a lack of urgency, breeding for this purpose has been slow (e.g. Nguyen *et al.*, 1997; Hirasawa *et al.*, 1998; Cooper, 1999; Cooper *et al.*, 1999; Fukai *et al.*, 1999). Although molecular biologists have tried hard to search for the genes (e.g. Courtois *et al.*, 2000; Saijo *et al.*, 2000; Tripathy *et al.*, 2000; Zhang *et al.*, 2001), it is still in the preliminary stage to transform the current cultivars for drought tolerance through molecular breeding. What we can do at the moment perhaps is to look at the current cultivars that have been used in this area. Rice is relatively sensitive in photoperiodism and to temperature changes. Cultivars therefore, are usually both latitude- and region-specific. We should focus on the region-specific cultivars first.

Third, we need to investigate whether some available irrigation methods are applicable with rice. Many different irrigation methods, such as deficit irrigation (e.g. Goodwin and Jerie, 1992; Boland *et al.*, 1993; Kang *et al.*, 2000b) and partial rootzone irrigation (Loveys *et al.*, 1997; Dry and Loveys, 1998; Zhang *et al.*, 1998; Kang *et al.*, 1998, 2000a; Zhang *et al.*, 2001), have been reported to improve WUE in many dry-land plants. We have also done some experiments on irrigated crops and found that not all the root system needs to be irrigated at any one time if we want to improve the WUE (Zhang *et al.*, 1998; Kang *et al.*, 1998, 2000a; Zhang *et al.*, 2001). It has been shown with many crops that substantial savings in water use can be made without major impact on the economic yield. Rice as a 'luxury water user' should not be an exception.

If we say that rice uses more irrigated water than any other crops, we believe that this is due mainly to surface evaporation and leakage into the ground from the paddy fields. Can water be saved through lower rates of evapotranspiration? Rice forms a condense canopy in the field. The total evapotranspiration from such a canopy is a long-distance diffusion process and determined by the total diffusion resistance and the vapor concentration difference. Factors that have influences on these two parameters include leaf temperature, stomatal resistance, vapor concentration inside the canopy and boundary resistance from inside to outside atmosphere. Since stomatal resistance is only a small fraction of the total diffusion resistance and a leaf heat-up may partially compensate for the increased stomatal resistance, the evapotranspiration is basically determined by how much light energy is intercepted by the canopy (see, for example, Monteith, 1975).

If water can be saved from evapotranspiration, very possibly we are talking about a trade-off of better water use efficiency for a smaller biomass yield. Total photosynthesis and biomass production may be reduced as a result of the better WUE. We may predict that when a shoot water deficit develops, leaf rolling or leaf wilting may reduce the light interception and therefore the photosynthesis. From this point of view, we should test whether such a loss may be partially compensated by, for example, higher planting density and/or better harvest index. Some interesting results have been obtained (see, for example, Yang et al., 2000, 2001a,b,c,d, 2002b). When soil drying was applied at the grain filling stage, a better utilization of pre-stored food enhanced the harvest index although early senescence shortened the duration of photosynthesis.

Rice is conventionally flood-irrigated. Can surface evaporation be reduced if some furrows are used for irrigation? At least with deficit irrigation, periodic dry-up of soil surface will surely reduce the surface evaporation and save water. In addition, periodic irrigation will also reduce possible percolation when compared to the conventional flood-irrigation. There is a lot to test here.

More evapotranspiration may be reduced if we can reduce the humidity inside the rice canopy. This may be achieved by 'small and pulse' irrigation so that soil surface is left dry periodically. In addition, furrow irrigation may also achieve the same purpose by wetting only part of the soil surface. In practice, less humidity inside the canopy should help control many humidity-sensitive diseases and insects. Frequently some of these pests, such as rice blast, rice leaf bacterial blight and rice hoppers, cause significant yield loss in humid areas. If these problems can be controlled better, it should bring the extra benefit of reduced irrigation.

8.3   A trade-off between grain yield and WUE?

Plants unavoidably lose a large quantity of water when they open their stomata for CO_2 uptake in less saturated air. Vapor defuses out from the substomatal cavity into the air while CO_2 goes in the opposite direction. Mathematical modeling of these two opposing diffusion processes has shown that water use efficiency (WUE) is largely a function of the CO_2 and vapour concentration differences between the inside and outside of the leaf (Jones, 1992). These two opposing fluxes are regulated by stomata, and stomatal behavior will therefore determine the WUE of a particular species.

Indeed WUE has been shown a conservative parameter with high inheritance. It is well known that C_4 plants have higher WUE than C_3 plants. Within C_3 plants, many reports have shown that genotypes can be selected for higher WUE according to their carbon isotope discrimination, a function of the CO_2 concentration gradient between the inside and outside of the leaf (see, for example, Craufurd et al., 1991; Ehdaie et al., 1991).

Why is WUE a conservative characteristic of plants? Jones (1992) explained that stomata operate in such a way that a reduction in carbon assimilation during stomatal closing or partial closing is responsible for only a small proportion of the total assimilation. This is particularly true when plant water requirements are well met. It is well known, however, that plants growing under water-limited conditions will have higher WUE. Indeed it has been predicted that plants generally have the capability to optimize their water use in the short term and maximize their chance of survival over a drought in the longer term. In the short term, for example a 24 h period, their carbon gained is maximized for limited amount of water lost (Cowan, 1977, 1982; Farquhar and Sharkey, 1982). In the longer term, for example in a season, their water loss should be regulated according to the amount of available water in the soil (Cowan, 1982, 1986; Jones, 1980). In a world where rainfall is unpredictable, long-term regulation means that plants must be able to 'detect' the soil drying and then 'respond' to it by regulating their stomata. Such a mechanism may be termed as a feed-forward mechanism, since the decreasing availability of water in the soil does not cause any significant water deficit (Davies and Zhang, 1991). Jones (1992) concluded that such a responsive pattern of stomatal behavior should be the best pattern for both plant survival and carbohydrate production.

Such optimized stomatal behavior means that plants should have a higher WUE with less water supply. If we hold the conventional view that plant biomass production is linearly coupled with the amount of water used, it is not surprising that higher WUE is a trade-off for lower biomass production. In agriculture, many ways of conserving water have been investigated and techniques such as partial irrigation, deficit irrigation or drip irrigation have shown that WUE can be enhanced (Stewart et al., 1981; Musick and Dusck,

1982; Hodges *et al.*, 1989; Graterol *et al.*, 1993; Stone and Nofziger, 1993; Kang *et al.*, 2000a). In general, these techniques are a trade-off: a lower yield for a higher WUE.

Can we increase WUE without much reduction of yield? It is well known that grain yield, a large proportion of the total biomass, shows a negative parabolic relationship with the amount of irrigation. This suggests that when water supply is sufficient, excessive vegetative growth may lead to less root activity, unhealthy canopy structure and a lower harvest index. That means that high biomass production, supported by high water supply, will not lead to high WUE if defined as the grain production per unit amount of water irrigated. Therefore the goal is to increase WUE under limited water supply and increase harvest index. Our recent research has shown that in many situations, grain yield can be improved while reducing the amount of water applied to the crop (Yang *et al.*, 2000, 2001c, 2002b), mainly via improved harvest index which has been shown as the key component in improving WUE of yield (see, for example, Ehdaie and Waines, 1993).

8.4 Grain filling: concurrent photosynthesis and remobilization of pre-stored carbon

Grain filling is the final stage of growth in cereals where fertilized ovaries develop into caryopses. At this stage, about 40–50% of total biomass is formed. Grain filling is contributed by both the concurrent photosynthesis during this period and remobilization of pre-stored carbon in sheath and stems.

There are many reports that grain filling is limited by photosynthesis during this period. For example, in our effort to reduce irrigation for wheat in Northern China where water shortage is already threatening sustainable wheat production, we reduced wheat irrigation from four to five times in a season to a single irrigation at the early flower-forming stage (Zhang *et al.*, 1998). We were able to enhance WUE, calculated as grain yield over total water consumption, by 24–30% with only a 15% reduction of grain yield from full irrigation. The reduction in grain yield was mainly a result of reduced grain size, due to early leaf senescence. Wheat production in many areas of Northern China usually suffers from hot and dry weather during grain-filling time (early senescence, for example), but this can be reduced to a minimum if irrigation is feasible. With the adoption of more restricted irrigation, we have found that up to a week was lost from the crucial grain-filling period when we had very hot and dry weather in 1995. Unlike other major cereal crops, rice has two hard and interlocked hulls that limit grain size; in wheat, grain size is much more variable and depends largely on the length of grain-filling period. In 1995 the grain size of single-irrigated wheat crop was greatly reduced as a result of early leaf senescence. As such, early leaf senescence and shortened grain filling period is a major limitation to wheat production in this area of China.

In field experiments we have shown that single-irrigated wheat plants were suffering from heat stress before senescence took place. Leaf temperature was substantially increased as a result of reduced leaf conductance. Heat stress causes damage in at least two ways. One is that photosynthesis is reduced so substantially and more quickly than respiration such that leaves may reach a compensation point when no net carbon is assimilated (Bjorkman *et al.*, 1980). The second is damage associated with the loss of thermal stability of membranes and proteins. At high temperatures, membrane lipids tend to be more fluid, and integral proteins tend to dissociate from hydrophilic regions and clump into hydrophobic areas so that the membrane may become leaky and denatured (Raison *et al.*, 1982; Bjorkman *et al.*, 1980).

It is possible that high yield breeding has selected cultivars that rely on high transpiration rates to keep leaf temperature stable, such that they are vulnerable to heat stress. Such high yield cultivars will yield substantially less when inadequate irrigation is supplied. Breeding for cultivars in areas where water shortage is accelerated should aim to produce genotypes that can tolerate heat stress and maintain green leaves during the grain-filling stage when irrigation is reduced. This is especially true for wheat production in areas with hot and dry weather conditions and limited water resources. An adequate grain-filling period is essential to maintain high yield.

8.5 Problems in grain filling

Monocarpic plants, such as rice and wheat, need the initiation of whole plant senescence so that stored carbohydrates in stems and leaf sheaths can be remobilized and transferred to their grains. Normally in these crops, pre-stored food contributes 25–33% to the final weight of a grain. Delayed whole plant senescence, leading to poorly filled grains and unused carbohydrate in straws, is a new problem increasingly recognized in rice and wheat production in recent years. Slow grain filling is always associated with delayed whole plant senescence. Although farmers can choose cultivars of early-maturation, there are still some situations that have made the delayed senescence a serious problem that needs attention:

1. It is well known that heavy use of nitrogen fertilizers can lead to delayed senescence and, in worst cases, canopy lodging. Although farmers are generally aware of this, the problem still occurs every year and everywhere, especially in highly productive areas (see Jiang *et al.*, 1997; Yue *et al.*, 1997). This is possibly related to the intensive nature of agriculture today: using less arable land to feed ever more people.

2. Selection of lodging-resistant cultivars to cope with the lodging problem has led to another problem in some cases, namely that stems are short and strong but stored carbohydrate is poorly used because the plants may stay 'green' for too long, particularly in some cases with short-grain rice cultivars. If weather conditions are favorable, and if such delayed senescence is also 'functionally' delayed, that is, photosynthesis and phloem translocation are functional, such delayed senescence may help achieve higher dry mass production and possibly higher grain yield (Thomas and Smart, 1993). However, in many cases, kernels and their connecting rachis or rachilla seem mature or senesce earlier than the stem and leaves. Much unused carbohydrate is left in the stem and sheath as a result (see, for example, Cao and Zhu, 1987; Liang *et al.*, 1994a; Ricciardi and Stelluti, 1995; Yang *et al.*, 1997).
3. Introduction of hybrid rice has been a fantastic success in China. Utilization of heterosis – a hybrid between the two subspecies (or ecotypes) of rice, the Japonica and Indica rices – has, however, also met the problem of delayed senescence. Grains are poorly filled or un-filled kernels (Yuan, 1987, 1990, 1996; Yuan *et al.*, 1992; Zhang, 1993; Liang *et al.*, 1994b; Zhu, 1995; Gu *et al.*, 1996). Such hybrid genotypes seem too vigorous in terms of keeping 'young'.

We may define cases described above as unfavorably delayed senescence, which means that no gain is obtainable from the extended grain-filling period. We should however, distinguish this situation from that in favorable conditions, in which early senescence should be avoided because it reduces the photosynthesis during the grain-filling period and therefore reduces grain weight (Zhang *et al.*, 1998).

Delayed senescence in 2 and 3 above is basically a genetically controlled characteristic and there are no practical measures to deal with this problem. Situation 1 is managed in practice with so called 'appropriate control of nitrogen fertilizers and canopy density', which is often too late when crops are in the grain-filling stage. In all of these three situations, a slow grain filling is a yield-limiting factor and any way in which the rate of grain filling can be enhanced should be beneficial to the final yield formation.

8.6   Controlled soil drying to promote whole plant senescence at grain filling

Our recent experience with field-grown wheat has found that soil drying during the grain-filling period can enhance early senescence (Table 8.1). While the grain-filling period was shortened by 10 days (from 41 to 31 days) in unwatered (during this period) plots, a faster rate of grain-filling and

Table 8.1 Comparison between wheat plots that were well-watered or unwatered during grain-filling stage. Fate of fed ^{14}C was measured on day 18 from anthesis.

Watering treatments	Duration from anthesis to maturation (days)	Fate of fed ^{14}C ($^{14}CO_2$ applied10 days early)		Total sugars left in stem (day 26 from anthesis)
		% in kernels	% in steman	
Well-watered	41	41.3	40.5	29%
Unwatered	31	81.3	9.6	8%

Reproduced from Zhang *et al.* (1998).

enhanced mobilization of stored carbohydrate minimized the effect on yield (Table 8.1). It seems possible that a controlled soil drying during the later stages of grain-filling may promote whole plant senescence, leading to increased re-translocation of pre-stored carbon reserve in the stem and sheath.

The mechanisms by which the utilization of pre-stored food was enhanced are not known. Many processes are likely to be involved, including storage carbohydrate hydrolysis, phloem loading, long distance translocation and phloem unloading into the kernels. Whatever the mechanisms, it is interesting to know whether such a treatment (soil drying) can be beneficial to grain filling in cases where filling is slow, apparently as a result of delayed senescence. The rationales behind such approach are:

1. A mild soil drying may not seriously disrupt the phloem function. It has been shown that phloem translocation may be less susceptible to drought than leaf photosynthesis (Boyer and McPherson, 1975; Kozlowski, 1978).
2. The duration of a developing grain to receive food, that is, the filling period, may be limited by the life span of its phloem link. A faster rate of filling will certainly have some advantages if a season is limited. It has been shown that 'stay green' (or delayed senescence in some sense) is not necessarily associated with the full function of photosynthesis (Thomas and Smart, 1993). It is also possible that the phloem link to the grains may lose its function earlier than the chlorophyll disappearance in the leaves in those 'stay-green' genotypes.
3. Even in weather conditions permitting delayed senescence and more photosynthetic assimilation (for example, on the functionally 'stay green' varieties), the gain from an accelerated grain filling from pre-anthesis carbohydrate reserve may outweigh any loss of photosynthesis as a result of imposed soil drying. Some delayed senescence seems genetically controlled – the cases of lodging-resistant cultivars that stay green too long or hybrid cultivars with too high heterosis. These cultivars always leave a substantial amount of food unused in their straws. In maize it is the fast reallocation of stem carbohydrate that is responsible for the high

grain weight, rather than the 'stay green' characteristics (Dwyer *et al.*, 1995).

A controlled soil-drying means that crops should not be soil-dried to a degree that overnight rehydration cannot be completed and photosynthesis is not too severely inhibited. It should be stressed that soil drying should be at the later stage of grain filling because early development of embryos (at the rapid cell division stage – the 'grain-setting' stage – is very susceptible to water deficit (Boyle *et al.*, 1991).

With the highly lodging-resistant rice cultivars, our results (Yang *et al.*, 2001c) showed that if a water deficit during grain filling of rice is controlled properly so that the plant can rehydrate overnight, photosynthesis should not be severely inhibited (Figures 8.1 and 8.2). A benefit from such a water deficit is that it can enhance plant senescence and lead to a fast and better remobilization of pre-stored carbon from vegetative tissues to the grains (Table 8.2). The early senescence induced by water deficit does not necessarily reduce grain yield even when plants are grown under normal N conditions. Furthermore, in cases where plant senescence is unfavorably delayed such as by heavy use of nitrogen, the gain from the enhanced remobilization and accelerated grain-filling rate may outweigh the loss of photosynthesis and shortened grain-filling period and increase the grain yield and harvest index (Table 8.3).

When senescence is unfavorably delayed, rice will show a prolonged and slow grain filling, for example under high N condition (Figure 8.3). Controlled soil drying increases the grain-filling rate and shortens the grain-filling period. The increased rate and shortened period are especially remarkable under high N conditions (Figures 8.3C and D). Our results (Yang *et al.*, 2001a) with lodging-resistant cultivars also showed that the final grain weight was not significantly different between well-watered and water-stressed treatments when normal amounts of N were applied. However, it was significantly increased under water-stressed plus high N treatment, implying that the gain from accelerated grain-filling rate outweighed the possible loss of photosynthesis as a result of a shortened grain-filling period when subjected to water stress during grain filling.

Similar results were also obtained with hybrid rice which shows very strong heterosis but a slow grain filling as a result of delayed whole plant senescence (Yang *et al.*, 2002b). The stronger the heterosis – the hybrid between japonica and indica subspecies – the higher the harvest index can be improved by a controlled soil drying at the grain-filling stage (Table 8.4). The grain-filling process and rate of the hybrid cultivars were substantially enhanced by the controlled soil drying. The grain yield was actually improved, rather than reduced, by the moderate soil drying the grain-filling stage in cases where heterosis is very strong, for example the japonica and indica hybrids (Table 8.5).

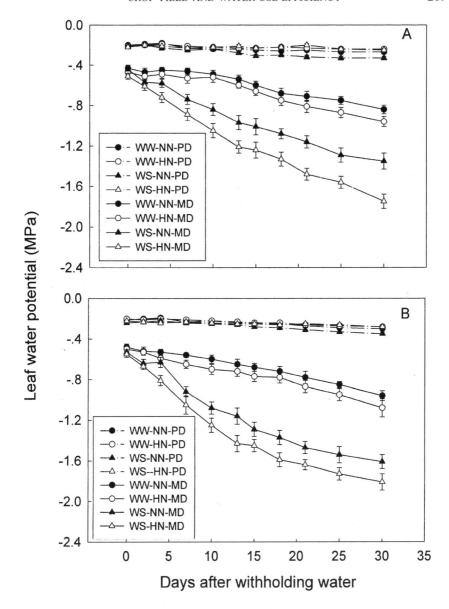

Figure 8.1 Changes of leaf water potentials of the japonica cultivar Wuyujing 3 (A) and indica cultivar Yangdao 4 (B) during the first 30–d after withholding water. The two cultivars used are highly lodging-resistant, i.e. staying green when grains are mature. NN and HN indicate normal and high levels of nitrogen application at heading time. WW and WS are well watered and water-deficit treatments during the grain filling. Measurements were made on the flag leaves at pre-dawn (PD, 0600 h) and at midday (MD, 1130 h). Vertical bars represent ±SE of the mean (n=6) where these exceed the size of the symbol. Reproduced from Yang *et al.* (2001c).

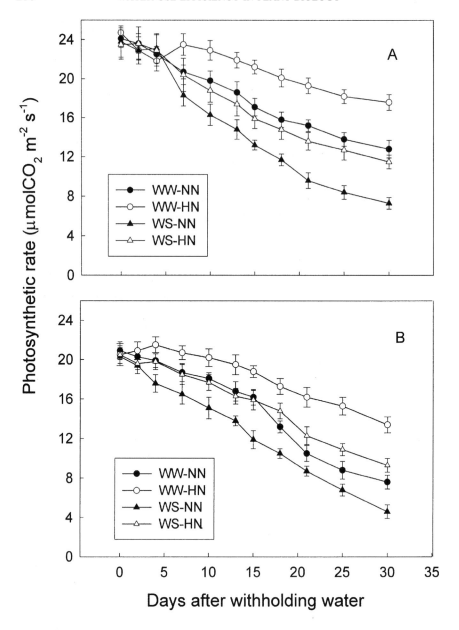

Figure 8.2 Photosynthetic rates of the flag leaves of the japonica cultivar Wuyujing 3 (A) and indica cultivar Yangdao 4 (B) during the first 30–d after withholding water. Treatment details are same as in Figure 8.1. Vertical bars represent ±SE of the mean (n=6) where these exceed the size of the symbol. Reproduced from Yang *et al.* (2001c).

Table 8.2 Remobilization of pre-stored assimilates in straws of rice subjected to various N and soil moisture treatments. The two cultivars used are highly lodging-resistant, i.e. staying green when grains are mature. NN and HN indicate normal and high levels of nitrogen application at heading time. WW and WS are well-watered and water-deficit treatments during the grain filling. Values are means of 20 plants. Letters indicate statistical significance at $P_{0.05}$ within the same cultivar. NSC stands for nonstructural carbohydrate in straws.

Cultivars	Water deficit treatment	Nitrogen applied	Remobilized C reserve %	Contribution to grain %	NSC residue mg g^{-1} DW	Total dry matter g m^{-2}	Harvest index
Wuyujing 3	WW	NN	47.5 c	14.4 c	142.3 b	1707.2 a	0.47 c
	WW	HN	24.5 d	7.6 d	218.5 a	1741.2 a	0.41 d
	WS	NN	74.6 a	28.5 a	64.5 d	1537.9 b	0.52 a
	WS	HN	61.2 b	21.5 b	103.5 c	1716.6 a	0.50 b
Yangdao 6	WW	NN	58.9 b	11.3 c	97.8 b	1788.0 a	0.51 c
	WW	HN	46.3 c	5.3 d	151.6 a	1743.6 a	0.47 d
	WS	NN	82.7 a	27.8 a	41.2 d	1578.8 b	0.56 a
	WS	HN	65.9 b	17.9 b	85.3 c	1768.3 a	0.53 b

Reproduced from Yang et al. (2001c).

Table 8.3 Grain-filling rate and grain yield of rice subjected to various N and soil moisture treatments (see Table 8.2). Active grain filling period and grain filling were calculated according to Richards (1959) equation. Values of total spikelets, grain weight and percentage of ripened grains were means 148–156 plants. Values of grain yield were means of all plants harvested from three plots of each treatment. Letters indicate statistical significance at $P_{0.05}$ within the same cultivar.

Cultivars	Water deficit treatment	Nitrogen applied	Active grain filling period d	Grain filling rate mg d⁻¹ grain⁻¹	Total spikelets 10^3 m⁻²	Ripened grains %	Grain weight mg grain⁻¹	Grain yield g m⁻²
Wuyujing 3	WW	NN	19.7 b	1.21 c	33.73 a	90.8 b	26.2 b	802.4 b
	WW	HN	24.8 a	0.91 d	33.78 a	84.2 c	25.1 c	713.9 c
	WS	NN	17.0 c	1.39 a	33.71 a	90.2 b	26.3 b	799.7 b
	WS	HN	19.1 b	1.28 b	33.62 a	94.2 a	27.1 a	858.3 a
Yangdao 6	WW	NN	23.9 b	1.02 c	41.80 a	80.5 ab	27.1 a	911.9 ab
	WW	HN	28.6 a	0.82 d	42.09 a	74.6 b	26.1 b	819.5 c
	WS	NN	18.4 d	1.31 a	41.81 a	78.9 b	26.8 a	884.1 b
	WS	HN	21.2 c	1.14 b	42.23 a	82.5 a	26.9 a	937.2 a

Reproduced from Yang et al. (2001c).

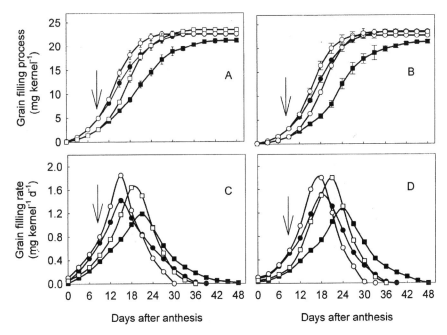

Figure 8.3   Grain filling process (A and B) and grain filling rate (C and D) of the japonica cultivar Wuyujing 3 (A and C) and indica cultivar Yangdao 4 (B and D) subjected to various nitrogen and soil moisture treatments. The treatments are: normal N (NN) +well watered (WW) (•), NN + water stressed (WS) (o), high N (HN) + WW ■), and HN + WS (□). Grain filling rate was calculated according to Richards (1959) equation. Arrows in the figure indicate the start of withholding water. Vertical bars in the figure A and B represent ±SE of the mean (n=2) where these exceed the size of the symbol. Reproduced from Yang *et al.* (2001a).

Heterosis of hybrid rice is usually a function of the genetic relations of the two parents. Hybrids from japonica and indica rice show stronger heterosis and also tend to stay green longer in later grain filling than the indica/indica rice (Figure 8.4, compare the A with B and C). When the whole plant stays green, its non-structural carbon (NSC), mainly the starch, in the culm also stays high for long (Figure 8.5). We found that concentration of NSC in the culm and sheath during grain filling was very different between indica/indica and japonica/indica hybrid(s) under WW treatments (Figure 8.5). NSC in the culm and sheath of the two japonica/indica hybrids showed a 'V' shape pattern – initially decreasing from 7 to 21 days after anthesis, but increasing thereafter. NSC concentrations at maturity for both japonica/indica hybrids were nearly the same as at anthesis. For the indica/indica hybrid, NSC in the culm and sheath decreased sharply from 7 to 32 days after anthesis and slowly thereafter. Water deficits substantially reduced NSC in the culm and sheath of all three hybrids. The more severe the water deficit, the more the NSC was reduced. Under MD and SD treatments the patterns of the NSC

Table 8.4 Remobilization of pre-stored carbon assimilates in straw of three hybrid rice cultivars subjected to various soil moisture treatments. The values in the table: † = [non structural carbohydrate (NSC) in culms and sheaths at anthesis – NSC in residue]/NSC in culms and sheaths at anthesis 100; ‡ = (NSC in culms and sheaths at anthesis NSC in residue)/grain weight × 100; § = transfer ratio of total assimilate: (panicle weight at maturity panicle weight at anthesis)/[NSC in culms and sheaths at anthesis + (dry weight of plant at maturity – dry weight of plant at anthesis)] × 100; ¶ = amount of NSC remaining in culms and sheaths at maturity; # = Total grain weight/total aboveground dry weight. All values are means of 20 plants. Means within a column with the same letter for a hybrid are not significant at the $P = 0.05$ level.

Hybrid	Water deficit treatment	Remobilized C reserve	Contribution to grain %	TRA §	NSC in residue mg g^{-1} DW	Harvest index #
Shanyou 63 (indica/indica)	Well-watered	64 c	19 c	71 c	87 a	0.48 b
	Moderate water-deficit	76 b	26 b	86 b	57 b	0.53 a
	Severe water-deficit	89 a	38 a	92 a	33 c	0.55 a
Ce 03/Yangdao 4 (japonica-indica)	Well-watered	14 c	6 c	47 c	201 a	0.41 c
	Moderate water-deficit	61 b	24 b	80 b	92 b	0.48 b
	Severe water-deficit	74 a	32 a	88 a	61 c	0.53 a
PC311/Zaoxian-dang 18 (japonica-indica)	Well-watered	7 c	2 c	23 c	215 c	0.37 c
	Moderate water-deficit	53 b	21 b	65 b	103 b	0.46 b
	Severe water-deficit	67 a	27 a	80 a	85 a	0.51 a

Reproduced from Yang *et al.* (2002b).

Table 8.5 Grain-filling rate and grain yield of three rice hybrids subjected to various soil moisture treatments. Active grain filling period and grain filling rate were calculated according to Richards (1959) equation. WW, MD, and SD are well-watered, moderate water-deficit, severe water-deficit during the grain filling. Values of total spikelets, grain weight and the percentage of ripened grains were means of 150 plants. Grain yields were means of all the plants harvested from three plots of each treatment. Means within a column with the same letter for a hybrid are not significant at the $P = 0.05$ level.

Hybrid	Water deficit treatment	Active grain filling period d	Grain filling rate mg d^{-1} grain^{-1}	Total spikelets $\times 10^3$ m^{-2}	Ripened grain %	Grain weight mg grain^{-1}	Grain yield g m^{-2}
Shanyou 63 (indica/indica)	WW	21.2 a	1.16 c	40.7 a	83.6 a	27.3 a	929 a
	MD	18.5 b	1.30 b	41.1 a	81.9 a	26.8 a	902 a
	SD	16.7 c	1.39 a	40.4 a	74.2 b	25.9 b	776 b
Ce 03/Yangdao 4	WW	24.2 a	0.94 c	46.9 a	76.8 b	25.3 a	911 b
	MD	20.7 b	1.11 b	46.6 a	80.1 a	25.5 a	952 a
	SD	18.4 c	1.23 a	47.1 a	77.5 b	25.1 a	916 b
PC311/Zaoxiandang 18 (japonica-indica)	WW	27.3 a	0.82 c	48.3 a	67.1 b	24.9 a	807 b
	MD	22.0 b	1.03 b	48.5 a	74.8 a	25.2 a	914 a
	SD	18.6 c	1.19 a	48.1 a	71.2 a	24.5 a	839 b

Reproduced from Yang et al. (2002b).

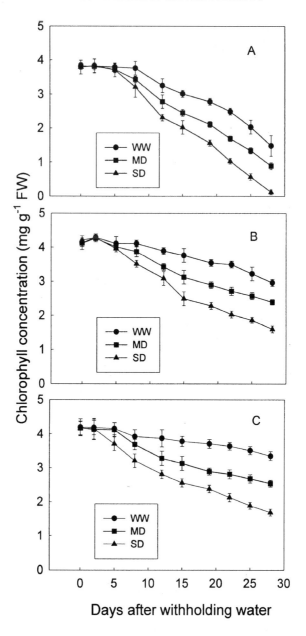

Figure 8.4   Changes in chlorophyll content in the flag leaves of an indica/indica hybrid rice Shanyou 63(A), and two japonica/indica hybrids, Ce 03/Yangdao 4 (B) and PC 311/Zaoxiandang 18 (C), during the first 28–d after withholding water. WW, MD, and SD are well-watered, moderate water-deficit, severe water-deficit during grain filling. Vertical bars represent ±SE of the mean (n=6) where these exceed the size of the symbol. Reproduced from Yang *et al.* (2002b).

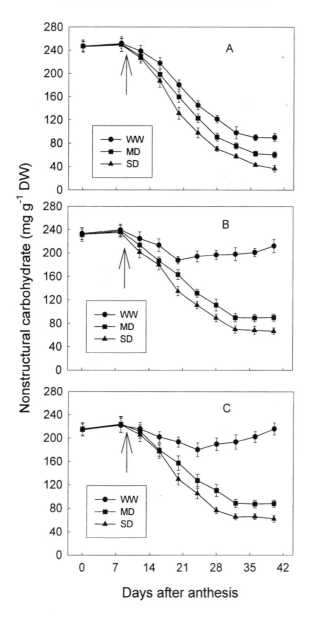

Figure 8.5 Nonstructural carbohydrate concentrations in the culm and sheath of an indica/indica hybrid rice Shanyou 63(A), and two japonica/indica hybrids, Ce 03/Yangdao 4 (B) and PC 311/ Zaoxiandang 18 (C), during the grain filling. WW, MD, and SD are well-watered, moderate water-deficit, severe water-deficit during grain filling. Arrows in the figure indicate the start of water deficit treatments. Vertical bars represent ±SE of the mean (n=6) where these exceed the size of the symbol. Reproduced from Yang *et al.* (2002b).

were similar for both indica/indica and japonica/indica hybrid(s) (Figure 8.5).

Wheat grain yield is more vulnerable to shortened grain-filling period than rice. As shown with wheat under reduced irrigation, grain size can be greatly reduced by less irrigation at this period (Zhang et al., 1998). However, in cases with high N and very strong lodging-resistant cultivars (Yang et al., 2000), moderate soil drying at later grain filling enhances the harvest index substantially with an improved WUE, if defined as the grain yield over the irrigated amount (Table 8.6). Soil drying at grain-filling stage can greatly shorten the filling period and reduce the grain size and yield (Table 8.7). However, in cases where staying green is a problem as a result of heavy use of nitrogen, the moderate soil drying may not necessarily reduce the grain yield of wheat (Table 8.7). An extra benefit from such practice is that less water may be required for wheat irrigation under these circumstances.

8.7 Plant hormones and key enzymes in the enhanced remobilization of pre-stored reserve

It is well known that plant hormones are involved in grain filling and seed development (Davies, 1987; Brenner and Cheikh, 1995). There are many reports that auxins, gibberellins (GAs) and abscisic acid (ABA) regulate the grain development (Karssen, 1982; Davies, 1987; Kende and Zeevaart, 1997; Hansen and Grossmann, 2000). We observed that cytokinins and indole-3–acetic acid contents in the rice grains transiently increases at early filling stage and coincides with the rapid increase of grain-filling rate (Yang et al., 2001a, 2002a,c). The heights and timing of such peaks are regulated by N nutrition status, that is, high N associated with lower peaks and their later appearance. Controlled soil drying only hastens the declines of cytokinin and indole-3-acetic acid contents at the late grain-filling stage.

Contents of gibberellins in the rice grains are also higher at early grain-filling stage than late. High N enhances while water stress substantially reduces the GA accumulation. ABA content in the rice grains is low at early grain-filling stage. Water stress increases the ABA accumulation greatly in the grains. The peak values of ABA in the grains are significantly correlated with the maximum grain filling rates. Our results suggest that an altered hormonal balance in the grains by water stress during grain filling, especially the decrease in GAs and increase in ABA, enhances the remobilization of pre-stored carbon to the grains (Yang et al., 2001a). Such a regulation may shorten the grain filling period but accelerate grain-filling rate that could be beneficial to cases where slow grain filling is a problem in rice production.

The main storage form of non-structural carbohydrate (NSC) in the straw of rice is starch (Murayama et al., 1961; Murata and Matsushima, 1975; Cao et al.,

Table 8.6 Remobilization of stored assimilates in straw of wheat subjected to various N and soil moisture treatments. The two cultivars are lodging-resistant wheat, ie. Staying green at later grain filling stage. Symbols NN and HN indicate normal and high levels of N application at heading time, and WW, MD, SD and WS are well-watered, mildly dried, severely dried during the grain filling. †= [non structural carbohydrate (NSC) in stems and sheaths at anthesis (a) − NSC residue]/a × 100. ‡ = (NSC of stems and sheaths at anthesis − NSC of stems and sheaths at maturity)/grain weight × 100. * = Transfer ratio of total assimilate: (spike weight at maturity spike weight at anthesis)/[NSC in stems and sheaths at anthesis + (dry weight of plant at maturity − dry weight of plant at anthesis)] × 100. ** = the amount of NSC remaining in stems and sheaths at maturity. Values are means of 15 plants. Letters indicate statistical significance at $P_{0.05}$ within the same cultivar and within the same N level.

Cultivars	Nitrogen applied	Soil drying treatments	Remobilized C reserve† %	Contribution to grain‡	TRA * %	NSC Residue ** mg g⁻¹ DW	Harvest index
Yangmai 158	NN	WW	47 c	18 c	65 c	188 a	0.39 b
	NN	MD	57 b	24 b	72 b	150 b	0.45 a
	NN	SD	70 a	40 a	80 a	106 c	0.44 a
	HN	WW	33 c	11 c	52 c	232 a	0.35 b
	HN	MD	54 b	18 b	69 b	162 b	0.44 a
	HN	SD	62 a	26 a	73 a	133 c	0.43 a
Yangmai 931	NN	WW	47 c	19 c	60 c	185 a	0.38 b
	NN	MD	57 b	26 b	67 b	147 b	0.40 a
	NN	SD	69 a	37 a	76 a	108 c	0.40 a
	HN	WW	30 c	12 c	49 c	222 a	0.31 b
	HN	MD	50 b	17 b	63 b	173 b	0.40 a
	HN	SD	66 a	28 a	72 a	120 c	0.40 a

Reproduced from Yang *et al.* (2000b).

Table 8.7 Grain-filling rate and grain yield of wheat subjected to various N and soil moisture treatments (see Table 8.6). Active grain filling period and grain filling were calculated according to Richards (1959) Equation (in Materials and Methods). Values of kernel weight and grain yield were means of 216 plants harvested from six pots of each treatment in the pot experiment, or means of 840 plants harvested from two plots of each treatment in the field experiment. Letters indicate statistical significance at $P_{0.05}$ within the same cultivar and within the same N level.

Cultivars Pot Exp.	N applied	Soil drying treatment	Active grain filling period d	Grain filling rate mg d^{-1}	Kernel weight mg grain^{-1}	Grain yield g per pot
Yangmai 158	NN	WW	27 a	1.3 b	40 a	55 a
	NN	MD	20 b	1.6 a	36 b	47 b
	NN	SD	17 c	1.3 b	25 c	32 c
	HN	WW	38 a	0.8 c	33 c	43 c
	HN	MD	22 b	1.7 b	40 a	57 a
	HN	SD	19 c	1.9 a	38 b	51 b
Yangmai 931	NN	WW	29 a	1.2 c	40 a	50 a
	NN	MD	21 b	1.6 b	37 b	46 b
	NN	SD	17 c	1.7 a	32 c	39 c
	HN	WW	40 a	0.7 c	31 c	39 c
	HN	MD	25 b	1.5 b	41 a	53 a
	HN	SD	21 c	1.6 a	37 b	47 b

Reproduced from Yang et al. (2000b).

1992). Starch needs to be degraded as glucose first and then sucrose is re-synthesized when the carbon is remobilized from the stem to the grains (Murayama *et al.*, 1961; Venkateswarlu and Visperas, 1987; Beck and Ziegler, 1989). Starch degradation can occur via hydrolytic and phosphorolytic reactions and probably the concerted action of several enzymes (Beck and Ziegler, 1989; Nielsen *et al.*, 1997). These enzymes include: α-amylase, β-amylase, α-glucosidase, and starch phosphorylase. Based on its high affinity, sucrose-phosphate synthase (SPS) is expected to play a major role in the re-synthesis of sucrose (Whittingham *et al.*, 1979; Wardlaw and Willenbrink, 1994), and sustain the assimilatory carbon flux from source to sink (Isopp *et al.*, 2000).

It has been reported that starch degradation in stolons of white clover is controlled by the α-amylase activity (Gallagher *et al.*, 1997). Water stress can induce α-amylase in the barley leaves (Jacobsen *et al.*, 1986) and enhance β-amylase activity in the cucumber cotyledons (Todaka *et al.*, 2000). We also investigated all the possible enzymes involved in the starch degradation and sucrose re-synthesis in rice stems during grain filling (Yang *et al.*, 2001d). Our results showed that both α- and β-amylase activities were enhanced by the soil drying, with the former enhanced more than the latter, and significantly correlated with the concentrations of soluble sugars in the stems. The other two possible starch-breaking enzymes, α-glucosidase and starch phosphorylase, showed no significant differences in the activities between the well-watered and water stress treatments. Water stress also increased the SPS activity that is responsible for sucrose production. All of these results suggest that an enhanced source-supplying capacity is promoted when controlled soil drying accelerates the carbon remobilization.

On the sink side and in the grains, grain filling means an active metabolism of carbohydrate. Over 33 enzymes have been reported to be involved in this process, although only four enzymes are considered to play a key or regulative role (Nakamura *et al.*, 1989; Preiss, 1991; Kato, 1995). In rice grains, sucrose synthase acts as the first step in the pathway (Kato, 1995) – the hydrolysis of sucrose – and its activity has accordingly been linked to sink strength in the developing grain (Liang *et al.*, 2001). The predominant pathway for starch biosynthesis in plants is composed of three reactions catalyzed by adenine diphosphoglucose pyrophosphorylase (AGPase), starch synthase, and starch branching enzyme (Nakamura *et al.*, 1989; Preiss *et al.*, 1991; Okita, 1992; Martin and Smith, 1995). The activities of the three enzymes are closely associated with the increase in starch content during the development of rice endosperm (Nakamura and Yuki, 1992).

In our study of rice grain filling under a controlled soil drying, activity of sucrose synthase, responsible for the hydrolysis of sucrose, was substantially enhanced and positively correlated with starch accumulation rate in the grains (Yang *et al.*, 2003). Both soluble and insoluble invertases, the other possible sucrose-breaking enzymes, were less enhanced by soil drying and showed no significant correlation with starch accumulation rate. Starch branching enzyme and soluble starch synthase activities were also enhanced by the soil drying, with the former enhanced more than the latter, and were significantly correlated with starch accumulation rate. AGPase activity was little affected by the water stress. These results suggest that the soil drying increased remobilization and grain-filling rate are attributed to enhanced sink strength by regulating sucrose synthase and starch branching enzyme activities in rice grains when subjected to water stress during the grain-filling period.

8.8   Conclusions

Water shortage has become a major threat to agriculture in the twenty-first century. As Borlaug has advocated, we should use every drop to produce more crops. Many developing countries will probably be the worst-hit in terms of water crisis. Many countermeasures have been proposed in recent years. We should adopt different approaches in areas with different rainfalls and water resources. In the near future, rain-fed agriculture should be encouraged and irrigated areas should be reduced rather than expanded. Rice crops use a significant amount of water in agricultural areas in the world. In areas where water resources are increasingly scarce, water-saving cultivation of rice should be researched and tested to a greater extent.

Water use efficiency, if defined as the biomass accumulation over water consumed, may be a highly inherited characteristic of a specific genotype and can be selected in plant breeding. WUE indeed can be enhanced by less irrigation, particularly via stomatal regulation. However, such enhancement is largely a trade-off between lower biomass production and higher WUE.

WUE, can be enhanced in agriculture in many ways. Many irrigation methods, such as drip irrigation and deficit irrigation, or cropping practices (such as mulching and surface harrowing), reduce soil surface evaporation and therefore increase WUE. Partial rootzone irrigation, as suggested recently (Loveys et al., 1997; Kang et al., 1997, 1998, 2000; Zhang et al., 2001), may maintain the partial stomatal opening for longer and therefore enhance WUE for biomass production. However, its contribution to long-term WUE improvement in the field crops remains to be confirmed and less surface evaporation may also save irrigated water as its merit.

We have presented a case here that we may enhance WUE through an improved harvest index. Harvest index has been shown as a variable factor in crop production, especially in cases where whole plant senescence of rice and wheat is unfavorably delayed. Such delayed senescence can delay the remobilization of pre-stored carbon reserves in the straws and results in lower harvest index. A controlled soil drying – the moderate drying such that overnight rehydration of plants is still possible – should enhance the whole plant senescence and therefore improve the remobilization of pre-stored carbon reserve. The gains from the improved harvest may outweigh any possible loss due to shortened photosynthetic period in grain filling, such as the cases with high N nutrition, lodging-resistant cultivars that stay green for too long, and hybrid cultivars with too high heterosis. One more advantage for this practice is that less water is used in the grain filling and a shortened growing season is required. This can be an extra benefit in areas where water shortage is a problem and a short growing season is favored.

Acknowledgement

We are grateful for research grants funded by the FRG of Hong Kong Baptist University, RGC of Hong Kong University Grants Council, Area of Excellence of Plant and Fungal Biotechnology in the Chinese University of Hong Kong, the National Natural Science Foundation of China (Project No. 39970424) and the State Key Basic Research and Development Plan (G 1999011704).

References

Baker, J.T., Allen, L.H., Boote, K.J. and Pickering, N.B. (1997) Rice responses to drought under carbon dioxide enrichment. 1. Growth and yield. *Global Change Biology*, **3**, 119–128.

Beck, E. and Ziegler, P. (1989) Biosynthesis and degradation of starch in higher plants. *Annual Review of Plant Physiology and Plant Molecular Biology*, **40**, 95–117.

Bjorkman, O., Badger, M.R. and Armond, P.A. (1980) Response and adaptation of photo-synthesis to high temperatures, in *Adaptation of Plants to Water and High Temperature Stress* (eds N.C. Turner and P.J. Kramer), Wiley, New York, pp. 233–249.

Boland, A.M., Mitchell, P.D., Jerie, P.H. and Goodwin, I. (1993) Effect of regulated deficit irrigation on tree water use and growth of peach. *Journal of Horticultural Science*, **68**, 261–274.

Boonjung, H. and Fukai, S. (1996) Effects of soil water deficit at different growth stages on rice growth and yield under upland conditions. 1. Growth during drought. *Field Crops Research*, **48**, 37–45.

Borlaug, N.E. and Dowswell, C. (2000) Global food security: Harnessing science in the 21st century, in *Gene Technology Forum*, Kasetsart University, Thailand. March 7, 2000.

Boyer, J.S. and McPherson, H.G. (1975) Physiology of water deficits in cereal crops. *Advances in Agronomy*, **27**, 1–23.

Boyle, M.G., Boyer, J.S. and Morgan, P.W. (1991) Stem infusion of liquid culture medium prevents reproductive failure of maize at low water potentials. *Crop Science*, **31**, 1246–1252.

Brenner, M.L. and Cheikh, N. (1995) The role of hormones in photosynthate partitioning and seed filling, in *Plant Hormones* (ed P.J. Davies), Kluwer Academic Publishers, The Netherlands, pp. 649–670.

Cabangon, R.J. and Tuong, T.P. (2000) Management of cracked soils for water saving during land preparation for rice cultivation. *Soil Tillage Research*, **56**, 105–116.

Cao, X., Zhu, Q. and Yang, J. (1992) Classification of source-sink types in rice varieties with corresponding cultivated ways, in *Prospects of Rice Farming for 2000* (ed S. Min), Zhejiang Publishing House of Science and Technology, Hanzhou, China, pp. 360–365.

Cao, X. and Zhu, Q. (1987) Study on characteristics of the relationship between source and sink in rice varieties and their classification. *Acta Agronomica Sinica*, **13**, 265–272.

Cooper, M. (1999) Concepts and strategies for plant adaptation research in rainfed lowland rice. *Field Crops Research,* **64**, 13–34.

Cooper, M., Fukai, S. and Wade, L.J. (1999) How can breeding contribute to more productive and sustainable rainfed lowland rice systems? *Field Crops Research* , **64**, 199–209.

Courtois, B., McLaren, G., Sinha, P.K., Prasad, K., Yadav, R. and Shen, L. (2000) Mapping QTLs associated with drought avoidance in upland rice. *Molecular Breeding*, **6**, 55–66.

Cowan, I.R. (1977) Stomatal behaviour and environment. *Advances in Botanical Research*, **4**, 117–228.

Cowan, I.R. (1982) Water use and optimization of carbon assimilation, in *Physiological Plant Ecology II, vol. II* (ed Lange *et al.*), Springer-Verlag, Berlin, pp. 589–613.

Cowan, I.R. (1986) Economics of carbon fixation in higher plants, in *On the economy of plant form and function* (ed T. J. Givnish), Cambridge University Press, Cambridge, pp. 133–170.

Craufurd, P.Q., Austin, R.B., Acevedo, E. and Hall, M.A. (1991) Carbon isotope discrimination and grain-yield in barley. *Field Crops Research*, **27**, 301–314.

Davies, P.J. (1987) The plant hormones: Their nature, occurrence, and functions, in *Plant Hormones and Their Role in Plant Growth and Development* (ed P.J. Davies), Martinus Nijhoff Publishers, The Netherlands, pp. 1–11.

Davies, W.J. and Zhang, J. (1991) Root signals and the regulation of growth and development of plants in drying soil. *Annual Review of Plant Physiology and Plant Molecular Biology*, **42**, 55–76.

Dry, P.R. and Loveys, B.R. (1998) Factors influencing grapevine vigour and the potential for control with partial rootzone drying. *Australian Journal of Grape Wine Research*, **4**, 140–148.

Dwyer, L.M., Andrews, C.J., Stewart, D.W., Ma, D.L. and Dugas, J.A. (1995) Carbohydrate levels in field-grown leafy and normal maize genotypes. *Crop Science*, **35**, 1020–1027.

Ehdaie, B., Hall, A.E., Farquhar, G.D., Nguyen, H.T. and Waines, J.G. (1991) Water-use efficiency and carbon isotope discrimination in wheat. *Crop Science*, **31**, 1282–1288.

Ehdaie, B. and Waines, J.G. (1993) Variation in water-use efficiency and its components in wheat: I. Well-watered pot experiment. *Crop Science*, **33**, 294–299.

Farquhar, G.D. and Sharkey, T.D. (1982) Stomatal conductance and photosynthesis. *Annual Review of Plant Physiology*, **7**, 315–345.

Fujii, M., Horie, T. (2001) Relative contributions of tolerance and avoidance to drought resistance in dry-matter production of different rice cultivars at different fertilization levels. *Japanese Journal of Crop Science*, **70**, 59–70.

Fukai, S., Pantuwan, G., Jongdee, B. and Cooper, M. (1999) Screening for drought resistance in rainfed lowland rice. *Field Crops Research,* **64**, 61–74.

Gallagher, J.A., Volence, J.J., Turner, L.B. and Pollock, C.J. (1997) Starch hydrolytic enzyme activities following defoliation of white clover. *Crop Science*, **37**, 1812–1818.

Goodwin, I. and Jerie, P. (1992) Regulated deficit irrigation: from concept to practice. *Australia and New Zealand Wine Industry Journal*, **5**, 131–133.

Graterol, Y.E., Eisenhauer, D.E. and Elmore, R.W. (1993) Alternate – furrow irrigation for soybean production. *Agricultural Water Management*, **24**, 133–145.

Gu, S. *et al.* (1996) Genetic analysis on the grain filling degree on japonica-indica hybrid rice. *Chinese Journal of Rice Science*, **10**, 129–137.

Hansen, H. and Grossmann, K. (2000) Auxin-induced ethylene triggers abscisic acid biosynthesis and growth inhibition. *Plant Physiology*, **124**, 1437–1448.

Hirasawa, H., Nemoto, H., Suga, R., Ishihara, M., Hirayama, M., Okamoto, K. and Miyamoto, M. (1998) Breeding of a new upland rice variety 'Yumenohatamochi' with high drought resistance and good eating quality. *Breeding Science,* **48**, 415–419.

Hodges, M.E., Stone, J.F., Garton, J.E. and Weeks, D.L. (1989) Variance of water advance in wide spaced furrow irrigation. *Agricultural Water Management*, **16**, 5–13.

Isopp, H., Frehner, M., Long, S.P. and Nosberger, J. (2000) Sucrose-phosphate synthase responds differently to source-sink relations and photosynthetic rates: *Lolium perenne* L. growing at elevated p_{CO2} in the field. *Plant, Cell and environment*, **23**, 597–607.

Jacobsen, J.V., Hanson, A.D. and Chandler, P.C. (1986) Water stress enhances expression of an α-amylase gene in barley leaves. *Plant Physiology*, **80**, 350–359.

Jearakongman, S., Rajatasereekul, S., Naklang, K., Romyen, P., Fukai, S., Skulkhu, E., Jumpaket, B. and Nathabutr, K. (1995) Growth and grain yield of contrasting rice cultivars grown under different conditions of water availability. *Field Crops Research,* **44**, 139–150.

Jiang, D. *et al.* (1997) Effects of top-dressing at different developmental stages on the senescence of root system of winter wheat. *Acta Agronomica Sinica,* **23**, 181–190.

Jones, H.G. (1980) Interaction and integration of adaptive responses to water stress: The implications of an unpredictable environment, in *Adaptation of plants to water and high temperature stress* (eds N. C. Turner and P. J. Kramer), Wiley, New York, pp. 353–365.

Jones, H.G. (1992) *Plants and Microclimate: A Quantitative Approach to Environmental Plant Physiology,* 2nd edn, Cambridge University Press, Cambridge.

Kang, S., Zhang, J., Liang, Z.S., Hu, X.T. and Cai, H.J. (1997) Controlled alternate partial rootzone irrigation: A new approach for water saving regulation in farmland. *Agricultural research in Arid and Semiarid Areas,* **15**(1), 1–6.

Kang, S., Liang, Z., Hu, W. and Zhang, J. (1998) Water use efficiency of controlled alternate irrigation on root-divided maize plants. *Agricultural Water Management,* **38**, 69–76.

Kang, S., Liang, Z.S., Pan, Y.H., Shi, P.Z. and Zhang, J. (2000a) Alternate furrow irrigation for maize production in arid areas. *Agricultural Water Management,* **45**, 267–274.

Kang, S., Shi, W. and Zhang, J. (2000b) An improved water-use efficiency for maize grown under regulated deficit irrigation. *Field Crops Research,* **67**, 207–214.

Karssen, C.M. (1982) The role of endogenous hormones during seed development and the onset of primary dormancy, in *Plant growth substances* (ed P.F. Wareing), Academic Press, London, pp. 623–632.

Kato, T. (1995) Change of sucrose synthase activity in developing endosperm of rice cultivars. *Crop Science,* **35**, 827–831.

Kende, H. and Zeevaart, J.A.D. (1997) The five 'classical' plant hormones. *Plant Cell,* **9**, 1197–1210.

Kobata, T., Okuno, T. and Yamamoto, T. (1996) Contributions of capacity for soil water extraction and water use efficiency to maintenance of dry matter production in rice subjected to drought. *Japanese Journal Crop Science,* **65**, 652–662.

Kozlowski, T.T. (1978) *Water Deficits and Plant Growth, Vol. 5.* Academic Press, New York.

Liang, J. *et al.* (1994) The changes of stem-sheath reserve contents of rice and affecting factors during grain filling. *Chinese Journal of Rice Science,* **8**, 151–156.

Liang, J., Zhang, J. and Cao, X. (2001) Grain sink strength may be related to the poor grain filling of *indica-japonica* rice (*Oryza sativa*) hybrids. *Physiologia Plantarum,* **112**, 470–477.

Lilley, J.M. and Ludlow, M.M. (1996) Expression of osmotic adjustment and dehydration tolerance in diverse rice lines. *Field Crops Research,* **48**, 185–197.

Loveys, B.R., Grant, J., Dry, P.R. and McCarthy, M.G. (1997) Progress in the development of partial rootzone drying. *The Australian Grapegrower and Winemaker,* **404**, 18–20.

Martin, C. and Smith, A.M. (1995) Starch biosynthesis. *Plant Cell,* **7**, 971–985.

Mitchell, J.H., Siamhan, D., Wamala, M.H., Risimeri, J.B., Chinyamakobvu, E., Henderson, S.A. and Fukai, S. (1998) The use of seedling leaf death score for evaluation of drought resistance of rice. *Field Crops Research,* **55**, 129–139.

Monteith, J.L. (1975) *Vegetation and the Atmosphere, Volume 1, Principles.* Academic Press, London.

Murata, Y. and Matsushima, S. (1975) Rice, in *Crop Physiology* (ed T. Evans), Cambridge University Press, Cambridge, pp. 73–79.

Murayama, N., Oshima, M. and Tsukahara, S. (1961) Studies on the dynamic status of substances during ripening processes in rice plant. *Science of Soil Manure,* **32**, 261–265.

Musick, J.T. and Dusck, D.A. (1982) Skip – row planting and irrigation of graded furrows. *Transactions of American Society of Agriculture and Engineering,* **25**, 82–87.

Nakamura, Y., Yuki, K. and Park, S.Y. (1989) carbohydrate metabolism in the developing endosperm of rice grains. *Plant and Cell Physiology,* **30**, 833–839.

Nakamura, Y. and Yuki, K. (1992) Changes in enzyme activities associated with carbohydrate metabolism during development of rice endosperm. *Plant Science,* **82**, 15–20.

Nguyen, H.T., Babu, R.C., Blum, A. (1997) Breeding for drought resistance in rice: Physiology and molecular genetics considerations. *Crop Science,* **37**, 1426–1434.

Nielsen, T.H., Deiting, U. and Stitt, M. (1997) A β-amylase in potato tubers is induced by storage at low temperature. *Plant Physiology,* **113**, 503–510.

Okita, T.W. (1992) Is there an alternative pathway for starch synthesis? *Plant Physiology*, **100**, 560–564.

PerezMolpheBalch, E., Gidekel, M., SeguraNieto, M., HerreraEstrella, L. and OchoaAlejo, N. (1996) Effects of water stress on plant growth and root proteins in three cultivars of rice (*Oryza sativa*) with different levels of drought tolerance. *Physiol Plant*, **96**, 284–290.

Preiss, J., Ball, K., Smith-White, B., Iglesias, A., Kakefuda, G. and Li, L. (1991) Starch biosynthesis and its regulation. *Biochemistry Society Transactions*, **19**, 539–547.

Raison, J.K., Pike, C.S.and Berry, J.A. (1982) Growth temperature-induced alterations in the thermotropic properties of *Nerium oleander* membrane lipids. *Plant Physiology,* **70**, 215–218.

Richards, F.J. (1959) A flexible growth function for empirical use. *Journal of Experimental Botany*, **10**, 290–300

Ricciardi, L. and Stelluti, M. (1995) The response of durum wheat cultivars and Rht1/rht1 near-isogenic lines to simulated photosynthetic stresses. *Journal of Genetics and Breeding*, **49**, 365–374.

Saijo, Y., Hata, S., Kyozuka, J., Shimamoto, K. and Izui, K. (2000) Over-expression of a single Ca2+-dependent protein kinase confers both cold and salt/drought tolerance on rice plants. *Plant Journal*, **23**, 319–327.

Scartazza, A., Lauteri, M., Guido, M.C. and Brugnoli, E. (1998) Carbon isotope discrimination in leaf and stem sugars, water-use efficiency and mesophyll conductance during different developmental stages in rice subjected to drought. *Australian Journal of Plant Physiology*, **25**, 489–498.

Stewart, B.A, Dusck, D.A. and Musick, J.T. (1981) A management system for conjunctive use of rainfall and limited irrigation of graded furrows. *Soil Science Society of America Journal*, **45**, 413–419.

Stone, J.F. and Nofziger, D.L. (1993) Water use and yields of cotton grown under wide-spaced furrow irrigation. *Agricultural Water Management* **24**, 27–38.

Thomas, H. and Smart, C.M. (1993) Crops that stay green. *Annals of Applied Biology*, **123**, 193–219.

Todaka, D., Matsushima, H. and Morohashi, Y. (2000) Water stress enhances β-amylase activity in cucumber cotyledons. *Journal of Experimental Botany*, **51**, 739–745.

Tripathy, J.N., Zhang, J., Robin, S., Nguyen, T.T. and Nguyen, H.T. (2000) QTLs for cell-membrane stability mapped in rice (*Oryza sativa* L.) under drought stress. *Theoretical and Applied Genetics*, **100**, 1197–1202.

Venkateswarlu, B. and Visperas, R.M. (1987) Source-sink relationships in crop plants. *International Rice Research Paper Series*, International Rice Research Institute 125, 1–19.

Wardlaw, I.F. and Willenbrink, J. (1994) Carbohydrate storage and mobilization by the culm of wheat between heading and grain maturity: the relation to sucrose synthase and sucrose-phosphate synthase. *Austrian Journal of Plant Physiology*, **21**, 255–271.

Whittingham, C.P., Keys, A.J. and Bird, I.F. (1979) The enzymology of sucrose synthesis in leaves, in *Encyclopedia of Plant Physiology Volume 6* (eds M. Gibbs and E. Latzko), Springer-Verlag, Berlin, pp. 313–315.

World Meteorological Organization. (1997) Comprehensive Assessment of the Freshwater Resources of the World.

Wopereis, M.C.S., Kropff, M.J., Maligaya, A.R. and Tuong, T.P. (1996) Drought-stress responses of two lowland rice cultivars to soil water status. *Field Crops Research,* **46**, 21–39.

Yang, J. (1997) Studies on photosynthetic characteristics, accumulation and transport of photoassimilate of japonica-indica hybrids. *Acta Agronomica Sinica*, **23**, 82–88.

Yang, J., Zhang, J., Huang, Z., Zhu, Q. and Wang, L. (2000) Remobilization of carbon reserves is improved by controlled soil-drying during grain filling of wheat. *Crop Science*, **40**, 1645–1655.

Yang, J., Zhang, J., Wang, Z., Zhu, Q. and Wang, W. (2001a) Hormonal changes in the grains of rice subjected to water stress during grain filling. *Plant Physiology*, **127**, 315–323.

Yang, J., Zhang, J., Huang, Z., Wang, Z., Zhu, Q. and Liu, L. (2001b) Water deficit-induced senescence and its relationship to the remobilization of pre-stored carbon in wheat during grain filling. *Agronomy Journal*, **93**, 196–206.

Yang, J., Zhang, J., Wang, Z., Zhu, Q. and Wang, W. (2001c) Remobilization of carbon reserves in response to water-deficit during grain filling of rice. *Field Crops Research*, **71**, 47–55.

Yang, J., Zhang, J., Wang, Z. and Zhu, Q. (2001d) Activities of starch hydrolytic enzymes and sucrose-phosphate synthase in the stems of rice subjected to water stress during grain filling. *Journal of Experimental Botany*, **52**, 2169–2179.

Yang, J., Zhang, J., Wang, Z., Zhu, Q. and Liu, L. (2002a) Abscisic acid and cytokinins in the root exudates and leaves and their relations with senescence and remobilization of carbon reserves in rice subjected to water stress during grain filling. *Planta*, **215**, 645–652.

Yang, J., Zhang, J., Liu, L., Wang, Z. and Zhu, Q. (2002b) Carbon remobilization and grain filling in japonica/indica hybrid rice subjected to post-anthesis water deficits. *Agronomy Journal*, **94**, 102–109.

Yang, J., Zhang, J., Huang, Z., Wang, Z., Zhu, Q. and Liu, L. (2002c) Correlation of cytokinin levels in the endosperms and roots with cell number and cell division activity during endosperm development in rice. *Annals of Botany*, **90**, 369–377.

Yang, J., Zhang, J., Wang, Z., Zhu, Q. and Liu, L. (2003) Activities of enzymes involved in sucrose-to-starch metabolism in rice grains subjected to water stress during grain filling. *Field Crops Research*, **81**, 69–81.

Yeo, M.E., Cuartero, J., Flowers, T.J. and Yeo, A.R. (1997) Gas exchange, water loss and biomass production in rice and wild *Oryza* species in well-watered and water-limiting growth conditions. *Botanica Acta*, **110**, 32–42.

Yuan, L. (1987) Strategies of hybrid rice breeding. *Hybrid Rice*, **1**, 1–3.

Yuan, L. (1990) Advances in two-line hybrid rice research. *Scientia Agricultra Sinica*, **23**, 1–6.

Yuan, L. (1996) Strategies of selection for hybrid combination between japonica and indica rice. *Hybrid Rice*, **2**, 1–3.

Yuan, L. (1992) *Current Status of Two Line Hybrid Rice Research*. Agriculture Publisher of China, Beijing.

Yue, S. (1997) Effects of top-dressing at different developmental stages on the senescence of flag leaves and grain weight of winter wheat. *Scientia Agricultra Sinica*, **29**, 59–69.

Zhang, B. (1993) Physiological and ecological characters of tow line hybrid rices. IV. characteristics of grain filling and its physiological basis. *Journal of Fujian Agricultural College*, **22**, 141–147.

Zhang, J., Sui, X., Li, B., Su, B., Li, J. and Zhou, D. (1998) An improved water-use efficiency for winter wheat grown under reduced irrigation. *Field Crops Research*, **59**, 91–98.

Zhang, J., Jia, W. and Kang, S. (2001) Partial rootzone irrigation: its physiological consequences and impact on water use efficiency. *Acta Botanica Boreali-Occidentalia Sinica*, **21**, 191–197.

Zhang, J., Zheng, H.G., Aarti, A., Pantuwan, G., Nguyen, T.T., Tripathy, J.N., Sarial, A.K., Robin, S., Babu, R.C., Nguyen, B.D., Sarkarung, S., Blum, A. and Nguyen, H.T. (2001) Locating genomic regions associated with components of drought resistance in rice: comparative mapping within and across species. *Theoretical and Applied Genetics*, **103**, 19–29.

Zhu, Q. (1995) Study on the index of grain filling degree of hybrid rice. *Journal of Jiangsu Agriculture*, **16**, 1–5.

9 Molecular approaches to unravel the genetic basis of water use efficiency

Roberto Tuberosa

9.1 Introduction

Increasing crop productivity while enhancing the sustainability of agricultural practices will be the major challenge faced by plant scientists during the twenty-first century. Improving the livelihood and well-being of billions of people now languishing in poverty and malnutrition will depend, among other numerous factors, on the possibility of markedly enhancing the yield of crops and stabilizing it from season to season. This challenge is even more daunting in view of the increasing vagaries of the seasonal trends in rainfall, possibly as a result of global warming. It is thus evident that improving water use efficiency (WUE) and yield under arid conditions will play an increasingly pivotal role for better exploiting the productivity potential of each crop. In rice, it has been estimated that to meet the food requirements of the fast growing population in Asia, the average yield should reach approximately 8 tonnes per hectare by 2010, that is, 25% above the present value (Gowda *et al.*, 2003). It should be noted that under optimal conditions, yields of up to 17 tonnes have been reported for hybrid rice in China (Xiao *et al.*, 1996), an indication that both biotic and abiotic stresses prevent reaching such high values in farmers' fields.

Clearly, a better knowledge of the genetic factors governing variability in seasonal WUE and harvestable yield will help the breeder to devise more effective strategies for improving yield potential while optimizing water harvest. The release of improved cultivars requiring lower amounts of water/ unit yield and characterized by high yield potential and stability is thus an essential prerequisite for more profitable and sustainable agricultural practices, particularly in rainfed, drought-prone areas. Indeed, WUE has now become an important target in many breeding programmes, also in consideration of the escalating costs of irrigation water, when this option is available. The emerging concept is that while in the past we have made significant efforts to adapt the environment to the plant, the newly released cultivars should be genetically tailored to improve their ability to withstand environmental setbacks and to optimize the use of water and nutrients. Molecular techniques hold great potential to dissect the genetic and functional basis of complex traits and, once such genes have been identified, to manipulate them more effectively, thus maximizing the returns in terms of response to selection of the target traits (Miflin, 2000; Tuberosa *et al.*, 2002a). Genomics and genetic engineering offer an array of innovative, integrative approaches to plant biochemistry, physiology and breeding that allow unprecedented opportunities for crossing reproductive barriers among species (Khush, 1999; Serageldin, 1999).

In this context, the main objectives of this chapter are (i) to critically review some of the information that genomic approaches have provided on the genetic basis of the traits affecting WUE in annual crops and (ii) to illustrate and discuss to what extent this knowledge may contribute to improve WUE and drought tolerance through genetic engineering and/or marker-assisted selection. Some basic information will be provided concerning the methodologies, genetic resources and technological platforms utilized to dissect the genetic basis of WUE and related traits, in the hope that a better understanding and appreciation of the possibilities offered by these novel platforms will further stimulate interdisciplinary research. A more 'holistic' approach is an essential prerequisite for harnessing the value of gene discovery and its applications to improve WUE and yield performance of crops, particularly under water-limited conditions. As to the terminology for categorizing the mechanisms conferring tolerance to drought, this article follows the nomenclature reported by Ludlow and Muchow (1990), which distinguishes between traits providing escape from drought and those providing resistance to drought, with the latter ones further categorized in terms of dehydration avoidance and dehydration tolerance. Dehydration avoidance depends on maintenance of turgor through an increase in water uptake and/or reduction in water loss, while dehydration tolerance relies on biochemical mechanisms allowing the cell to tolerate water loss. This notwithstanding, from a breeding standpoint increasing emphasis is being placed on the mechanisms contributing to increase yield *per se*, rather than the characteristics enhancing plant survival under extreme drought, in the notion that these may have a negative trade off under less severe circumstances (Passioura, 2002). A number of reviews have extensively analysed and discussed the mechanisms and the traits underlying drought tolerance in a wide range of crops (Boyer, 1982; Ceccarelli, 1984; Morgan, 1984; Blum 1988, 1996; Quarrie, 1991; Passioura, 1996, 2002; Richards, 1996, 2000; Turner, 1997; Quarrie *et al.*, 1999b; Mitra, 2001; Araus *et al.*, 2002; Richards *et al.*, 2002).

9.2 Target traits influencing WUE

Throughout this chapter, the term WUE will prevalently be defined as the amount of dry matter produced (grain yield in case of grain crops when considering seasonal WUE) per unit of water lost through evapotranspiration. Other chapters in this volume have exhaustively dealt with the issues related to the physiological bases of WUE as well as their implications in terms of dry matter production and partitioning. Because WUE is to a varying degree influenced by any trait affecting the CO_2 fixation and water status of the plant, its genetic basis virtually encompasses all the genes governing the bio-chemical and morpho-physiological characteristics regulating the carbon and water balance of the plant (for example, chlorophyll content, PEPcarboxylase activity, root size and architecture, osmotic adjustment, ABA concentration, stomatal conductance, leaf thickness and angle, etc.). In grain crops grown in water-stressed environments, a meaningful formula that highlights the critical role of WUE in determining grain yield (GY) was suggested by Passioura (1977):

$$GY = W \times WUE \times HI$$

where W is the total amount of water used by the crop and evaporated from the field, WUE is the water use efficiency and HI is the harvest index, i.e. the ratio between GY and total biomass. When considering this formula, we should be aware of the possible interdependence of these variables, with the result that selection for improving WUE in order to increase GY may be partially counterbalanced by a reduction in the amount of water extracted from the soil. Some traits have been shown to influence both W and WUE. The most important factor is matching the phenological development pattern of the crop and the seasonal rainfall pattern (Richards, 1996; Turner, 1997; Villegas et al., 2000). Early vigour potentially increases both W and WUE, while deep roots or osmoregulation may also help the crop to use more water (Ludlow and Muchow, 1990). The greatest genetic gains in bread wheat yields have been attributed to changes in HI, with negligible increases in total biomass production (Austin et al., 1989; Siddique et al., 1989; Duvick and Cassman, 1999; Tollenaar and Wu, 1999). However, the scenario in other crops may be different (for durum wheat, see Pfeiffer et al., 2000) also in relation to the availability of water before and after flowering (Richards et al., 2002).

From an agronomic standpoint, a valuable formula to properly address the factors influencing WUE in field-grown crops has been proposed by Richards (1991):

$$WUE \text{ (biomass)} = TE/(1 + Es/T)$$

where TE is the transpiration efficiency (above ground dry weight/transpired water), Es is the water lost by evaporation from the soil surface and T is water lost through transpiration by the crop. The analysis of the variables of this formula provides a useful framework to identify the agronomic and breeding strategies most suitable to optimize WUE and maximize yield in target environments differing in terms of rainfall distribution during the crop vegetative and reproductive phases. At the leaf level, 'intrinsic WUE' indicates the ratio of the instantaneous rates of CO_2 assimilation and stomatal transpiration. Of course, the interpretation of this formula in the context of final yield requires the integration of many other factors acting at levels of increasing morpho-physiological complexity moving from a single leaf to the crop community. The factors influencing intrinsic WUE have been exhaustively analysed by Condon *et al.* (2002) who pointed out how an increased intrinsic WUE can be achieved either through lower stomatal conductance or higher photosynthetic capacity or both. The same authors cautioned about the possible penalties in terms of yield upon manipulation of each variable and concluded that to achieve more widespread gains in cereal yield derived from greater intrinsic WUE, it is necessary to decouple intrinsic WUE and low crop growth rate.

The physiological traits to be considered as potential selection targets for improving yield must be genetically and, preferably, causally correlated with yield and should have a greater heritability than yield itself. Ideal traits are those whose measurement is fast, accurate and inexpensive. Additionally, the trait should integrate environmental and physiological cues experienced by the plant for an extended period of time preceding data collection, rather than providing information only for the time of data collection (Araus *et al.*, 2002). From these considerations it is clear that none of the physiological traits so far targeted meets all these ideal requirements. However, according to the characteristics of the target environment, on a case-by-case basis it may be more convenient to select for a trait known to be genetically associated with WUE and yield, rather than attempting a direct selection of the former traits. Because this chapter focuses on the identification of quantitative trait loci (QTLs) more directly affecting WUE and related traits, some general background information is provided on those traits for which QTLs have been reported. Clearly, these traits represent only a fraction of all the morpho-physiological features that have been shown to influence WUE and yield under drought (Blum, 1988; Ludlow and Muchow, 1990; Turner, 1997; Richards *et al.*, 2002).

9.2.1 Carbon isotope discrimination

Carbon isotope discrimination ($\Delta^{13}C$) is a measure of the ratio of stable carbon isotopes ($^{13}C/^{12}C$) in the plant dry matter compared to the value of the same ratio in the atmosphere (Farquar *et al.*, 1989). Because of the differences in leaf anatomy and the mechanisms of carbon fixation in species with the C_3 or C_4 pathway, studies on $\Delta^{13}C$ have wider implications for C_3 species where, as compared to C_4 species, the variation in $\Delta^{13}C$ is larger and has a greater impact on crop yield (Farquar et al., 1989). $\Delta^{13}C$ is negatively associated with transpiration efficiency over the period during which the dry mass accumulates (Condon et al., 1990; Turner, 1997). Under drought stress, $\Delta^{13}C$ is a good predictor of stomatal conductance (Condon et al., 1990, 2002) and WUE in different crops (Araus et al., 1993; Matus et al., 1996; Turner, 1997). A number of studies conducted under varying conditions of water availability have shown that in bread wheat the correlation between $\Delta^{13}C$ and final grain yield varies from positive, when ample water is available to the crop, to negative in drought conditions, with no correlation at all in intermediate conditions (Condon et al., 1993). These results can be interpreted based on the influence of both stomatal conductance and photosynthetic activity on $\Delta^{13}C$ and on the fact that biomass production is limited in wet years by a lower stomatal conductance, which becomes an advantage under drought conditions. Additional complexity is added when other physiological traits are considered, such as leaf temperature. In the field, higher canopy temperatures were exhibited by wheat genotypes characterized by high transpiration efficiency consequent to a lower stomatal conductance (Turner, 1997). This effect partially counterbalanced, but did not eliminate, the differences in transpiration efficiency (Richards and Condon, 1993). When determined in grains, $\Delta^{13}C$ correlates positively with growth cycle duration (Araus et al., 1997) and negatively with leaf temperature (Acevedo, 1993; Richards et al., 2002). Therefore, the relationships between $\Delta^{13}C$ and grain yield depend on the environmental conditions, the phenology of the crop and the plant organ (e.g., leaf or grain) from which the samples are collected (Araus et al., 1997; Merah et al., 2001; Royo et al., 2001). Substantial genetic variation for grain $\Delta^{13}C$ has been reported in several C_3 species (Turner, 1997), with high values for broad-sense heritabilities (e.g., from 0.76 to 0.85 in durum wheat; Merah et al., 2001) and a low genotype $\times$ environment $(G \times E)$ interaction (Richards, 1996). For these characteristics, $\Delta^{13}C$ is an attractive target for improving WUE and yield. However, the high cost required to measure each sample limits its potential applications and makes it an interesting candidate for marker-assisted selection.

9.2.2   Stomatal conductance

Stomatal conductance plays a pivotal role in determining $\Delta^{13}C$ and WUE (Farquar et al., 1989; Condon et al., 2002; Richards et al., 2002). Other chapters in this book provide ample evidence of its integrative nature and how this allows stomata to respond to important environmental and metabolic signals to optimize the CO_2/water balance. From a breeding standpoint, a retrospective study conducted by Fischer et al. (1998) on an historical series (from 1962 to 1988) of eight succesful bread wheat cultivars released by CIMMYT provides interesting clues on how stomatal conductance has changed as a consequence of direct selection aimed at improving yield. When the eight cultivars were tested under optimal conditions during three seasons, a strong positive correlation was observed ($r = 0.94$; Fischer *et al.*, 1998). Also in cotton, high yields were associated with high stomatal conductance and reduced leaf temperature (Radin *et al.*, 1994; Lu *et al.*, 1998; Ulloa *et al.*, 2000). Altogether, these results indicate the possibility of raising the yield potential, hence water use, of these two species through an indirect selection for stomatal conductance and/or leaf temperature and suggest the value of identifying the corresponding QTLs in order to implement marker-assisted selection, which would eliminate the time-consuming procedures required to properly measure stomatal conductance. In fact, it is difficult to measure accurately stomatal conductance in a reasonably large number of plants and to properly account for all the fluctuating environmental factors known to affect stomatal conductance during the day (e.g., wind, solar radiation, humidity, etc.). From a technical standpoint, encouraging results have recently been reported in wheat using a viscous air-flow porometer that measures resistance to mass flow through a leaf to provide rapid estimates of leaf conductance (Rebetzke *et al.*, 2000, 2003). A shortcoming in the use of the viscous air-flow porometer is that the values are prevalently determined by the leaf surface with the lowest conductance. Rebetzke *et al.* (2003) reported significant genetic differences between generation means for conductance measured on different days in three wheat crosses and detected a large variability in family-mean heritabilities (from 6 up to 70%), depending on the investigated cross and time of sampling. The same authors suggested that in order to maximize the genetic gain for altered leaf conductance, it is advisable to delay screening of populations until later in the day and repeat measurements for at least two days. As to studies aimed at identifying QTLs, large populations of highly inbred families (e.g., recombinant inbred lines: RILs; double-haploids: DHs) should preferably be evaluated in order to minimize the confounding effects of dominance and allow for proper replications in time and space. Because of the difficulty in measuring stomatal conductance in a large number of plants, only a handful of studies have reported QTLs for this trait (Lebreton *et al.*, 1995; Price *et al.*, 1997, 2002a; Sanguineti *et al.*, 1999; Ulloa *et al.*, 2000; Herve *et al.*, 2001). A more attractive and integrative way to monitor stomatal conductance indirectly through an extended period of time is based on the measurement of the natural oxygen isotope composition ($\Delta^{18}O$) in leaf and grain materials (Barbour *et al.*, 2000). As compared to stomatal conductance, measuring $\Delta^{18}O$ in plant material offers three advantages.

1. It provides an integrative measure in terms of stomatal conductance and leaf temperature over the period that the analysed tissue was formed.
2. It avoids a number of experimental problems typical of measuring stomatal conductance.

3. It is possiblle to collect a large number of samples and requires very little labour in the field.

In the historical series of CIMMYT wheat cultivars tested under irrigated conditions, leaf $\Delta^{18}O$ was strongly correlated with stomatal conductance ($r = -0.93$; Barbour et al., 2000). In this case, grain yield was more strongly correlated with leaf $\Delta^{18}O$ ($r = -0.90$) as compared to leaf $\Delta^{13}C$ ($r = -0.71$). However, the authors cautioned that in situations where stomatal conductance and grain yield are not strongly correlated, the value of $\Delta^{18}O$ as a yield predictor, hence also as an indirect selection criterion, may be questionable. A project funded by the EU (http://137.204.42.130/iduwue/) is in progress to identify QTLs for $\Delta^{18}O$, other WUE-related traits and yield in durum wheat grown under different water regimes in Mediterranean countries.

9.2.3   Canopy temperature

Canopy temperature is an integrative measure of several biochemical and physiological traits acting at the stomatal, leaf and whole-plant level. In the field, genotypes with a cooler canopy temperature under drought stress, or a higher canopy temperature depression (CTD), will use more of the available water in the soil, thus avoiding excessive dehydration and mitigating the negative effect of water stress on grain yield (Blum, 1988; Ludlow and Muchow, 1990). New thermal imaging technology can now report subtle differences in leaf temperature in both laboratory and field conditions (Jones et al., 2003). CTD is mainly useful in hot and dry environments, typical of countries with a Mediterranean-type climate, with measurements preferably made on recently irrigated crops in cloudless and windless days at high vapour pressure deficits. Under these circumstances and provided that data are collected when the canopy is sufficiently expanded to cover the soil, the CTD can be a good predictor of wheat grain yield ($r = 0.6-0.8$; Reynolds and Pfeiffer, 2000). In bread wheat, yield progress was found to be associated with cooler canopies (Fischer et al., 1998); additionally, genetic gains in yield have also been reported in response to direct selection for CTD (Reynolds et al., 1999). One of the factors influencing leaf temperature via an effect on transpiration through stomatal conductance is the concentration of abscisic acid in the xylem sap and in the leaf.

9.2.4   Abscisic acid

Abscisic acid (ABA) is a fundamental component of the complex mechanisms allowing the plant to match the water supply with the water demand and to optimize growth and survival in response to daily and more long-term environmental fluctuations (Passioura, 2002). Indeed, a universal response observed in plants subjected to drought and other abiotic stresses is an increase in ABA concentration (Zeevaart and Creelman, 1988; Quarrie, 1991; Sharp, 1996; Bray, 1997). ABA has been shown to affect many traits influencing the water balance of the plant through mechanisms of dehydration avoidance and dehydration tolerance. In maize seedlings subjected to artificially-induced conditions of water deprivation, an increased ABA concentration enhanced the root/shoot ratio (Sharp, 1996), an adaptive change involving an interaction with ethylene production (Spollen *et al.*, 2000; Sharp, 2002) which could be beneficial for avoiding dehydration under certain conditions (e.g., availability of water in deeper layers of the soil profile). It has also been shown that ABA facilitates water uptake into roots as the soil begins to dry, particularly under non-transpiring conditions, when the apoplastic path of water transport is largely excluded (Hose *et al.*, 2000, 2001).

In droughted barley plants, an interaction between osmotic stress and ABA may represent an important mechanism accounting for leaf growth inhibition (Dodd and Davies, 1996). Additionally, ABA modulates the expression of a large number of genes whose products protect the cell from the harmful effects of dehydration (Bray, 1997, 2002a; Close, 1997). Due to its pivotal role in influencing WUE and adaptation to drought and because it is reasonably quick to collect data in many samples exploiting monoclonal antibodies (Walker-Simmons *et al.*, 1991), a number of studies have been devoted to the identification of QTLs controlling ABA concentration and the analysis of their associated effects on other drought-related traits and yield (see section 9.3.3). Given the effects of ABA on gene expression (Hoth *et al.*, 2002), root elongation (Sharp, 1996), leaf expansion (Dodd and Davies, 1996; Bacon, 1999; Reymond *et al.*, 2003), stomatal conductance (Tardieu *et al.*, 1992, 1993; Tuberosa *et al.*, 1994), seed abortion (Setter, 1997) and seed growth (Tuberosa *et al.*, 1992), the identification and manipulation of QTLs for sensitivity to ABA could be exploited to optimize the water balance of crops and their yield under arid conditions. In maize, significant genetic variability among inbred lines has been detected for stomatal sensitivity to ABA concentration of detached leaves (Conti *et al.*, 1994) and for the elongation rate of pollen tubes growing in substrates supplemented with increasing doses of ABA (Frascaroli and Tuberosa, 1993). Genetic variability in stomatal sensitivity to ABA has also been reported in barley (Borel *et al.*, 1997). These preliminary findings indicate the feasibility of identifying suitable lines for a QTL study aimed at dissecting the genetic basis of sensitivity to ABA in these two species. It should also be mentioned that an increase in ABA concentration is usually associated with a decreased activity of the majority of the photosynthetic enzymes, although an increased activity of carbonic anydrase has been reported (Popova *et al.*, 1996), a finding which may imply a higher fixation through the PEPcarboxylase during stomatal closure.

9.2.5 Water potential and relative water content of the leaf

Sustained turgor is the primary reason for sustained function under drought (Blum, 2002). Indeed, cases for sustained function at low water status as the main reason for drought tolerance are comparatively less numerous. Therefore, it is important to monitor the water status of the genotypes being compared under field conditions. Monitoring the water status of plants in the field provides information related to their WUE as well as their degree of water stress which can be used as selection criteria. Leaf water potential is recognized as an indicator of whole plant water status (Turner, 1990). Maintenance of high leaf water potential under dry conditions indicates dehydration avoidance (Blum, 1988, 2002; Ludlow and Muchow, 1990). Similarly, the leaf relative water content also provides important information on the water status of the plant, offering the advantage of collecting a high number of samples in a short time, an important prerequisite for QTL studies trying to link variation in physiological parameters to variation in yield. Both leaf water potential and relative water content offer a highly integrated measure of the interaction among all the factors involved in maintaining the flow of water through the plant, i.e. water uptake from the roots, its flow through the xylem to the leaf and its evapo-transpiration to the atmosphere regulated by stomatal conductance, presence of cuticle waxes and trichomes, leaf temperature, canopy size and duration, etc. Similarly to stomatal conductance, only a few studies have reported QTLs for leaf water potential and/or relative water content.

9.2.6 Osmotic adjustment

Osmotic adjustment is a metabolic process entailing a net increase in intercellular solutes in response to water stress (Morgan, 1984). As soil moisture declines, osmotic adjustment favours maintenance of turgor, hence the integrity of metabolic functions. Osmotic adjustment has been implicated in sustaining yield under conditions of water deficit in oilseed *Brassica* species (Kumar and Singh, 1998), cotton (Saranga *et al.*, 2001), rice (Jongdee *et al.*, 2002), sorghum (Tangpremsri *et al.*, 1995) and wheat (El-Hafid *et al.*, 1998). It should be noted that the value of osmotic adjustment as a desirable selection target from a breeding standpoint has been questioned (Munns, 1988), based on the notion that drought-tolerant genotypes endowed with a higher capacity to adjust osmotically are likely to be characterized by slow growth, hence biomass production, due to the metabolic requirements of osmolyte biosynthesis. However, in cases of very limited water availability, a higher capacity to accumulate osmolytes may allow the plant to better survive a prolonged and severe drought spell and a more prompt recovery upon establishment of more favourable conditions. Even though the interpretation of osmotic relations in genetically engineered plants can be cumbersome (Blum *et al.*, 1996), transformation experiments have shed light on the mechanisms by which plants may benefit from an altered capacity to accumulate osmolytes (see section 9.6.3.2). It seems plausible that the tradeoff between the metabolic requirements of osmotic adjustment and the potential benefits for the crop varies on a case-by-case basis as a function of the crop, the dynamic and severity of the drought episode(s).

9.2.7   Root traits

Among all plant organs, the root exhibits an astounding level of morphological plasticity. This peculiarity allows the plant to better respond and adapt to the chemical and physical properties of the soil biota, particularly under drought conditions (Sharp and Davies, 1985; McDonald and Davies, 1996; Maggio *et al.*, 2001; Bacon *et al.*, 2003). Ample genetic variation for root traits has been reported in several crops (O'Toole and Bland, 1987). The concept of root ideotype should be elaborated upon a detailed knowledge of the factors that limit the availability of nutrients and stored moisture in the soil as well as on the awareness of the metabolic cost sustained by the plant to develop a more vigorous root system. In this respect, it is notable that a successful recurrent selection for an increased grain yield in drought-stressed tropical maize was associated with a decrease in root mass (Bolaños *et al.*, 1993), counterintuitively from the widespread notion that under drought conditions a more vigorous root system is advantageous. The effects of root size and architecture on final yield will thus depend on the water distribution in the soil and the level of competition for water resources within the plant community. In this respect, when additional stored moisture is available in deeper soil layers, selection for faster growing and deeper roots could improve water use and stabilize yield under drought conditions, particularly in crops with a limited capacity to adjust osmotically.

In *Phaseolus acutifolius*, high yield in drought conditions was associated with deep root penetration (Mohamed *et al.*, 2002). In rainfed rice, the importance of a deep root system has also been repeatedly emphasized (Nguyen *et al.*, 1997; Ito *et al.*, 1999; Blum, 2002). The difficulty of measuring root characteristics in a large number of plants has so far limited the number of studies to determine the genetic basis of variability in root architecture and size, and its effects on yield. Additionally, studies on roots of field-grown plants often entail destructive approaches and are greatly complicated by heterogeneity in soil structure and composition. An alternative to studies investigating roots in field-grown plants, hydroponics allows for a rapid and low-cost screening of root characteristics in a large number of plants at an early growth stage. The addition of polyethilenglycol (PEG) to the hydroponics solution offers an interesting opportunity for evaluating the adaptive response of different genotypes to dehydration (Nagy *et al.*, 1995). Unlike in field conditions where sampled genotypes of a mapping population are likely to experience different stress intensities, the addition of PEG allows us to expose plants to predetermined and uniform conditions of water stress, a condition that facilitates the correct interpretation of the cause-effect relationships of traits' association. A major shortcoming of hydroponics is the unnatural environment where roots grow, a condition that limits the extrapolation of the results to field-grown plants.

In maize, a significant, albeit weak, positive association has been reported between seminal root traits in hydroponics and root pulling resistance in the field (Landi *et al.*, 2001). A reasonable compromise to avoid both the unnatural conditions present in hydroponics and the difficulty of studying roots in the field is offered by growing plants in pots and/or chambers filled with soil, an approach that has been extensively utilized for rice (Azhiri-Sigari *et al.*, 2000; Wade *et al.*, 2000). These experiments are particularly suited for discovering QTLs prevalently expressed in a constitutive manner, which, as such, are more likely to influence variability for root traits also among field-grown genotypes.

Additionally, pot experiments allow us to measure with great precision the amount of water provided to each plant, its water use and WUE. In rice, the lines of a mapping population (Bala × Azucena) were screened for root growth in thin glass-sided soil-filled chambers adopting two water regimes: an early water deficit in which seeds were sown in wet soil but received no additional water, and a late water deficit in which plants were watered for seven weeks before withholding water for the final week (Price *et al.*, 2002b; see also section 9.3.3.1). In the early water deficit treatment, although shoot growth slowed markedly, maximum root length increased, indicating major changes in partitioning. This change in partitioning was also revealed as major differences in root mass and root/shoot ratio between treatments. Another rice mapping population (IR64 × Azucena) was studied to ascertain the associations between root traits measured in soil-filled columns and grain yield under low-moisture stress and non-stress conditions. The overall correlation between maximum root length and grain yield was positive under water-stressed conditions and negative in non-stress conditions, thus confirming the influence of the water regime on the association between morpho-physiological traits and yield.

9.2.8   Early vigour and flowering time

An early and vigorous establishment under conditions of low evapotranspiration, may allow annual crops to optimize the efficiency of water use in the soil and limit the loss of water due to direct evaporation from the soil surface, thus leaving more stored moisture available for later developmental stages when water may become limiting (Lopez-Castaneda *et al.*, 1996). An early establishment also reduces the potential of an inhibition of stomatal conductance as a consequence of root-borne signalling (e.g., ABA) derived by a shallow and superficial root system (Blum, 1996). As trade-off, an excessively vigorous canopy development may deplete soil moisture too soon and/or limit rainfall interception. The desired degree of vigour will thus depend on the environmental characteristics of the target environment. Early vigour has recently been exploited to obtain more water use efficient and higher yielding wheat lines (Asseng *et al.*, 2003). Recent work in maize has reported a number of QTLs governing the rate of leaf expansion (Reymond *et al.*, 2003), opening up new possibilities for a more direct manipulation of this important trait.

Another phenological trait of paramount importance in breeding for an optimal seasonal WUE is flowering time, actually recognized as the most critical factor to optimize adaptation, hence yield, to environments differing for water availability and distribution through the growing season (Richards, 1991). A recent study on the model plant Arabidopsis has evidenced a genetic association between $\Delta^{13}C$ and early flowering based on pleiotropic effects (McKay *et al.*, 2003), highlighting the central role of genetic variability in phenology in optimizing the length of the vegetative and reproductive phases to the available water resources. For annual crops grown in temperate climates, an important additional factor to be considered is the tolerance to low temperature either during the germination phase when planting occurs at the end of the cold season or during the early vegetative growth when planting occurs in the autumn. Indeed, one way to ensure an increased WUE is to plant annual crops before winter, which allows the crop to grow when more water is available and when the evapotranspirative demand is lower. From an agronomic standpoint, an example of the validity of this approach is offered by the shift occurring in the sowing date (from late winter to late fall) of sunflower in Spain, which enables the crop to better escape drought and to obtain higher yields (Gimeno *et al.*, 1989).

As a more general rule, less water is required for growth when temperatures are cooler, provided of course that they are high enough to allow for significant carbon assimilation and plant growth. A vast number of studies have investigated the genetic basis of flowering time, in view of the economic importance of this trait. In annual crops, the genetic basis of flowering time is more complex in temperate species (e.g., barley and wheat) as compared to tropical species (e.g., rice, sorghum and maize), due to the presence in the former group of vernalization genes influencing flowering time in response to cold temperatures. In cereals, the switch from the vegetative to the reproductive phase is controlled, according to the species, by several genes responsive to vernalization and/or length of the day as well as by earliness *per se* genes (Tuberosa *et al.*, 1986; Laurie *et al.*, 1995; Laurie, 1997; Yano *et al.*, 2000;

Snape *et al.*, 2001; Takahashi *et al.*, 2001; Kojima *et al.*, 2002; Salvi *et al.*, 2002; Kulwal *et al.*, 2003). It should be mentioned that QTLs for flowering time have already been cloned in rice (Yano *et al.*, 2000; Takahashi *et al.*, 2001; Kojima *et al.*, 2002) while in maize the positional cloning of *Vgt1* is well advanced (Salvi *et al.*, 2002, 2003).

From this quick review it is evident that genetic variability in WUE can be traced to the interaction of a multitude of quantitatively inherited morpho-physiological traits, whose effects on yield may vary to a considerable extent both in terms of magnitude and direction according to the prevailing environmental conditions. Identifying the QTLs underpinning such traits and understanding their cause-effect relationships is one way to simplify this complexity and make it amenable to a more direct and effective manipulation by the breeder. The following section reviews a number of studies that have reported QTLs for the above-mentioned traits.

9.3   QTLs for WUE and related traits in crops

Studies for identifying QTLs influencing WUE and related traits, including yield, should be conducted considering at least two water regimes (for example, well-watered and water-stressed), thus allowing one to distinguish between the constitutive and adaptive nature of QTL effects. In other words, testing over a wider range of environmental conditions will provide the opportunity to sort out QTLs which show a more limited interaction with the environment from those, the vast majority, whose effects are to a greater extent influenced by variation in water availability and/or other environmental factors. This type of information would be of great value for applying, at a later stage, marker-assisted selection more effectively, also according to the level of drought stress expected in each target environment. Evaluating at higher levels of yield potential also increases traits' heritability thus increasing the probability of detecting QTLs.

9.3.1   *The QTL approach: where genetics, crop physiology and breeding meet*

The traditional methods (for example, diallel analysis, generation means analysis, factorial design, etc.) used to investigate the genetic control of quantitative traits (Falconer, 1981; Hallauer and Miranda Fo, 1988) provide no information on the chromosome regions governing naturally occurring variability in WUE and on the genetic causes (linkage or pleiotropy) of its association with yield. With the advent of molecular marker technologies, these constraints can now be partially overcome through the identification of the quantitative trait loci (QTLs), that is, the relevant loci where the presence of functionally polymorphyc alleles influences genetic variability of quantitatively inherited traits (Stuber *et al.*, 1987, 1999; Tanksley, 1993; Lee, 1995; Asins, 2002; Morgante and Salamini, 2003).

9.3.1.1 QTL discovery

QTL analysis allows us to dissect the complex inheritance of quantitative traits into single factors which can then be selected based on the polymorphic molecular markers flanking the QTL region. The dissection of a quantitative trait into its discrete genetic components is possible through the production of an experimental population (for example, commonly from approximately 100–200 genotypes consisting of F_2 plants, F_3 families, recombinant inbred lines, doubled haploids, etc.) derived from a cross between two inbred lines differing for the trait(s) of interest (Lee, 1995; Quarrie, 1996; Prioul *et al.*, 1997). A QTL study includes an accurate phenotypic evaluation and molecular profiling of a mapping population coupled with a statistical analysis to test the level of significance of the differences between the phenotypic values of the alternative marker alleles, averaged across all individuals in the mapping population. The genetic maps required to identify the QTLs are assembled using different categories of molecular markers, more commonly RFLPs (restriction fragment length polymorphisms), AFLPs (amplified fragment length polymorphisms) and SSRs (simple sequence repeats), with each category presenting advantages and shortcomings. A number of reviews have analysed in detail the issues related to the construction of linkage maps based on the use of molecular markers (Tanksley, 1993; Lee, 1995; Liu, 1998). Herein, it is important to point out that the centiMorgan (cM) is the measuring unit of genetic maps and that marker pairs separated by more than 50 cM segregate independently, while values lower than 50 cM and approaching zero indicate an increasing degree of association (linkage). Because a marker distance of approximately 20 cM is usually sufficient to detect the presence of QTLs having a major effect on the phenotype, the number of markers required varies according to the genome size of the species from as little as 40–50 as in the case of the model species Arabidopsis, to well over 300 as in the case of large-genome species such as durum and bread wheat. Another important point relates to the ratio between the genetic and physical distance (i.e., cM *vs.* basepair), which varies greatly (up to 200–300 fold) according to the frequency of crossing-over events (i.e., recombination rate) along the chromosome (Boyko *et al.*, 2002). This factor is of paramount importance for the level of genetic resolution achievable in a particular chromosome region and will influence the results obtained applying marker-assisted selection and the possibility of positionally cloning the gene(s) underlying a QTL.

Statistical approaches based on information of multiple markers (Zeng, 1994; Liu, 1998) identify the chromosome portion which is more likely to contain the gene(s) underlying the QTL. The most frequently used output statistic to describe the results is the LOD (logarithm of the odds ratio) score, computed as the $\log_{10}$ of the ratio between the chance of a real QTL being present given the effect measured at that position divided by the chance of having a similar effect with no QTL being present. The graphical output provides us with a confidence interval around the QTL peak, thus delineating the range of the most likely QTL position moving away from the peak. In order to avoid declaring 'false-positive' QTLs (i.e., declaring the presence of a 'ghost' QTL when the QTL is actually absent), a reasonably high threshold value for the LOD score should be set (usually > 2.5). Iterative softwares applying bootstrapping procedures allow us to set more precise threshold levels to the mapping population being considered. Several reviews have discussed the statistical approaches and features of QTL analysis (Tanksley, 1993; Churchill and Doerge, 1994; Jansen and Stam, 1994; Lee, 1995; Beavis, 1998; Stuber et al., 1999; Hackett, 2002; Broman et al., 2003). In order to better appreciate the value of QTL analysis and its application in the context of advancing our understanding of the genetic and physiological bases of WUE, a number of relevant issues are addressed briefly below.

When QTL data are available for two or more mapping populations of the same species, it is possible to compare the position of QTLs by using 'anchor markers' (usually RFLPs and/or SSRs) common to the linkage maps being compared and/or indirectly through a comparison with a reference map when the number of markers in common to the mapping populations is limited. Reference maps are available for many crops (for an example on maize, see Davis et al., 1999). A number of these maps have been subdivided into sectors (bins) flanked by reliable RFLP and/or SSR markers, i.e. markers whose map information can be accurately transferred from one segregating population to another. A reference map also integrating information on mutants provides the additional advantage to compare the map position of mutants with that of QTLs, thus contributing relevant information for the identification of possible candidate genes for the investigated trait. On this line, Robertson (1985) suggested that a mutant phenotype at a particular locus may simply be caused by an allele with a much more drastic effect in comparison to that of the QTL alleles at the same locus. In maize, Robertson's hypothesis has been validated for a QTL for plant height co-localized with the mutant *dwarf3* (Touzet et al., 1995; Winkler and Helentjaris, 1995) but not for QTLs for ABA concentration in the leaf when their position was compared to that of mutants impaired in ABA biosynthesis (Tuberosa et al., 1998b).

9.3.1.2   Interpreting traits' association
The relationships between physiological traits, including WUE, have usually been investigated analysing correlated changes in time, the effects of treatments known to influence the target traits and/or with the analysis of

mutants (Quarrie and Jones, 1977; Zeevaart and Creelman, 1988). The presence of a genetic association between traits can be ascertained through the evaluation of (i) related families divergently selected from a population segregating for the target trait (Innes *et al.*, 1984; Tuberosa *et al.*, 1986; Bolaños and Edmeades, 1993; Sanguineti *et al.*, 1996; Ulloa *et al.*, 2000; Landi *et al.*, 2001; Rebetzke *et al.*, 2002) and/or (ii) a number of unrelated genotypes differing in their mean value for the traits of interest (Condon *et al.*, 1993; Tuberosa *et al.*, 1994; Araus *et al.*, 1997; El Hafid *et al.*, 1998). Each approach has flaws and rarely allows us to determine to what extent the association is due to pleiotropy and/or linkage. Alternatively, QTL analysis enables us to investigate with greater precision the genetic basis of trait association merely by looking for co-location on the genetic map of the corresponding QTLs and comparing their effects on the investigated traits (Tanksley, 1993). Consequently, QTL studies contribute to elucidating the causal pathways linking two or more traits, as is the case with morpho-physiological traits and yield under drought (Lebreton *et al.*, 1995; Prioul *et al.*, 1997; Simko *et al.*, 1997; Sanguineti *et al.*, 1999). Linkage can be distin-guished from pleiotropy whenever an improved genetic resolution in the region of interest leads to the identification of recombinant progenies for the associated traits, i.e. the parental alleles of the linked genes controlling the two traits separate in a small number of recombinant progenies. The opposite finding (i.e., no recombinant progeny) cannot prove with certainty that pleio-tropy is the cause of the association; however, pleiotropy becomes increas-ingly more plausible when co-segregation of the two traits is maintained even when additional individuals of the mapping population are evaluated.

An empirical way for ascertaining the cause of trait association at overlapping QTLs is to analyse the sign of the association between QTL effects on each trait. In fact, linkage becomes an unlikely cause when the sign of the association between the genetic effects remains the same at all QTLs affecting the associated traits. However, the presence of contrasting trait associations at regions showing co-location of QTLs does not disprove pleiotropy, since the sign of the association could vary according to the environmental conditions present when the gene(s) underlying that particular QTL was/were expressed (Sanguineti *et al.*, 1999). The final proof for pleiotropy can be obtained through the cloning and direct manipulation (e.g., genetic engineering and/or mutagenesis) of the gene(s) underlying the QTL in question. In this context, a combination of mutants and near isogenic lines (NILs) were deployed in Arabidopsis to unveil a positive genetic correlation between flowering time and $\Delta^{13}C$ and to demonstrate the presence of pleiotropic genes controlling both traits (McKay *et al.*, 2003). These results suggested that the correlated evolution of $\Delta^{13}C$ and flowering time is explained in part by the fixation of pleiotropic alleles that alter both $\Delta^{13}C$ and time to flowering.

An opportunity for investigating in greater detail whether the overlap of QTLs for associated traits is due to linkage and/or pleiotropy is offered by the analysis of the traits starting from an adequately large (> 300–400 F$_2$ plants) mapping population obtained from the cross of NILs contrasted for the parental chromosome regions present at the QTL of interest. NILs can be obtained through repeated selfings (preferably, at least 6–7) of one or more individuals heterozygous at the QTL region prior to identifying the individuals homozygous for each one of the two parental segments. Alternatively, each parental line of the original mapping population evaluated for discovering the QTL can be used as recurrent parent in a backcross scheme in which a single genotype heterozygous at the QTL in question is utilized as donor of the alternative QTL regions; in this case, the congenic lines are identified as back-crossed derived lines (BDLs). With NILs it is thus possible to 'Mendelize' quantitative traits, which, in turn, allows for a more precise evaluation of the effects of the QTL in question, excluding possible epistatic interaction with a genetically heterogeneous background, the condition typical of all mapping populations. From a physiological standpoint, NILs offer the opportunity of using a small number of genotypes (two as a minimum) to compare the effect associated with functionally different parental alleles at a QTL of interest, contrary to the usual QTL studies where several hundred plants are usually evaluated. However, it should be appreciated that the results of these comparative studies can to a different extent be biased by the action of a number of linked genes affecting the investigated traits. In such circumstances, only the evaluation of an adequately large mapping population will improve genetic resolution.

9.3.2   Case studies in dicots

One of the first papers to report QTLs in plants searched for chromosome regions influencing the composition of stable isotopes of carbon as an indirect measure of WUE in tomato. Martin *et al.* (1989) analysed crosses of two drought-sensitive *Lycopersicon esculentum* cultivars and one drought-tolerant *L. pennellii* entry and showed that Δ^{13}C could be satisfactorily predicted with three independent RFLPs linked to QTLs mainly characterized by an additive type of gene action. Despite these encouraging results, further developments and/or applications of this interesting pioneering work have not been reported. A different interspecific cross in tomato (*L. esculentum* × *L. hirsutum*) was used to identify markers associated with QTLs influencing the anatomy of the vascular tissue in the transition zone between the root and the shoot. Measuring stem cross sections in an inbred backcross population having *L. esculentum* as the recurrent parent and *L. hirsutum* as the donor parent for large primary vascular bundles, a QTL region was identified on the distal portion of chromosome 2 affecting the size of primary vascular bundles, the shape of the vascular system and the thickness of the secondary vascular tissue. QTLs influencing the diameter of vascular bundles have also been reported in rice (Sasahara *et al.*, 1999).

Among dicots, soybean has been investigated extensively by Mian and co-workers (1996, 1998) for identifying chromosome regions affecting WUE. Because WUE has been shown to be associated with leaf ash (LASH), both traits were investigated in a population of 120 F_4 lines derived from the cross Young × PI416937. The data collected from leaf tissue sampled from five-week-old plants grown in the greenhouse showed significant differences among lines for both traits (Mian *et al.*, 1996). A total of four and six independent RFLP markers were associated with significant effects for WUE and LASH, respectively. When combined, each group of markers accounted for 38% and 53% of the variability among lines observed in WUE and LASH, respectively. In particular, one marker locus (*cr497–1*), on linkage Group J, explained 13% of the variability in WUE. Two QTL regions influenced both WUE and LASH. For each of these QTLs, the allele for increased WUE was associated with reduced LASH, thus accounting for the negative association between the two traits ($r = -0.40$). In a subsequent study, an F_2 population derived from S100 × Tokyo was tested in a greenhouse to determine the consistency of WUE QTLs with those previously described in Young × PI416937 (Mian *et al.*, 1998). In S100 × Tokyo, two independent markers were associated with WUE, one of which (*A063E*) was also associated with WUE in the other population. The second marker (*A489H*) on linkage group L, was unique to the S100 × Tokyo population and explained 14% of the variation in WUE. Collectively, the findings of these two studies indicate the polygenic nature of WUE and the possibility of applying marker-assisted selection to improve WUE in soybean. However, the effects of these QTLs under field conditions, particularly in terms of yield, are yet to be ascertained.

In this respect, Specht *et al.* (2001) measured grain yield in 236 RILs of a Minsoy × Noir-1 cross evaluated in the field at six water treatments (0, 20, 40, 60, 80 and 100% evapotranspiration) for two years (1994 and 1995). For each RIL, yield was regressed on the amount of water supplemented to replenish (from 0 to 100%) the seasonal evapotranspiration (ET), which allowed for the computation of a regression value (beta) that was used as an estimate of season-specific WUE. As an additional measure of transpiration efficiency, $\Delta^{13}C$ was also determined. The 1995 average yield-to-water regression was highly linear ($28\,\mathrm{kg\,ha^{-1}\,cm^{-1}}$). Significant variability among RILs was detected for $\Delta^{13}C$ and for the 'RIL × water treatment' interaction. The low correlation between $\Delta^{13}C$ and beta ($r = 0.26$) suggested that selection for enhanced leaf transpiration efficiency may not improve crop WUE. Conversely, the rather high correlation between the values of beta and yield ($r = 0.71$) underlined the difficulty of selecting for low beta (lower sensitivity to drought) while maintaining an acceptable yield potential. The major QTLs affecting $\Delta^{13}C$ and yield coincided with QTLs governing maturity and/or determinancy QTLs, thus confirming the important role of phenological traits in influencing WUE. Specht *et al.* (2001) concluded that improvement of soybean yield under drought would be better achieved by coupling a high-yield grand mean with a high value for yield beta.

Causal relationships among traits related to WUE and drought resistance were investigated in *Stylosanthes scabra* using QTL analysis (Thumma *et al.*, 2001). The phenotypic evaluation was carried out using cuttings of the 120 F_2 plants derived from a cross between two genotypes differing for transpiration efficiency, $\Delta^{13}C$ and specific leaf area (SLA). The plants were subjected to water stress in the field and were evaluated for biomass productivity (BP) traits (total, shoot and root dry matter), transpiration, transpiration efficiency, specific leaf area, $\Delta^{13}C$, relative water content and the osmoprotectant trans-4–hydroxy-N-methyl proline (MHP), which accumulates under water stress conditions. BP traits were positively correlated with transpiration efficiency and negatively correlated with $\Delta^{13}C$ and SLA, whereas transpiration was positively correlated with BP traits and negatively correlated with SLA. QTLs were detected for all investigated traits and several markers showed significant effects associated to more than one trait. At the chromosome regions where QTLs for transpiration efficiency overlapped with those for $\Delta^{13}C$, alleles affecting positively transpiration efficiency had a negative effect on $\Delta^{13}C$. Additionally, at other regions with overlapping QTLs, a negative association was also found between QTL effects of SLA *vs.* BP traits and transpiration, and between the QTL effects of MHP and relative water content.

Cotton is an important crop prevalently grown in hot climates characterized by a very high evapotranspirative demand. Saranga *et al.* (2001) tested mapping populations under well-watered and water-limited conditions and showed that productivity and fibre quality were partly accounted for by different independent QTLs, indicating that it is possible to combine adaptation to both arid and favourable conditions in the same genotype. QTL mapping was also used to investigate the association between fibre productivity and quality under dry conditions with a number of physiological traits often found to differ between genotypes adapted to arid *vs.* well-watered conditions. A reduced plant osmotic potential was the only feature clearly associated with cotton productivity under arid conditions, with a reduction in osmotic potential being associated with higher yield. Because cotton is subjected to extended periods of high temperature which can reduce lint yield even when an adequate water supply is available, a high stomatal conductance may be beneficial for cooling the leaf and avoiding excessively high tempera-tures. Two putative QTLs for stomatal conductance have been reported in a population derived from the cross NM24016 $\times$ TM1 (Ulloa *et al.*, 2000). Subsequently, ten selected F_4 families with high stomatal conductance and ten with low stomatal conductance were tested in two locations; in one site, significant differences between the two groups were detected (543 *vs.* 472 mmol $H_2O\,m^{-2}\,s^{-1}$). Lint yield was significantly affected by selection for stomatal conductance in one location, with the group with high stomatal conductance outyielding by 11% the family with low stomatal conductance. These two groups of families could be subjected to bulk-segregant analysis (Michelmore *et al.*, 1991) in order to identify polymorphic markers closely associated with major QTLs possibly affecting the observed difference in stomatal conductance.

Testing a population of RILs of sunflower under greenhouse conditions, Herve *et al.* (2001) identified QTLs for several physiological traits associated with water status (stomatal conductance, transpiration, pre-dawn leaf water potential and relative water content) and photosynthesis (leaf chlorophyll concentration, net photosynthesis and internal CO_2 concentration). A total of 19 QTLs were detected explaining from 9% to 63% of the phenotypic variance for each trait. Among these, two major QTLs for net photosynthesis were identified on linkage group IX. One QTL co-location was found on linkage group VIII for stomatal conductance and water status, while overlap of QTLs regulating photosynthesis, transpiration and leaf water potential occurred on linkage group XIV.

Wild plant species are often better adapted to stressful environments than their cultivated relatives. Cultivated lettuce (*Lactuca sativa*) and the wild relative *L. serriola* differ widely in both shoot and root characteristics. Approximately 100 $F_{2:3}$ families derived from an interspecific cross were evaluated by Johnson *et al.* (2000) in greenhouse and field experiments. In the greenhouse, root traits (taproot length, number of laterals emerging from the taproot and biomass) and shoot biomass were measured four weeks after planting. In the field, plants were grown for nine weeks and were then exposed to a mild drought stress induced by withholding water for one week. Thirteen QTLs were identified for root architectural traits and/or water acquisition, with each QTL accounting for 28–83% of the phenotypic variability. The QTLs for taproot length and those for the ability to extract water from deeper soil layers co-localized, with the *L. serriola* alleles providing deeper roots and higher water-harvest capacity. For this reason, the wild *L. serriola* was indicated as a potential source of agronomically useful alleles to optimize resource acquisition by cultivated lettuce and minimizing water inputs.

9.3.3   Case studies in cereals

Of all crops, rice, maize, wheat and barley provide the majority of all the grain required for feeding mankind. In most of the least developed countries, cereals are grown under rainfed conditions in environments characterized by relatively low and uncertain rainfall. While intermittent precipitation causes large inter-annual fluctuations in cereal production, terminal drought and heat usually represent the most critical and common constraints to cereal production, particularly in the tropics. Additionally, drought stress also limits the uptake of nitrogen and other nutrients, thus impairing kernel development and negatively affecting its nutritional value. Because the improvement of yield stability under arid conditions can in principle be accelerated by a proper use of molecular markers linked to major QTLs (Ribaut *et al.*, 2002), discovery of QTLs for WUE and related traits in cereals is particularly important for future productive gains through molecular breeding.

9.3.3.1   Rice
Upland rice cultivation relies exclusively on rainfall for its water supply, while paddy rice requires a high amount of water for each unit of carbon fixed. As compared to the other small-grain cereals, rice is a very wasteful crop in terms of water demand, requiring up to 5,000 kg of water for each kg of rice grain produced (Shen *et al.*, 2001). It has been estimated that more than half of the total 40 million hectares of rainfed lowland rice is affected annually by drought (Sarkarung and Pantuwan, 1999) and that the annual drought losses in rice production of the 24 million hectares cultivated in eastern India represented 22% of all losses from technical constraints (Widawsky and O'Toole, 1990). Through a number of international projects, considerable effort has been devoted to identify QTLs influencing water use and yield under drought and to verify the feasibility of using marker-assisted selection to improve drought tolerance in rice (Ito *et al.*, 1999; Mackill *et al.*, 1999; Atlin and Lafitte, 2002).

QTLs for the adaxial and abaxial stomatal frequencies were studied in a backcross population derived from Nipponbare × Kasalath (Ishimaru *et al.*, 2001). Four QTLs influenced adaxial and abaxial stomatal frequencies, suggesting that the same genes may pleiotropically control the stomatal frequency on both surfaces of the leaf. QTLs for leaf rolling as related to osmotic adjustment were also mapped and were found not to overlap with the QTLs for stomatal frequency.

Because a deep, thick root system positively impacts the yield of upland rice under water stress conditions (Nguyen *et al.*, 1997), great interest has been devoted to QTL discovery for root architecture and related traits (Lafitte *et al.*, 2002). Indeed, rice is the species for which the largest set of QTL data for root traits is presently available. Due to the difficulty of properly scoring root traits, marker-assisted selection could be conveniently deployed for improving root morphology (Shen *et al.*, 2001). For the sake of conciseness, only the results of a limited number of these studies are herein reviewed. Champoux *et al.* (1995) investigated the overlap of QTLs associated with root morphology and QTLs associated with drought avoidance/tolerance in a RIL population grown in three greenhouse experiments. The RILs derived from a cross between Moroberekan, a traditional *japonica* upland cultivar, and Co39, an *indica* cultivar of lowland adaptation. The RILs were also evaluated at the seedling, early vegetative and late-vegetative growth stage for their degree of drought avoidance/tolerance (visually rated based on a leaf rolling scoring system) in field experiments. In total, 14 chromosomal regions were found to significantly affect field drought avoidance/tolerance. Interestingly, 12 of these QTL regions also influenced root morphology. Based on these results, a selection for the Moroberekan alleles at marker loci flanking the putative root QTLs was advocated as an effective strategy for altering the root phenotype of rice towards that commonly associated with drought-resistant cultivars (Champoux *et al.*, 1995). In a subsequent study, a sample of 52 RILs derived from C039 × Moroberekan was investigated for the presence of QTLs associated with osmotic adjustment and dehydration tolerance (Lilley *et al.*, 1996). One major QTL for osmotic adjustment and two of the five QTLs influencing dehydration tolerance were found to partially overlap with QTLs for root morphology, thus indicating a genetic association among these traits. Both osmotic adjustment and dehydration tolerance were negatively correlated with root morphological characters associated with drought avoidance. A more vigorous root system was associated with Moroberekan alleles, while a high osmotic adjustment capacity and dehydration tolerance were associated with Co39 alleles. Therefore, the combination of a high capacity to osmotically adjust with an extensive root system requires breaking the linkage between these traits.

QTLs for seminal root length under different water-supply conditions were sought by Zhang *et al.* (2001a) in a RIL population derived from a cross between IR1552, an indica lowland rice, and Azucena, a tropical japonica upland rice. Lowland and upland growing conditions were mimicked using hydroponics and a paper roll culture system, respectively. One major QTL for seminal root length in solution culture (SRLS) and one for seminal root length in paper culture (SRLP) were detected on chromosomes 8 and 1, respectively. Significant epistatic interactions were evidenced for five pairs of loci for SRLS and one for SRLP, which accounted for about 60% and 20% of the total variability in SRLS and SRLP, respectively, indicating that in this rice population epistasis is a major genetic basis for seminal root length, and that different genetic systems control seminal root growth under different water regimes.

The rice mapping population that has been most extensively investigated for root traits and other WUE-related traits has been derived from Bala × Azucena (Price and Tomos, 1997; Price *et al.*, 1997, 2000, 2002a,b,c). Azucena has root traits that potentially contribute to drought resistance, while Bala has a number of shoot-related characteristics for a better adaptation to drought-prone environments. As compared to Azucena, Bala is shorter and with a different canopy structure, has highly sensitive stomata and a greater ability to adjust osmotically, does not roll its leaves readily, slows growth more rapidly when droughted and has a lower WUE (Price and Tomos, 1997). In a series of studies conducted under a wide range of experimental conditions, Price and co-workers identified and compared several QTLs for root characteristics, stomatal conductance, leaf rolling, leaf drying, leaf relative water content, etc. (Price and Tomos, 1997; Price et al., 1997, 2000, 2002a,b,c). The QTLs for the drought avoidance traits did not map to the same locations as those for root traits as would be expected if the traits all contributed to drought resistance. Therefore, in this population root-growth QTLs contributed only marginally to drought avoidance. Price et al. (2002c) suggested a number of reasons for this lack of co-location, such as shoot-related mechanisms of drought resistance and the difficulty of collecting precise data from field trials because of variability in soils and rainfall. Because QTLs for grain yield under well-watered conditions did not overlap with QTLs for root morphological traits under low-moisture, Price et al. (2002c) suggested the possibility of using marker-assisted selection to combine QTL alleles for higher grain yield and QTL alleles for root morphological traits desirable for low-moisture conditions but causing no yield penalty under favourable conditions.

Results on root QTLs obtained with other mapping populations and their utilization to improve drought tolerance have been discussed in a number of other studies (Yadav et al., 1997; Mackill et al., 1999; Courtois et al., 2000; Kamoshita et al., 2002; Lafitte et al., 2002). This extensive body of data on different mapping populations profiled with a set of common RFLP and SSR markers has made possible a detailed comparison of the QTL position for root traits across populations, thus leading to the identification of a number of key QTL regions with a more substantial and consistent effect in controlling variation in roots and other traits influencing WUE (Courtois et al., 2000; Zhang et al., 2001a,b; Price et al., 2002b,c) and other traits (Ali et al., 2000; Courtois et al., 2000; Tripathy et al., 2000; Price 2002b,c; Mei et al., 2003). In particular, some of the QTLs detected by Courtois et al. (2000) for leaf rolling, leaf drying and relative water content overlapped with QTLs influencing root morphology in the same population and QTLs for leaf rolling as reported from other populations. Based on the results of these comparative analyses carried out in rice, a number of congenic strains (e.g., NILs and BDLs) have been derived using MAS in order to more accurately test the effects of the single QTLs and the presence of epistatic interactions (Shen et al., 2001; Cortuois et al., 2000; Price et al., 2002c). A more recent study on a DH population from the cross CT9993-5-10-1-M × IR62266-42-6-2 subjected to water stress before anthesis identified 47 QTLs for various plant water stress indicators, phenology and production traits (Babu et al., 2003). Root traits were positively associated with yield under drought stress and the region *RG939–RG214* on chromosome 4 strongly affected root traits and yield.

9.3.3.2 Sorghum and maize

Even though sorghum is considered a crop well-adapted to hot, dry environments, drought is a major constraint in sorghum production worldwide. More than 80% of commercial sorghum hybrids in the United States are grown under non-irrigated conditions and although most of them have pre-flowering drought resistance, many do not show any significant post-flowering drought resistance, thus making breeding for improving post-flowering drought tolerance and seasonal WUE an important priority. Drought stress during grain filling causes rapid and premature leaf death. Because evaluation of stay-green is difficult and unreliable under field conditions, progress in improving stay-green in sorghum by conventional breeding methods has been quite slow. Consequently, increasing attention has been devoted to the discovery of QTLs affecting stay-green, a post-flowering drought resistance trait that contributes to normal grain filling and, indirectly, to an increase in seasonal WUE. Studies on drought-related traits in sorghum are also valuable for maize, in consideration of the similarity of these two species in carbon metabolism (C_4 metabolism) and the high level of synteny of their genomes (Gale and Devos, 1998); additionally, the smaller size of the sorghum genome makes this species particularly attractive for QTL studies and cloning of agronomically valuable genes. Significant progress in genome mapping of this crop has also been recently reported (Menz *et al.*, 2002). The physiological trait associated to drought resistance that has received most attention is leaf stay-green, i.e. delayed leaf senescence. The important role of stay-green traits under dry conditions has been reported in other species (Akhtar *et al.*, 1999; Bolaños and Edmeades, 1996; Thomas and Howarth, 2000) and a number of underlying mechanisms have been identified (Thomas and Howarth, 2000).

The evaluation of 98 RILs obtained from a cross between TX7078 (pre-flowering tolerant but post-flowering susceptible) and B35 (pre-flowering susceptible but post-flowering tolerant) allowed for the identification of 13 QTLs affecting one or more post-flowering drought tolerance traits (Tuinstra *et al.*, 1997). Two QTLs characterized by major effects on yield and stay-green under post-flowering drought were shown to affect yield also under fully irrigated conditions. Following these encouraging results, NILs were developed to test the phenotypic effects of three major QTLs affecting agronomic performance in drought and/or well-watered environments (Tuinstra *et al.*, 1998). In most cases, NILs contrasting for a specific QTL allele differed in phenotype as predicted by the previous QTL study. NILs contrasting at QTL marker *tM5/75* differed considerably in yield across environments. Further analysis indicated that differences in agronomic performance may be associated with a drought tolerance mechanism that also affects heat tolerance. Also, NILs contrasting at QTL marker *tH19/50* differed in yield under both water regimes. The analysis of these NILs indicated that these differences may be influenced by a drought tolerance mechanism that conditions plant water status and the expression of stay-green. NILs contrasted at QTL marker *t329/132* differed in yield and seed weight. In this case, differences appeared to be caused by two QTLs that are closely linked in repulsion. Stay green was also investigated in a RIL population developed from the cross B35 × Tx7000 and tested in seven environments (Xu *et al.*, 2000; Subudhi *et al.*, 2000). The comparison of the four stay green QTLs (*Stg1*, *Stg2*, *Stg3* and *Stg4*) discovered in this population with those described in earlier reports (Tuinstra *et al.*, 1997; Crasta *et al.*, 1999) indicated their consistency across genetic backgrounds (Sanchez *et al.*, 2002). A significant epistatic interaction for stay-green and chlorophyll content involving *Stg2* and a region on linkage group C was also detected, leading Subudhi *et al.* (2000) to conclude that *Stg2* is the most important QTL for stay-green.

Among the traits that can influence WUE through stomatal conductance in maize, extensive work has been carried out to identify QTLs for the concentration of ABA in the leaf (L-ABA) and their associated effects on other drought-related traits and yield (Lebreton *et al.*, 1995; Tuberosa *et al.*, 1998a,b; Sanguineti *et al.*, 1999). Among the 16 QTLs identified in the Os420 × IABO78 background (Tuberosa *et al.*, 1998b), the most important and consistent one mapped near *csu133*, on bin 2.04. This QTL has been validated through a divergent selection for L-ABA started from 480 (Os420 × IABO78) F_2 plants (Sanguineti *et al.*, 1996; Salvi *et al.*, 1997). For a more targeted manipulation of the QTLs controlling ABA concentration, it is desirable to dissect the biochemical and physiological bases governing variability for this trait. To this end, Tuberosa *et al.* (1998b) verified whether mapped mutants affecting ABA biosynthesis might be possible candidates for the QTLs controlling L-ABA. The major rate-determining step for ABA biosynthesis is controlled by *vp14* (Schwartz *et al.*, 1997; Milborrow, 2001) which has been mapped to bin 1.08 (Tan *et al.*, 1997). Based on the RFLP information of markers common to the reference UMC map and the Os420 × IABO78 map, it was shown that the map position of mutants impaired in ABA biosynthesis was not within the support intervals of the QTLs influencing L-ABA (Tuberosa *et al.*, 1998b). These results, leave the question open as to what sort of genes may underlie these QTLs. Feasible candidates include the genes involved in the intensity of the transduction signal associated with turgor loss, a major determinant in the regulation of ABA concentration (Jensen *et al.*, 1996) and/or genes controlling morpho-physiological traits (e.g., root size and architecture, osmotic adjustment, leaf angle, etc.) affecting the water balance of the plant, hence its turgor. Quarrie *et al.* (1999a) reported that recurrent selection for grain yield under drought conditions significantly changed allele frequencies at *csu133* in two populations ('Tuxpeño Sequia' and 'Drought Tolerant Population') developed at CIMMYT (Bolaños and Edmeades, 1993; Bolaños *et al.*, 1993). These results further substantiate the importance of the QTL region near *csu133* in controlling drought-related traits and yield in maize. A high number of QTLs were also found to influence ABA concentration in samples of leaf tissue and xylem sap collected from drought-stressed plants in the cross Polj17 × F2 (Lebreton *et al.*, 1995). All chromosomes, with the exception of chromosome 8, harboured QTLs influencing concentration of ABA. Analogously to what was reported in Os420 × IABO78 (Tuberosa *et al.*, 1998b), also in Polj17 × F2 a major QTL affected L-ABA concentration near *csu133* on bin 2.04. The QTLs for xylem ABA showed a poor overlap with those for L-ABA, a result in keeping with the low correlation reported for these two traits in maize (Zhang and Davies, 1990; Tuberosa *et al.*, 1994). Only two of the four QTLs which influenced leaf water potential were located in regions near QTLs for xylem ABA or L-ABA, a result which led Lebreton *et al.* (1995) to suggest that it was unlikely that variation in the degree of drought stress was the cause of variability in L-ABA.

In the Polj17 × F2 population investigated for the concentration of ABA in the leaf and in the xylem, Lebreton *et al.* (1995) searched also for QTLs affecting root traits. Four QTLs affected both seminal root number and root number at the base of the stem and seven QTLs influenced root pulling force (RPF). QTLs for root traits in hydroponics were identified in a mapping population derived from Lo964 × Lo1016 (Tuberosa *et al.*, 2002b). Several QTLs affected primary root length (R1L), primary root diameter (R1D), primary root weight (R1W) and/or the weight of the adventitious seminal roots (R2W). The QTL with the most sizeable effects was mapped on chromosome 1 (bin 1.06). In order to verify to what extent some of the QTL regions influencing root traits in hydroponics also modulate root characteristics in the field, the same mapping population was tested for root pulling force (RPF) in three field experiments (Landi *et al.*, 2002b). Among the 19 bins harbouring QTLs for RPF, 11 also harboured a QTL for one or more root traits measured in hydroponics. The most noticeable overlap for QTLs influencing root traits in hydroponics and in the field occurred on bin 1.06. The same bin also harboured QTLs for root traits in Polj17 × F2 (Lebreton *et al.*, 1995), in F288 × F271 (Barriere *et al.*, 2001) and in Ac7729 × Ac7643/TZSRW (Tuberosa *et al.*, unpublished results). Additionally, QTLs for the anthesis-silking interval and for grain yield under both well-watered and drought-stressed conditions were identified in the same bin in Lo964 × Lo1016 (Tuberosa *et al.*, 2002b), thus indicating the importance of producing NILs for this QTL region in order to better assess its role in regulating root traits and drought response in maize. In Lo964 × Lo1016, several overlaps occurred between the QTLs for root traits and those affecting grain yield

(Tuberosa et al., 2002b). Among the investigated root traits, R2W most frequently and consistently overlapped with QTLs for grain yield. In particular, an increase in R2W was positively associated with grain yield at four QTL regions (bins 1.06, 1.08, 10.04 and 10.07).

A novel approach integrating QTL analysis and an ecophysiological model based on previous studies (Ben Haj Salah and Tardieu, 1995, 1996) has been applied by Reymond et al. (2003) to identify QTLs affecting leaf elongation rate as a function of meristem temperature, water vapor pressure difference and soil water status recorded in six different experiments. The phenotypic trait subjected to QTL analysis was the slope of the regression of the elongation rate on the values of the environmental variables. Consistently with the model originally proposed, all the investigated responses were found to be linear and common to different experiments. ·Several QTLs were identified most of which were specific for the response to one variable only. Interestingly, a model based on the combined QTL effects predicted 74% of the variability for leaf elongation rate among a random sample of RILs (Reymond et al., 2003).

9.3.3.3 Barley and wheat

In barley, $\Delta^{13}C$ of the shoot tissue has been reported to be more heritable than other seedling traits (Ellis et al., 2002), similarly to what was previously found in wheat (Ehdaie and Waines, 1994), thus indicating that a substantial fraction of the phenotypic variation in plant $\Delta^{13}C$ can be genetically manipulated. However, early attempts to develop $\Delta^{13}C$ as a direct assay for yield in barley were unsuccessful (Acevedo, 1993). The reasons for this are suggested by the results of Ellis et al. (2002), showing that in the spring barley cross Derkado × B83–12/21/5, only one of the three primary QTLs for yield was associated with a primary QTL for stem $\Delta^{13}C$. Extensive work has been carried out to identify QTLs for a number of drought-related traits, grain yield and its components in a population of 167 RILs developed from the cross Tadmor and Er/Apm (Teulat et al., 1997, 1998, 2001a,b), two lines differing for traits associated with drought tolerance (for example, plant architecture, growth habit and chlorophyll content, osmotic adjustment, etc.). Despite the large heterogeneity in water availability among the six Mediterranean environments considered by Teulat et al. (2001a), a number of QTLs were consistently identified in several environments, particularly for plant height and kernel weight, thus indicating a prevalently constitutive mode of action of the genes underlying these QTLs. In total, the multiple-environment analysis revealed 24 consistent QTLs, 18 of which were common to other published work, while the six QTLs found to be unique suggested the likely presence of a functional polymorphism conferring specific adaptation to Mediterranean conditions or specific to the Tadmor and Er/Apm genetic background. A separate analysis of the irrigated and dry environments indicated that among the 16 QTLs showing main effects over the two water conditions, five were influenced by the water regime.

Similarly to other crops, the level of genetic variability present in cultivated barley, as compared to its wild progenitor, has been strongly reduced by domestication. To overcome this limitation, Tanksley and Nelson (1996) have advocated the use of advanced backcross quantitative trait analysis (ABQA), an approach that offers the opportunity to quickly discover and exploit beneficial QTL alleles present in wild germplasm. The ABQA is based on the evaluation of backcross (BC) families between an elite variety used as recurrent parent and a donor accession, more commonly a wild species sexually-compatible with the cultivated species. QTL analysis is usually delayed until the BC_2 generation, after selecting in BC_1 against characteristics with a negative effect on the agronomic performance (e.g., ear shattering in barley). ABQA has already proven its validity for the exploitation of exotic germplasm in tomato (Tanksley et al., 1996; Bernacchi et al., 1998) and rice (Xiao et al., 1998; Moncada et al., 2001). Therefore, wild barley (Hordeum spontaneum) is a source of valuable alleles for traits influencing resistance to abiotic stresses (Ellis et al., 2000; Forster et al., 2000; Ivandic et al., 2000; Robinson et al., 2000; Turpeinen et al., 2001). To this end, 123 DH lines derived from an H. vulgare × H. spontaneum backcross were investigated in three Mediterranean countries under rainfed conditions to identify agronomically favorable QTL alleles contributed by the wild parent (Talamè et al., 2004). Among the 81 putative QTLs found to influence growth habitus, heading date, plant height, ear length, ear extrusion, grain yield and/or 1000–grain weight, in 43 cases (53%) the wild parental line contributed the alleles with favorable effects (Talamè et al., 2004). As to grain yield, although the majority (65%) of the favourable QTL alleles were those of H. vulgare, at six QTLs the alleles increasing grain yield were contributed by H. spontaneum. These results indicate that ABQA is a useful germplasm enhancement strategy for identifying wild progenitor alleles capable of improving yield of the related crop cultivated under arid conditions.

Extensive work has been carried out by Quarrie and co workers to discover bread wheat QTLs influencing WUE and yield in a mapping population of 95 DH lines derived from the cross Chinese Spring × SQ1 and tested under different water regimes. A $\Delta^{13}C$ analysis carried out on kernels harvested for two consecutive years in rainfed trials identified QTLs on chromosomes 2A (near Xpsr375.1), 2B (near Xcbd453) and 4D (near Xpsr575.1) (Quarrie et al., 1995). No significant effect on $\Delta^{13}C$ was found associated with rht-B, which has been reported to affect WUE in Australian wheats (Richards and Condon, 1993). In a subsequent study, the DH lines were tested in 1999 and 2000 under irrigated and non-irrigated field conditions (Quarrie et al., 2004). A number of highly significant QTLs for yield were found under both droughted and irrigated conditions. QTLs for grain yield coincident between water treatments were identified on chromosomes 1A and 7A. Other yield QTLs were unique to the rainfed or irrigated treatment. In particular, a major QTL for yield only under droughted conditions was present distal on chromosome 1AL. Each of the major yield QTLs under both rainfed and irrigated conditions was associated with one of the yield components. The major yield QTL present in both treatments on chr. 7A was coincident with a QTL for grains per ear. These results confirm the presence of a major QTL for yield on chromosome 7A found previously in other field studies with the same population (Quarrie *et al.*, 1995). The major dwarfing gene *rht-B1* present in this mapping population on chromosome 4BS had no obvious effect on plant height in 2000, in contrast to the results for 1999, which was a much more favourable year for plant growth. The results for 2000 showed that the QTLs governing variability in grain yield under severe drought differed from those acting under irrigated conditions.

An RFLP linkage map of hexaploid wheat derived from W7984 × Opata was used to identify QTLs for WUE and other morpho-physiological traits (Zhang and Xu, 2002). Several QTL clusters were shown to influence the investigated traits. Among them, the centromeric region of 1A and 1B influenced the photosynthetic rate and root traits and chromosome 2 affected WUE and root traits. Other QTL clusters controlling root traits were located on the 6A and 6B chromosomes. The main QTL for WUE and related traits was identified near the centromeric region of the 6D chromosome.

A recent review has compiled and discussed a list of QTLs, chromosome regions and stress-related sequences involved in resistance to abiotic stresses in *Triticeae* (Cattivelli *et al.*, 2002).

9.4   What can we learn through the 'omics' approach?

Mapping QTLs for WUE and other traits associated with yield is an essential prerequisite for improving crops' performance through marker-assisted selection (MAS). Once major QTLs are mapped, the next logical step is to identify suitable candidate genes accounting for the QTL effects, validate their role and proceed with a more direct manipulation using the gene itself as marker for MAS within the same species and/or through genetic engineering in other species. The identification of candidate genes and the elucidation of their role can be facilitated combining different sources of information and technological platforms with suitable genetic resources (Pflieger *et al.*, 2001; Wayne and McIntyre, 2002). The spectacular progress recently achieved in the high throughput profiling of the transcriptome, proteome and metabolome (i.e., the so-called 'omics') provides new opportunities and challenges to analyse the changes in the concerted expression of thousands of genes and, as a valid example for this volume, to evaluate the effects of such changes in response to a water deficit.

9.4.1   Transcriptome analysis

A number of factors have so far limited a more widespread application of transcriptome profiling to studies aimed at dissecting the molecular bases of the response to drought and other abiotic stresses: (i) the high cost associated to these studies; (ii) the correlation between the expression level of genes and their final products can be lowered by translational and/or post-translational modifications, thus weakening the association between the changes at the level of the genes and the phenotype; (iii) it is difficult to properly account for biological and sampling variability; (iv) when using microarrays, the array may not contain genes encoding for low-abundant mRNAs and/or other key mRNAs, except of course when all the genes of one particular species are available (e.g., in Arabidopsis and rice); (v) even in such cases, differences in low-abundant mRNAs may not be detected because they are below a critical threshold level; (vi) it is difficult to profile gene expression in small samples. Due to the high cost, transcriptome profiling is better suited for surveying a limited number of samples, such as in the case with transgenics, parental lines of a mapping population, mutants and other congenic strains as well as samples obtained through bulk segregant analysis. Among the different platforms available for the mass-scale profiling of the transcriptome of drought-stressed plants, microarrays have been most frequently utilized (Hazen et al., 2003). As an alternative to microarrays, transcriptome profiling can be pursued also with the so-called 'open-ended' (i.e., not relying on a predetermined number of genes) methods such as differential display (Liang and Pardee, 1992), cDNA-AFLP (Bachem et al., 1996), SAGE (Serial Analysis of Gene Expression; Velculescu et al., 1995; Matsumura et al., 1999) MPSS (Massively Parallel Signature Sequencing; Brenner et al., 2000) and GeneCalling (Bruce et al., 2001). As compared to microarrays including only a portion of the genes of a species, these methods allow for an unbiased genome-wide survey.

Monitoring genome-wide changes in transcript abundance during a drought episode or any other environmental insult, provides important clues for recognizing transcript networks accounting for the adaptive response of the plant (Kreps et al., 2002; Seki et al., 2001, 2002; Hazen et al., 2003). An interesting application of transcriptome analysis is the identification of DNA motifs in the promoter region of genes that respond similarly to a particular transcription factor. Cluster analysis of the changes occurring under abiotic stress allows us to pinpoint the genes with a similar response to environmental cues, an indirect indication of the presence of cis-regulatory elements putatively responding to a common regulatory circuitry. However, a more robust approach to identify promoter motifs interacting with the same transcription factor relies on chromatin immunoprecipitation with DNA microarray (Nal et al., 2001; Weinmann et al., 2002).

Two microarray studies carried out in Arabidopsis showed that a number of novel genes previously unknown to respond to dehydration were transcriptionally regulated following an artificially-induced water deficit (Seki et al., 2001, 2002). Among the surveyed 1,300 full-length cDNAs which had been isolated from libraries prepared from stressed plants, Seki et al. (2001) described 44 genes that appeared to be induced by the water-deficit treatment. Of these genes, 30 which had not been previously identified as water-deficit-induced genes shared homology with putative cold-acclimation, LEA, nonspecific lipid tansfer and water channel proteins. In a subsequent experiment, a microarray including 7,000 full length cDNAs was used to monitor the response to cold, drought and salinity (Seki et al., 2002). In the drought treatment, 277 genes showed a 5–fold (or greater) increase in expression. The promoter region of the majority of these genes had the DRE-related core motif or the ABA-responsive element (ABRE). Interestingly, 11% of all stress-inducible genes encoded for TFs, suggesting that numerous transcriptional regulatory mechanisms underpin the drought, cold and/or high-salinity stress signal transduction pathways (Seki et al., 2002). A detailed analysis of the functional processes regulated by 130 genes that have been shown to be up-regulated by water deficit has been presented (Bray, 2002b).

Genes encoding for transcription factors have been shown to play a pivotal role in the adaptive response of plants to environmental stresses (Shinozaki and Yamaguchi-Shinozaki, 1996, 2000; Stockinger et al., 1997; Thomashow, 1999, 2001). It is therefore important to acquire information on the changes in their expression during the onset of the stress and also during the recovery phase. The total number of Arabidopsis genes encoding for transcription factors (TFs) have been estimated to be ca. 1,500 (i.e. ca 7% of all genes), the majority of which belong to gene families with many members. One of the greatest challenges is to understand and discriminate the function of closely related members within each gene family. For this purpose, the comparative analysis of expression profiles of different plant organs subjected to different stresses provides important information for assigning possible functions to the various members of each gene family. This approach was applied in Arabidopsis by Chen et al. (2002) who constructed a two-dimensional transcription factor matrix (genes vs. treatments or developmental stages/ tissues) reporting the changes in the mRNA levels of 402 genes coding for known and putative TFs. Groups of genes that respond similarly to different stresses were identified by cluster analysis of the fold-change in expression of each gene in the different treatments. Some of the results showed a partial overlap of the signalling pathways known to be activated by pathogen infection with pathways activated by leaf senescence, a useful indicator of the level of drought tolerance. The study conducted in Arabidopsis by Oono *et al.* (2003) during recovery from dehydration using a cDNA microarray containing ca. 7,000 full-length cDNAs evidenced 152 rehydration-inducible genes and their association with proline-inducible and water treatment-inducible genes. Of the 152 rehydration-inducible genes, 58 contained in their promoter regions the ACTCAT sequence involved in proline- and hypoosmolarity-inducible gene expression, suggesting that ACTCAT sequence is a major cis-acting element involved in rehydration-inducible gene expression, and that some novel cis-acting elements are involved in rehydration-inducible gene expression. A study carried out in Arabidopsis applying MPSS (massively parallel signature sequencing) provided information concerning the function of *abi1-1*, a gene controlling sensitivity to ABA (Hoth *et al.*, 2002). An ABA treatment of wild-type plants altered the expression of 1,354 genes, many of which encode signal transduction components and proteins involved in regulated proteolysis. In the *abi1-1* mutant, disrupting ABA sensitivity impaired or strongly reduced the regulation of 84% and 7% of the 1,354 genes, respectively, thus confirming the crucial role of this gene in coordinating gene expression following changes in ABA concentration. Additionally, because 9% of the 1,354 genes were normally regulated in the *abi1-1* mutant, the presence of two distinct ABA signalling pathways was suggested (Hoth *et al.*, 2002).

A targeted and a non-targeted gene expression profiling was applied in maize to dissect the stress sensitivity of reproductive development following a reduction in the photosynthates to the ear (Zinselmeier *et al.*, 2002), a condition typically caused by drought (Zinselmeier *et al.*, 1995). A four- to six-day drought episode reduces maize photosynthesis to near zero and can disrupt kernel growth, thus significantly reducing grain yield (Westgate and Boyer, 1985; Zinselmeier *et al.*, 1999). In the targeted approach, the 384 maize genes included in the microarray represented four metabolic pathways important for kernel growth, such as ABA signalling and starch biosynthesis. The genes controlling the starch biosynthetic pathway were found to be coordinately regulated under stress and the loss of starch was related to a decreased expression of these genes (Zinselmeier *et al.*, 2002). The non-targeted approach considered 1,502 maize genes, including over 300 unknown ESTs (Expressed Sequence Tags), that were annotated into 27 unique metabolic pathways. Transcriptome profiling during ear and kernel development revealed a set of genes that were affected by water stress regardless of tissue type, although in some tissues gene expression was more responsive to stress than other tissues; additionally, a set of genes unknown to respond to water stress was identified (Zinselmeier *et al.*, 2002). In maize, the early post-pollination phase is particularly sensitive to water deprivation (Bolaños and Edmeades, 1993; Zinselmeier *et al.*, 1999). Yu and Setter (2003) profiled mRNA extracted from endosperm and placenta/pedicel tissues at nine days after pollination in kernels of water-stressed plants. Of the 79 genes that were significantly affected by stress in placenta/pedicel tissue, 89% were up-regulated, whereas 82% of the 56 genes significantly affected in the endosperm were down-regulated. Only nine of the stress-regulated genes were in common between these tissues, thus pointing out that the transcriptional responses of placenta/pedicel and endosperm differ considerably. Yu and Setter (2003) postulated that the responsiveness of placenta to whole-plant stress factors (e.g., water potential, ABA and sugar flux) and of endosperm to indirect factors may play key roles in determining the threshold for kernel abortion in maize.

Dehydrated leaf and root samples of young barley plants were investigated with a microarray containing 1,463 transcripts derived from cDNA libraries of water-stressed barley plants (Ozturk et al., 2002). The same study also monitored the response to salinity. The two abiotic stresses affected largely diverse sets of transcripts and the differences were often in isoforms of transcripts for similar functions, an indication that the same function may be required by the plant to adapt to more than one abiotic stress. These differences involved different activation circuits either through alternative signal transduction, separate transcription factors and/or altered promoter structures. When focusing on the drought-induced transcripts, Ozturk et al. (2002) listed approximately 100 strongly up-regulated sequences, half of which were functionally unknown; furthermore, a number of these transcripts have not been reported as drought-inducible. From an applicative standpoint, Ozturk et al. (2002) questioned whether arrays of the size similar to that used in their study are sufficient for providing meaningful information from a breeding standpoint. In this respect, it should also be emphasized that the results of experiments carried out adopting stress treatments whose dynamic and intensity differ greatly from the conditions present in the field, may only be partially applicable to a breeding context, particularly in relation to the acclimation phase. A recent microarray study in barley has indeed shown a low, albeit significant, correlation between the fold-changes in gene expression under conditions of dehydration induced through a 'shock' treatment and the fold-changes observed following a more gradual and natural water loss in a soil glasshouse (Talamè et al., unpublished results). The set of genes responding similarly to both types of experimental conditions deserves further attention, also in consideration that for these genes it may be more feasible to extrapolate to field conditions the findings of expression studies carried out under controlled conditions.

9.4.2   Proteomics and metabolomics

A major shortcoming of expression profiling is that DNA sequence alone is unable to predict the structure, function and abundance of proteins in an organism. Additionally, other factors may influence the final level and role of key metabolites regulating growth and final yield. Recent work in yeast and mammalian cells has shown that abiotic stresses can profoundly alter the translational machinery (Patel *et al.*, 2002; Uesono and Toh, 2002), hence weakening the association between the level of mRNAs and their final products. It is thus clear that deciphering gene functions can be facilitated through the profiling of the proteome and metabolome, which, as compared to the transcriptome, are functionally 'closer' to the observed phenotypic traits evaluated by breeders and crop physiologists.

The past decade has witnessed a rapid progress in the enabling platforms that allow for a high-throughput analysis and accurate determination of the protein profiles of a particular tissue. In particular, improved sample preparation, narrow range immobilized pH gradient (IPG) strips, image analysis and the use of tryptic fingerprinting with mass spectrometry (e.g., MALDI-TOF) allow for the quantitative profiling of up to ca. 2,000–2,500 proteins in each sample (Fiehn, 2002; Roberts, 2002; Salekdeh *et al.*, 2002). The objectives of proteomics are to investigate the changes in structure and abundance of the proteome in response to developmental and environmental cues. The proteome profiling of the genotypes of a mapping population allows for the identification of QTLs influencing protein quantity, the so-called PQLs, i.e. Protein Quantity Loci (Damerval *et al.*, 1994; Touzet *et al.*, 1995; de Vienne *et al.*, 1999; Thiellement *et al.*, 1999, 2002; Zivy and de Vienne, 2000; Consoli *et al.*, 2002). The coincidence on the map of a PQL peak with its protein-coding locus would indicate that allelic differences at that locus influence the expression of the protein, whereas co-localization between a PQL and a QTL for a different trait would suggest an association between variability of the level of the candidate protein and the variability of the trait in question. A detailed description of this approach together with examples of its application in order to identify suitable candidate genes for QTLs influencing drought resistance in maize have been presented by de Vienne *et al.* (1999) and Pelleschi *et al.* (1999).

In rice, more than 2,000 proteins were detected reproducibly in drought-stressed and well-watered leaves of two lines contrasted for traits (e.g., osmotic adjustment, hardpan penetration capacity and depth of root system) relevant for adaptation to dry conditions (Salekdeh *et al.*, 2002). Among the approximately 1,000 proteins that were reliably quantified, 42 showed significant changes in abundance and/or position on the two-dimensional gel caused by drought and re-watering, with many of the proteins that were apparently up- or down-regulated under stress showing the reverse behaviour upon re-watering. Among the enzymes of carbon metabolism, cytosolic triose phosphate isomerase and chloroplastic fructose-1,6-biphosphate aldolase showed a 30–40% up-regulation under drought stress, confirming previous results reported in drought-stressed maize (Riccardi *et al.*, 1998). Given the central role of these enzymes in carbon metabolism and energy transduction, Salekdeh *et al.* (2002) suggested that their increase during the drought episode could reflect altered patterns of carbon flux in response to reduced photosynthesis and an increased need for osmotic adjustment in leaf tissue. A marked increase also characterized the actin depolymerizing factor. This factor promotes the disaggregation of actin filaments regulating cell shape, including the ABA-mediated changes in guard cells during stomatal closure (Hwang and Lee, 2001), thus suggesting that its induction may reflect the need for leaf cells to adjust their cytoskeletons to the cell shrinkage caused by dehydration (Salekdeh *et al.*, 2002). Another up-regulated protein whose role may be relevant for the adaptive response to drought, was the chloroplastic glutathione-dependent dehydroascorbate reductase, an enzyme which is part of the ascorbate-GSH pathway of oxidative stress tolerance. Interestingly, the increase in the level of the reductase was observed only in the drought-tolerant line. The surge of antioxidant activity observed in rice and reported also in other species exposed to drought could partially mitigate the damaging cellular effect associated to the increased production of free radicals during drought, thus contributing to the maintenance of an adequate redox potential (Smirnoff and Pallanca; 1996; Foyer and Noctor, 2000; Bartels, 2001). The relevance of the different changes in protein profiles observed in the two lines tested by Salekdeh *et al.* (2002) can now be determined through more detailed studies based on QTL mapping and/or reverse genetic approaches.

Another promising avenue in the realm of functional genomics is the metabolic profiling of tissue samples. Metabolome profiling aims at the identification and quantitation of all metabolites in a given biological sample. Although the development of metabolomic methods and tools is progressing rapidly, an understanding of the resulting data is hindered by a fundamental lack of biochemical and physiological knowledge about network organization in plants (Weckwerth and Fiehn, 2002; Steuer *et al.*, 2003). To this end, metabolic databases provide a valuable framework to predict biochemical pathways and products given a certain genotype and to reveal the phenotype of silent mutations (Fell, 2001), thus contributing to a more comprehensive view of the functional characteristics under investigation. The present technology permits the profiling of approximately 2,000 different metabolites in a single sample (Fiehn, 2002). Metabolome profiling can be used to monitor the metabolic changes occurring during exposure to an abiotic stress in one or more genotypes and, when applied to a mapping population, to identify the QTLs regulating the level of key metabolites and verify their coincidence with QTLs for yield and/or genes involved in metabolic pathways. In maize, a rather extensive literature describes the changes occurring during a drought episode in the level of a limited number of key metabolites such as sugars and starch in the reproductive organs and in the growing kernel (Zinselmeier *et al.*, 1995,

1999). In drought-stressed maize, it has been suggested that the observed reduction in acid invertase activity could impair sink strength because photosynthates cannot be converted rapidly to starch (Zinselmeier *et al.*, 1995). Nonetheless, limited work has been completed to link this information to QTLs for the same metabolites. QTLs for invertase activity have been described in a maize population subjected to drought stress (Pelleschi *et al.*, 1999). The number of QTLs for invertase activity detected under drought (nine in total) was more than twice the number detected under well-watered conditions (four in total), an indirect indication of the important role of this enzyme under drought conditions. One QTL common to both treatments was located near *Ivr2* on bin 5.03. A number of QTLs for invertase activity were mapped in close proximity to carbohydrate QTLs with the two main clusters mapping on bins 1.03 and 5.03. Drought produced an early stimulation of acid-soluble invertase activity in adult leaves, whereas cell wall invertase activity was unaffected. This response was closely related to the mRNA level for only one (*Ivr2*) of the invertase genes.

Despite the encouraging results herein reported, it should be appreciated that, at present, proteomics and metabolomics can indirectly report changes occurring in only a fraction of the genome and are often unable to detect appreciable changes in genes expressed at low levels, some of which (e.g., transcription factors) may actually play a very important role in determining QTLs for tolerance to drought and other abiotic stresses. Nonetheless, the information obtained through proteomics and metabolomics, when appropriately combined with transcriptome and phenotypic data, greatly enhances our capacity to identify suitable candidate genes, as illustrated in the next section.

9.5   From QTLs to genes for WUE and related traits

The final genetic dissection of a QTL is completed only if the observed phenotypic variability is directly connected with a DNA polymorphism (i.e., sequence variation). Several options are available to proceed from a supporting interval delimiting the QTL to the actual gene(s) responsible for the QTL effect. According to the magnitude of the QTL effect and the mapping accuracy, the support interval of the QTL may include several hundred genes. Additionally, also non-coding regions may be responsible for QTLs through a cis-acting effect on the promoter region of nearby genes. It is evident that to identify the right 'needle' in this 'genetic haystack' is a rather daunting undertaking, although one well worth pursuing.

9.5.1 *Positional cloning*

The main strategy so far adopted for identifying the gene(s) underlying a QTL relies on positional cloning. All the prerequisites necessary to positionally clone Mendelian genes (Tanksley *et al.*, 1995) are also required for the positional cloning of a QTL. However, a much larger effort is needed for the phenotypic scoring of the segregating progenies, particularly if the heritability of the trait is not high. The main prerequisites for attempting the cloning of a QTL for which congenic strains have been obtained are as follows: (i) the confining of the QTL to a 2–3 cM interval (or even shorter) coupled with the availability of at least one pair of NILs carrying contrasting parental alleles at the target region; (ii) the availability of a large mapping population (> 1,500 plants) derived from the cross of the two NILs; (iii) the presence of a high level of polymorphism between the NILs in order to allow for a level of marker density (most frequently AFLPs) in the target region suitable to chromosome walking or, preferably, chromosome landing; (iv) a high recombination rate (i.e., a high ratio between genetic and physical distances) in the target chromosome region; (v) the availability of a contiged BAC (Bacterial Artificial Chromosome) and/or YAC (Yeast Artificial Chromosome) genomic library spanning the QTL region; (vi) a system (e.g., transformation) for validating the identity and testing the effects of the selected candidate gene. It is evident that the routine implementation of the positional cloning approach is possible only for a very limited portion of the QTLs so far described, particularly in view of the large amount of resources required to successfully complete a QTL cloning project. Assigning a QTL peak with the accuracy required for a positional cloning effort is possible only when the QTL effect is substantial and highly heritable (Salvi *et al.*, 2002). Even in this case, several hundred of individual progenies have to be scored molecularly and phenotyped to achieve the required level of genetic resolution. An additional complexity is added in species with large genomes (e.g., cereals), rich in heterochromatin and redundant genes, two peculiarities that will inevitably complicate steps (iv) and (v) of the cloning process.

9.5.2 *The candidate gene approach*

During the gene discovery process, it is possible to reduce the number of possible candidates for a particular QTL adopting the so-called 'candidate gene approach' (Causse *et al.*, 1995; Pflieger *et al.*, 2001; Wayne and McIntyre, 2002). At its simplest, the candidate gene approach exploits all possible sources of information concerning the role and function of a particular coding sequence and verifies whether it may represent a feasible candidate for the QTL in question. If a suitable cause-effect relationship can be established between a candidate gene and a QTL, then validation of its role could be attempted directly through genetic engineering and/or the screening of knockout mutants, avoiding the tedious procedures of the positional cloning approach. The 'candidate gene approach' has already been applied to maize in order to identify candidate genes for tolerance to drought (de Vienne *et al.*, 1999) and low nitrogen (Hirel *et al.*, 2001). For QTLs influencing drought tolerance, two categories of genes which seem particularly promising as candidates are those encoding for proteins involved in the signalling pathways activated by the early stages of water stress (e.g., kinases) and those encoding for transcription factors (e.g., DREB-like proteins) which, in turn, modulate the expression of gene clusters responsive to such factors (see section 9.6.3.2).

The availability of large sets of cereal ESTs (Expressed Sequence Tags) together with the 'omics' data collected under drought, provide interesting opportunities to streamline the identification of candidate genes at target QTLs. It is also technically feasible to assemble physical maps of crops' genomes by ordering overlapping genomic clones (BACs or YACs), and exploiting the ESTs to cross-link the physical map to the genetic map (Kurata *et al.*, 1997; Yim *et al.*, 2002). At that stage, QTL cloning will be reduced to fine mapping at the genomic clone level, followed by the sequencing of the genomic clone most likely to contain the gene(s) responsible for the QTL. This procedure has already been successfully exploited for QTL cloning in rice (Yano *et al.*, 2000; Takahashi *et al.*, 2001; Kojima *et al.*, 2002).

A novel application of transcriptome profiling that may facilitate the identification of candidate genes on a genome-wide basis is offered by the so-called eQTLs, namely QTLs influencing variability for the level of expression (hence the 'e') of a particular gene. A quantitative microarray-based analysis of gene expression carried out on all the progenies of a mapping population will eventually lead to identify eQTLs influencing the observed variability among progenies in the mRNA level of single genes. In turn, this information may facilitate the identification of candidate genes for QTLs, particularly when it is possible to establish a plausible cause-effect relationship between the target trait (e.g., tolerance to dehydration) and the expression level of the putative candidate gene, hence its product, such as a transcription factor regulating a suite of downstream genes encoding for proteins protecting the cell from the harmful effects of dehydration). In plants, eQTLs through genome-wide microarray profiling have only been reported by a recent study in maize (Schadt *et al.*, 2003). The potential of the eQTL approach to identify candidate genes for QTLs influencing tolerance to abiotic stresses has recently been validated in wheat for *Fr-A2*, a QTL affecting freezing tolerance (Vagujfalvi *et al.*, 2003). When plants of the frost-tolerant and frost-susceptible parental lines of the mapping population were grown at 15 °C, they differed in the transcription level of the cold-induced gene *Cor14b*. Transcript levels of this gene were determined in each line of the mapping population, allowing Vagujfalvi *et al.* (2003) to identify a QTL coincident with the QTL for frost survival at the *Fr-A2* locus, thus providing strong circumstantial evidence that differential regulation of the expression of the *Cor14b* gene mediated frost tolerance. Clearly, the costs for profiling the high number of RNA samples required to identify eQTLs are still too high to conceive a more routine application of this approach, particularly at the whole-genome level. A similar strategy can be devised also deploying proteomics and metabolomics to identify QTLs controlling the level of particular proteins (e.g., PQLs) and/ or metabolites. Also in this case, the overlap of the profile of a QTL for the trait with the map position of the candidate gene and the corresponding 'omics' QTL provides circumstantial evidence on the importance of the gene itself. More compelling evidence on the role of the gene can then be obtained by altering its expression level by genetic engineering, through the analysis of mutants (e.g., knockout mutants, TILLING, etc.) and/or through linkage-disequilibrium analysis targeting the region of interest in a sufficiently large sample of unrelated genotypes where the degree of linkage disequilibrium is sufficiently low to provide a level of genetic resolution suitable for validating the role of the candidate sequence (Buckler and Thornsberry, 2002; Rafalski, 2002).

Among the few QTLs that so far have been cloned in crop species, those controlling transition from the vegetative to the reproductive stage in rice (Yano *et al.*, 2000; Takahashi *et al.*, 2001; Kojima *et al.*, 2002) are of applicative interest for improving seasonal WUE and drought resistance through the manipulation of flowering time, a key trait for drought escape. In maize, a similar approach is well under way for the cloning of *Vgt1* (Salvi *et al.*, 2003), a QTL on bin 8.05 near *umc89a* controlling flowering time (Salvi *et al.*, 2002). Interestingly, the chromosome segment corresponding to *Vgt1* does not coincide with any of the nearby annotated genes, thus indicating the importance of non-coding regions for the control of variation in quantitative traits. Under this respect, an approach solely based on the analysis of knock-out mutants at known candidate genes would have not reached the same result. In fact, when a polymorphism in a non-coding region is the underlying cause of a QTL, the analysis of knock-out mutants of candidate genes may lead us to identify the coding sequence ultimately affecting the phenotype, but it will fail to identify the cis-acting sequence governing its regulation. In maize, a positional cloning effort is also under way to attempt the positional cloning of *root-ABA1*, a QTL on bin 2.04 influencing root architecture, ABA concentration and other drought-related traits (Tuberosa *et al.*, 1998b; Sanguineti *et al.*, 1999; Landi *et al.* 2002a).

An increasing amount of information is being obtained on the colinearity relationships between the sequenced genomes of model species like Arabidopsis and rice and the genomes of the main crops (Paterson *et al.*, 1996; Bowers *et al.*, 2003). Although it has been shown that colinearity between Arabidopsis and rice (Devos *et al.*, 1999), and between Arabidopsis and maize (van Buuren *et al.*, 2002) has been eroded to such a degree that the map information of Arabidopsis does not seem to help the identification of related genes in cereals, the level of synteny between rice and the other grasses (Gale and Devos, 1998; Goff *et al.*, 2002; Ware *et al.*, 2002) provides opportunities to exploit high resolution colinearity maps in order to facilitate the positional cloning of QTLs in cereals. However, a recent study comparing rice and wheat sequences has revealed numerous chromosomal rearrangements that will significantly complicate using rice as a model for cross-species transfer of information in evolutionarily non-conserved regions (Sorrells *et al.*, 2003). Similar opportunities will be available for dicot crops more closely related to Arabidopsis (for example, rapeseed and cotton). Although the major emphasis throughout this paper has been on crop species, the contribution to gene discovery of model species such as Arabidopsis should not be disregarded. This is particularly true for the isolation of genes controlling traits influencing WUE (Nienhuis *et al.*, 1994) and drought tolerance (Meyre *et al.*, 2001) in this species. The small size of the genome and the adult plant, its short cycle, coupled with an efficient transformation system, and the availability of the genome sequence, mutant collections and other genomic platforms (Koumproglou *et al.*, 2002; Greene *et al.*, 2003; Schmid *et al.*, 2003) are all factors contributing to making this small weed a valuable model to identify QTLs and clone the corresponding genes. Surprisingly, little work has been carried out in Arabidopsis to identify QTLs for WUE and related traits; analogously, no QTLs for ABA accumulation nor root characteristics have been described in Arabidopsis. The presence of genetic variability among ecotypes and recombinant inbred lines of Arabidopsis for $\Delta^{13}C$ and seasonal WUE has been reported (Nienhuis *et al.*, 1994), although its heritability was rather low (18%).

9.6 Conventional and molecular approaches to improve WUE and drought tolerance

Improving WUE to enhance survival under rather extreme arid conditions disregarding implications on yield under more favourable conditions can be a counterproductive approach for the majority of agricultural conditions. Presently, the focus has shifted more on the improvement of yield per se, even if this may be associated with a lower intrinsic WUE of the selected materials. Whichever is the preferred ideotype, the integration of conventional breeding approaches with the emerging molecular platforms herein described offers new opportunities to speed up the yearly production gains necessary to keep pace with the food and fiber requirements of a burgeoning world population.

9.6.1 Empirical vs. analytical breeding

Yield improvements in drought-stressed conditions obtained with conventional breeding approaches have for the most been achieved with limited knowledge of the genetic basis of WUE and drought tolerance. According to Blum (2002), the complexity of drought tolerance can be greatly reduced if two major points are considered: 1) a number of plant traits crucial for the control of plant water status and yield under drought are constitutive and not stress adaptive; 2) plant water status, more than plant function, controls crops' performance under drought. Empirical breeding relies on the phenotypic evaluation of large numbers of genotypes tested in field trials representative of the target environments. The main obstacle to the success of this approach is represented by the low heritability of yield, particularly under arid conditions, as a consequence of large G × E interactions (Ceccarelli, 1984; Blum, 1988). Compared to empirical breeding, the analytical approach tries to identify heritable attributes whose selection improves field performance. In this case, additional genetic variability can be acquired through adequate crossing schemes deploying germplasm accessions contributing new alleles for the trait of interest. If sufficient genetic variability is available for the target trait, other factors to be considered before using it as a selection criterion are its heritability and its genetic correlation with yield, which should both be high. Selection for a trait characterized by high heritability is feasible at early generations, thus reducing the high costs associated with a breeding scheme where selection for yield is delayed until later generations. From a practical standpoint, the ideal genotype will be determined by the prevailing conditions of the target environment. Turner (1997) pointed out that doubling grain production, hence grain WUE, in Australian wheats during the past 120 years was prevalently achieved through an increase in harvest index (Perry and D'Antuono, 1989; Turner et al., 1989) and better matching of water use to water supply (Siddique et al., 1990). This notwithstanding, the low grain yield per unit of water of the modern wheat genotypes warrants further efforts to improve seasonal WUE. Harvest index is a valid example of a trait selected in favourable environments through the use of dwarfing genes (Austin, 1999) that can also be beneficial when the selected genotypes are grown under unfavourable conditions. For those traits with a fairly consistent effect on yield irrespectively of the level of water deficit, it is more convenient to carry out the selection under favourable environments in order to maximize heritability, hence response to selection. A field evaluation under more favourable conditions (e.g., under irrigation) allows the breeder to reduce the high level of unpredictability for yield usually associated with field trials carried out under rainfed conditions. Nonetheless, testing the selected elite materials under dry environments remains an essential prerequisite to determine their real adaptive value.

An example of a physiological trait advantageous under drought and carrying no yield penalty in favourable conditions has been provided by the work of Richards and Passioura (1981a,b) to investigate the effects of xylem vessel diameter of seminal roots on final grain yield in wheat. In this case, locally adapted cultivars were backcrossed with a Turkish landrace with a small xylem vessel diameter, a positive characteristic under growing conditions prone to terminal drought; in these circumstances, a thriftier use of water by the crop prior to anthesis translates in a higher soil moisture availability during grain filling. The backcrossed lines, when tested under dry conditions, outyielded by 3–11% the recurrrent parents. Importantly, the same lines showed no yield penalty in wet seasons, because under these circumstances the water was prevalently absorbed by the nodal roots from the more superficial layers of the soil (Richards and Passioura, 1989).

When a large number of genotypes are compared under drought for a physiological trait influenced by developmental factors and/or the water status of the plant, it is important to account for variability in flowering time and fluctuating environmental factors that may affect the trait according to the time and day of sampling. In this case, interaction of the genotype with variation in phenology and the dynamics of the drought episode(s), if not accounted for appropriately, can bias the interpretation of the results. In this context, an interesting approach is to consider morpho-physiological traits characterized by a low interaction with measurable environmental variables, such as the elongation rate at the base of the leaf (Reymond et al., 2003). In maize, elongation rate has been shown to be genotype-specific and linearly influenced by environmental variables (e.g., water availability, temperature, etc.), thus providing a way to properly account for variation due to environmental conditions when comparing data of a large set of genotypes sampled at different times and/or in different experiments (Reymond et al., 2003).

9.6.2   Marker-assisted selection

Until now, from a strictly breeding standpoint, the application of marker-assisted selection (MAS) for the improvement of WUE and/or drought tolerance has been attempted in a rather limited number of cases, mainly due to the difficulty in identifying major QTLs with a sufficiently large and stable effect for justifying the monetary investment required by MAS (Atlin and Lafitte, 2002). Conversely, MAS has been successfully utilized to transfer disease resistance genes (Young, 1999) and to speed up the recovery of the recurrent parental genome in backcrossing schemes (Hospital and Charcosset, 1997; Ribaut et al., 1997a).

Several morpho-physiological characteristics influencing WUE are valid targets for programs of MAS aimed at improving yield. Traits characterized by low heritability and/or requiring a great deal of resources for an accurate phenotyping are ideal candidates for a MAS approach. For these reasons, substantial efforts have been devoted to the identification of molecular markers linked to QTLs influencing root characteristics, undoubtedly one of the most resource-demanding and difficult category of traits to be assessed, particularly in field conditions. Once major QTLs have been identified, MAS can be utilized to derive congenic strains and/or to introgress the desired QTL alleles in different genetic backgrounds. The evaluation of testcrosses obtained by crossing a number of tester lines with pairs of NILs contrasted at a QTL region allows us to more accurately determine the magnitude and consistency of the QTL effect, and to evaluate its real agronomic value. Additionally, once pairs of congenic strains have been derived for different QTLs, crossing schemes can be devised to more precisely test for epistatic interactions between the isogenized QTLs, avoiding the shortcomings pertaining to the evaluation of epistasis using an entire mapping population. In fact, a reliable assessment for the presence of epistatic QTL effects in a mapping population usually requires the evaluation of a number of progenies much larger than those commonly utilized (from ca. 100 to 250) for QTL discovery. Although the derivation of congenic strains through MAS does not directly lead to short-term applications, it is an essential step to 'Mendelize' single QTLs, an important prerequisite for their positional cloning. A number of NILs have been obtained for QTLs of traits relevant to WUE and drought tolerance (Tuinstra et al., 1998; Shen et al., 2001; Landi et al., 2002a; Price et al., 2002c; Sanchez et al., 2002) and several others are presently being derived. In maize, MAS has been deployed to derive pairs of backcross-derived lines (BDLs) differing for the parental alleles at a major QTL (root-ABA1) on bin 2.04 near csu133 that affects ABA concentration and root traits (Tuberosa et al., 1998b; Sanguineti et al., 1999; Landi et al., 2002a). A field evaluation conducted under well-watered and water-stressed conditions during two consecutive seasons indicated that each pair of BDLs differed significantly and markedly for L-ABA, thus confirming the effectiveness of the MAS (Landi et al., unpublished results). Furthermore, each pair of BDLs showed significant differences also for root lodging, a finding that could depend on differences of the two parental lines for root architecture, consistently with the hypothesis postulated by Sanguineti et al. (1999).

As to the utilization of MAS to improve yield under drought conditions, major integrated breeding programs have been undertaken in cereals at the CGIAR centres, namely at IRRI for rice (reviewed in Saxena and O'Toole, 2002), ICRISAT for pearl millet and sorghum (Crouch and Serraj, 2002; Hash et al., 2003) and CIMMYT for maize (Ribaut et al., 2002). Due to the key role played under rainfed conditions by roots in determining rice yield (Nguyen et al., 1997), MAS for root depth has been deployed at IRRI to more specifically tailor new genotypes to the range of environments where rice is cultivated (Mackill et al., 1999). After identifying QTLs affecting root parameters in a

DH population derived from the cross IR64 × Azucena, Shen *et al.* (2001) used marker-assisted backcrossing to transfer the Azucena allele at four QTLs for deeper roots (on chromosomes 1, 2, 7 and 9) from selected DH lines into IR64. Three backcross progenies were selected solely on the basis of their genotypes at the marker loci in the target regions up to the BC_3F_2 generation. Twenty-nine selected BC_3F_3 NILs were developed and compared to IR64 for the target root traits and three non-target traits. The phenotypic evaluation of root traits indicated that the backcross MAS was successful for QTL regions on chromosomes 1, 7 and 9, but not for the QTL on chromosome 2 where none of five NILs differed from IR64 for root traits. This result prompted Shen *et al.* (2001) to re-analyse the initial data with a more sophisticated software package that revealed two linked QTLs with opposite effects in this area. Some of the NILs were taller than IR64 and all had a decreased tiller number because of a likely co-introgression of linked QTLs. A review on QTL mapping in rice, including also QTLs affecting drought resistance, detected epistatic interactions for many QTLs (Li *et al.*, 2001). Based on these findings, a new strategy of molecular breeding has been developed to facilitate simultaneous QTL identification and introgression from germplasm accessions (Li *et al.*, 1999).

In maize, MAS has been deployed to introgress QTL alleles for reducing the anthesis-silking interval (ASI; i.e., the interval between the extrusion of the anthers and the silks), a trait found to be consistently associated with grain yield under conditions of water deficits (Bolaños *et al.*, 1993; Bolaños and Edmeades, 1996; Ribaut *et al.*, 1997b; Ribaut *et al.*, 2002). Drought usually delays silk extrusion in maize, the most critical stage in terms of deleterious effects of water deprivation on yield (Westgate and Boyer, 1985), thus leading to a negative association of ASI and grain yield (Bolaños and Edmeades, 1996). The availability of molecular markers linked to the QTLs for ASI allows for a more effective selection under drought as well as when drought fails to occur at flowering (Ribaut *et al.*, 1997a, 2002). A backcross marker-assisted selection (BC-MAS) project based on the manipulation of five QTLs affecting ASI was started a decade ago (Ribaut *et al.*, 1997a, 2002). CML247 was used as the recurrent parent and the line Ac7643 was the drought-tolerant donor. CML247, an elite line with high yield *per se* under well-watered conditions, is drought susceptible and shows long ASI under drought. The QTL regions with alleles for short ASI were introgressed through MAS from Ac7643 into CML247. A number of lines (ca. 70) derived through BC-MAS were crossed with two testers and were evaluated for three consecutive years under several water regimes. Under severe stress conditions reducing yield by at least 80%, the mean of the selected lines was higher than that of the unselected control. However, this advantage decreased at a lower stress intensity, and disappeared for a stress-reduced yield less than 40%. Across the water-limited trials, a few genotypes consistently outperformed the controls. Interestingly, under well-watered conditions the selected lines did not show any yield penalty when compared to the control lines. Notwithstanding the success of this BC-MAS experiment, Ribaut *et al.* (2002) pointed out that QTL manipulation to improve germplasm for polygenic traits has a number of limitations, the most distinct being the inability to predict the phenotype of any given genotype based on its allelic composition, a constraint particularly important when epistatic interactions regulate the expression of the target trait. Another limitation of MAS pertains to the high costs associated to QTL discovery and validation as well as the release of superior lines. New strategies to overcome these limitations have been implemented at CIMMYT that are aimed at improving the cost effectiveness of MAS and its capacity to deliver improved germplasm (Ribaut and Betran, 2000).

In chickpea, MAS is being deployed at ICRISAT to introgress QTL alleles associated with a large root size into elite germplasm (Saxena *et al.*, 2002). More than 90% of the global chickpea-growing area is rainfed, for the most part characterized by soil moisture conserved from the preceding rainy season. Terminal drought can curtail chickpea yield from 20% to more than 50%. Hence, a deep root system capable of extracting additional moisture should positively impact yield in drought-prone areas. At ICRISAT, similar projects for using MAS to improve drought tolerance are under way also for groundnut and pigeonpea (Crouch and Serraj, 2002).

Among major crops, large acreages of cotton are grown under both irrigated and rainfed conditions, thus making $G \times E$ interactions of even greater importance than usual in designing crop-improvement strategies. In a recent study carried out at different water regimes, Paterson *et al.* (2003) described 17 QTLs affecting fiber quality only in the water-limited treatment, while only two were specific to the well-watered treatment. These results suggest that improvement of fiber quality in cotton under water stress may be even more difficult than improvement of this already complex trait under well-watered conditions, thus reducing the expected rate of genetic gain. To overcome these difficulties, Paterson *et al.* (2003) suggested deploying MAS based on a number of markers found to be associated to the QTLs discovered in their study.

A more routine application of MAS will largely depend on the cost of the molecular analyses. In a recent study, Morris *et al.* (2003) compared the cost-effectiveness of marker-assisted and conventional maize breeding for the introgression of an elite allele at a single dominant gene into an elite maize line. In this particular case, neither method showed clear superiority in terms of both cost and speed: conventional breeding schemes are less expensive, but MAS-based breeding schemes are faster. These findings may question to some extent the validity of integrating MAS with the conventional breeding program. Nonetheless, as new major QTLs with large effects on the phenotype are discovered, DNA-assisted selection will offer opportunities to accelerate the pace of improving yield under arid conditions. High-throughput genotyping based on the scoring of markers that do not require the use of gels (Salvi *et al.*, 2001) coupled with quick DNA extraction protocols are needed to streamline MAS, lower the associated costs and to make this approach more widely applicable.

9.6.3   Genetic engineering

Recombinant DNA approaches offer additional opportunities to investigate and understand the role of candidate genes and to attempt their direct manipulation in order to improve crops' WUE and resilience to water stress. A number of model species have provided a wealth of information on the molecular mechanisms by which the plant cell withstands dehydration (Zhu *et al.*, 1997; Bohnert and Cushman, 2000; Bartels and Salamini, 2001). The real challenge is now to ascertain to what extent the manipulation of these mechanisms may lead to increased yields in the field. A number of reviews have exhaustively analysed and discussed how crops grown in arid conditions may benefit from the direct manipulation of the genome (Holmberg and Bülow, 1998; Bajaj *et al.*, 1999; Bohnert and Shen, 1999; Bohnert and Bressan, 2001; Mitra, 2001; Chen and Murata, 2002). Among the vast literature available on this topic, this chapter will present a few examples where genetic engineering has been deployed to manipulate some of the physiological factors more likely to affect intrinsic WUE and/or confer tolerance to dehydration. Although pertinent to the topic herein discussed, the results obtained with the direct manipulation of traits conferring escape from drought, such as flowering time, or other prevalently constitutive traits affecting seasonal WUE (e.g., partitioning of photosynthates) have not been surveyed. An important category of genes affecting the water balance of the plant and still little explored as to their impact on WUE and yield is represented by the aquaporins, hydrophobic proteins which regulate water flow across plasma membranes, with an increasingly important role in regulating hydraulic conductivity and water flow as water availability declines (Tyerman *et al.*, 1999, 2002; Maurel and Chrispeels, 2001). However, the high genomic redundancy (i.e., high number of members) of this gene family hinders the application of genomics and/or genetic engineering to more clearly ascertain the role of aquaporines in regulating WUE.

9.6.3.1   Genes affecting intrinsic WUE
Concentrating CO_2 at the site of Rubisco should allow metabolically engineered C_3 crops to decrease stomatal conductance under dry conditions without a substantial decrease in CO_2 assimilation. Because of the inherently higher WUE of C_4 vs. C_3 species, the past decade has witnessed a renewed interest for transferring the characteristics of the C_4-like metabolism from C_4

to C_3 crops through genetic engineering (Miyao, 2003). Earlier attempts to use conventional breeding approaches to transfer via interspecific cross the Krantz anatomy from C_4 to C_3 species of the genus *Atriplex* (Björkman *et al.*, 1971) were only marginally successful and did not lead to any appreciable applicative result. A decade of efforts aimed at genetically engineering the C_4 pathway in C_3 species has provided crucial information to devise more effective strategies to improve the photosynthetic efficiency in tobacco, potato and rice (Hausler *et al.*, 2002; Miyao, 2003). It is now possible to overexpress the enzymes of the C_4 pathway in desired locations in the leaves of C_3 plants, otherwise present al lower concentration (Ku *et al.*, 1996, 1999; Matsuoka *et al.*, 2000, 2001; Jeanneau *et al.*, 2002b). From these experiments, it appears that the overproduction of a single C_4 enzyme can alter the carbon metabolism of C_3 plants, although this does not lead to positive effects on photosynthesis. The emerging paradigm is that improving the photosynthetic performance of C_3 transgenic plants requires the concerted overproduction of multiple enzymes of the C_4 pathway. Despite this progress, considerable controversy remains as to what extent this strategy might be successful in achieving the desired goals (Leegood, 2002; Miyao, 2003). As an example, it may be possible that the C_4-like pathway could support C_3 photosynthesis under conditions of limited CO_2 fixation, such as drought (Miyao, 2003).

Encouraging results have been reported by Jeanneau *et al.* (2002a), who transformed maize with two objectives: first, to assess the role of *Asr1*, a candidate gene encoding for a putative transcription factor, as the underlying cause of a genetically linked QTL for drought tolerance, and second, to ectopically alter the rate of CO_2 fixation in leaves through changes of C_4 phosphoenolpyruvate carboxylase (C_4-PEPC) activity using a sorghum C_4-PEPC cDNA. Interestingly, the maize line with the highest C_4-PEPC expression level exhibited a substantial increase (+30%) in intrinsic WUE accompanied by a dry weight increase (+20%) under moderate drought conditions in the field. Reverse effects were measured for transgenic plants under-expressing the corresponding proteins. These encouraging results suggest the feasibility of increasing the level of endogenous biochemical activities related to water economy and/or drought tolerance. Further field testing under different water regimes will indicate if this strategy will allow for the development and release of commercial high-yielding maize hybrids more tolerant to drought.

One way to alter WUE is through the manipulation of genes affecting stomatal conductance. Guard cells of a T-DNA insertion mutant for the Arabidopsis ABC-transporter AtMRP5 (mrp5-1) were found to be insensitive to external calcium and ABA, and caused a reduced transpiration rate of mrp5-1 in the light (Klein *et al.*, 2003). Excised leaves of mrp5-1 plants exhibited reduced water loss and an increased WUE, and after witholding water they survived much longer due to reduced water use. Wild-type characteristics were shown by transgenic mutant plants overexpressing AtMRP5 under the control of the CaMV 35S. Borel *et al.* (2001a,b) transformed *Nicotiana plumbaginifolia* with an antisensed zeaxanthin epoxidase gene (ZEP) to alter ABA accumulation in roots and xylem sap. ZEP is a key enzyme for the regulation of ABA concentration in plant tissues (Milborrow, 2001). In a number of greenhouse and growth chamber experiments, all the transgenic tobacco lines showed a linear relationship between predawn leaf water potential and ABA concentration in the xylem sap, which allowed Borel *et al.* (2001a) to use a quantitative approach to assist the interpretation of the behaviour of transgenic lines in terms of ABA accumulation and stomatal conductance, thus providing valuable insights in the drought-avoidance mechanisms. Another approach to enhance drought avoidance through a decreased loss of water from the stomata has focused on the concentration of malate in the guard cells via the targeted manipulation of NADP-malic enzyme (ME; Laporte *et al.*, 2002). The accumulation of potassium, chloride and/or malate during stomatal opening increases the turgor pressure of the guard cells, which increases stomatal conductance. In tobacco, the ectopic expression of maize ME, which converts malate and NADP to pyruvate, NADPH and CO_2, altered stomatal behaviour and water relations. The ME-transformed plants had lower stomatal conductance and were more water-use efficient than the wild type, despite having the same growth rate. Adding chloride via the transpiration stream partially reversed the effects of ME expression on stomatal aperture size, consistently with the interpretation that ME expression altered malate metabolism in guard cells. These results suggest that it is possible to devise effective strategies to genetically manipulate stomatal conductance to alter plant water use.

9.6.3.2 Genes affecting dehydration tolerance

A variety of genes are involved in the intricate mechanisms enabling the plant to perceive dehydration and signal the right cues to the genes whose products may mitigate the negative effects of drought (Bohnert and Bressan, 2001; Zhu, 2002). These genes encode for proteins or enzymes with vital roles in reducing water loss, protecting the cellular machinery, repairing cellular damage and restoring a new cellular homeostasis more compatible with the stressful conditions (Bray, 2002a; Ramanjulu and Bartels, 2002). According to their role, the genes involved in the response to water stress can be categorized into three hierarchically connected groups:

(1) signalling genes controlling the biochemical cross-talk in response to metabolic and environmental cues,
(2) regulatory genes coordinating the response of
(3) functional genes (i.e. the real 'effectors') responsible for the phenotypic changes at the biochemical, physiological and/or morphological level.

From an applicative standpoint, a strategy based on the ectopic expression of genes that are closer to the top of this hierarchical cascade (i.e., categories 1 and 2) is more likely to affect the phenotype, with all the obvious advantages and shortcomings.

Functional genes. In order to avoid an excessive loss of water under arid conditions, plant cells respond by raising osmotic pressure through the accumulation of inorganic ions (e.g., K^+, Na^+ and Cl^-) and/or organic solutes such as glycine-betaine, proline, polyols (e.g., inositol and mannitol) and sugars (e.g., sucrose, fructans and trehalose), whose accumulation does not interfere with the cellular metabolism (Bohnert et al., 1995; Bohnert and Sheveleva, 1998). Genes coding for crucial steps in the biosynthesis of osmolytes have been isolated from plants and microorganisms which share some osmoprotective mechanisms with plants (Nuccio et al., 1999; Bohnert and Shen, 1999). The accumulation of osmolytes through the engineering of biosynthetic pathways has been the focus of intense research (Blum et al., 1996; Chen and Murata, 2002; Rontein et al., 2002). Transgenic tobacco plants able to synthesize and accumulate mannitol were obtained deploying a bacterial gene encoding for mannitol-1-phosphate dehydrogenase. Plants producing mannitol showed increased drought and salt tolerance (Tarczynski et al., 1993; Thomas et al., 1995). Similarly, a good degree of drought tolerance was obtained in tobacco and sugar beet plants engineered using microbial fructosyltransferase genes that lead to the accumulation of fructans, high-soluble polyfructose molecules that are produced by many vascular plants and bacteria (Pilon-Smits et al., 1995, 1999). In these studies, the same *SacB* gene from *Bacillus subtilis* was used to produce bacterial fructans in sugar beet. Transgenic tobacco and sugar beet plants accumulated fructans to low levels (up to approximately 0.5% of dry weight) in both roots and shoots and showed significantly better growth under drought than untransformed plants, while no significant differences were observed under control conditions between the transgenic and wild type.

The effects of an overproduction of trehalose have been investigated in tobacco engineered with the yeast gene *TPS1* (trehalose phosphate synthase) encoding trehalose synthase (Holmstrom et al., 1996; Lee et al., 2003) or with bacterial trehalose synthesizing enzymes (trehalose-6-phosphate synthase and trehalose-6-phosphate phosphatase; Pilon-Smits et al., 1998). The leaves of these tobacco plants, as compared to the controls, showed a better water retention, a higher photosynthetic efficiency and a higher dry weight accumulation under drought stress (Pilon-Smits et al., 1998), confirming the role of osmolyte-synthesizing genes in conferring adaptation to dehydration (Rontein et al., 2002). Since most of the osmolytes did not accumulate in amounts large enough to play a role in osmotic adjustment, the protective mechanism remains presently unclear (Hong et al., 2000). It has been suggested that osmolytes may protect macromolecules during dehydration. An osmoprotectant effect of trehalose accumulation on biological membranes has recently been reported by Lee *et al.* (2003) after engineering the tobacco chloroplast or nuclear genomes with the yeast *TPS1* gene. As compared to nuclear transformation, chloroplast transformation prevents the escape of the engineered trait to other sexually-compatible species. The chloroplast transformant showed a much higher (169-fold) *TPS1* transcript and trehalose accumulation (25-fold higher) than the best surviving nuclear transgenic plant. When grown in PEG, transgenic chloroplast thylakoid membranes showed high integrity, whereas chloroplasts in untransformed plants were bleached. Following intense drying, chloroplast transgenic seedlings successfully rehydrated while control plants died. Although during dehydration there was no difference between control and transgenic plants in water loss, dehydrated leaves from transgenic plants recovered upon rehydration while control leaves dried out (Lee *et al.*, 2003). Additional evidence for an osmoprotectant role of osmolytes in transgenic plants has been provided by the work of Shen *et al.* (1997), who reported a positive effect of mannitol in the chloroplast as radical scavenger of reactive-oxygen species (ROS) that rapidly accumulate in response to different environmental stresses, causing widespread oxidative cellular damage (Pastori and Foyer, 2002). Indeed, one strategy to enhance tolerance to drought and other environmental stresses is to ectopically increase the capacity to scavange ROS (McKersie *et al.*, 1996; Van Camp *et al.*, 1996; Roxas *et al.*, 1997; Oberschall *et al.*, 2000). Although more work is needed to understand the actions and the interactions of the different detoxification enzymes during stress, it is conceivable that decreasing oxidative stress offers an additional strategy for providing protection against drought and other environmental stresses. Finally, an improved biomass productivity and WUE under water deficit conditions were reported in transgenic wheat grown under greenhouse conditions and constitutively expressing the barley *HVA1* gene, an ABA-responsive barley gene, a member of group 3 late embryogenesis abundant (LEA) protein genes (Sivamani *et al.*, 2000).

Signalling and regulatory genes. The overexpression of single genes acting at the end of the hierarchical cascade providing adaptation to drought is often insufficient to reach the required level of tolerance due to the multifactorial nature of the response to abiotic stresses. An alternative strategy to enhance stress tolerance relies on the simultaneous transcriptional activation of a subset of all the stress-induced genes. This approach can be pursued through the engineering of genes involved in the signal transduction pathway (Saijo *et al.*, 2000). Protein kinases and phosphatases play a key role in signalling processes (Smith and Walker, 1996; Mizoguchi *et al.*, 1997, 2000; Luan, 1998) through phosphorylation/dephosphorylation of proteins, such as transcription factors. In this context, the identification and characterization of kinases and phosphatases acting in response to specific stimuli is considered a very promising approach. In tobacco, the overexpression of the *At-DBF2* gene encoding a serine-threonine kinase enhanced the level of tolerance to drought, cold and heat tolerance due to a constitutive expression of several stress-responsive genes (Lee *et al.*, 1999). In rice, when a stress-induced gene encoding a calcium-dependent protein kinase (OsCDPK7) was expressed ectopically, an enhanced level of stress-responsive genes was reported after exposure to drought and salt but not to cold (Saijo *et al.*, 2000), an indication that the signalling mechanisms of salt/drought tolerance and cold tolerance differ, sharing OsCDPK7 as a common component. Importantly, no significant effects of overexpression on the phenotype were observed (Saijo *et al.*, 2000), an encouraging result from an applicative standpoint since the overexpression of regulatory genes has often caused severe growth retardation (Liu *et al.*, 1998; Gilmour *et al.*, 2000; Vannini *et al.*, 2003a). Due to the lack of noticeable pleiotropic effects of the overexpression of CDPK7, Saijo *et al.* (2000) suggested that this enzyme is normally kept inactive and only upon interaction with stress stimuli the downstream signalling is triggered. Subsequent work showed that in wild-type rice plants exposed to drought, OsCDPK7 was expressed predominantly in vascular tissues of crowns and roots, where water stress occurs most severely (Saijo *et al.*, 2001).

A promising avenue for a coordinated ectopic expression of genes influencing the level of tolerance to abiotic stresses is offered by the manipulation of the *CBF/DREB* genes (Shinozaki and Yamaguchi-Shinozaki, 1996; Thomashow, 1999; Seki *et al.*, 2003). These genes encode for transcription factors which, in turn, bind to DNA motifs present in the promoter region of downstream stress-induced genes (e.g., LEA genes) whose product mitigate the negative effects of dehydration. Although the engineering of *CBF/DREB* genes was originally targeted to increase freezing tolerance, it was observed that also drought tolerance can be similarly enhanced (Shinozaki and Yamaguchi-Shinozaki, 2000). This is not surprising, since under both conditions dehydration is one of negative effects that the plant cell has to endure. Following freezing, ice formation in the extracellular spaces drives water from inside the cell to the apoplast, causing severe osmotic and dehydration stresses (Gilmour *et al.*, 2000; Liu *et al.*, 1998). The constitutive overexpression of *CFB1* (Jaglo-Ottosen *et al.*, 1998) or *DREB1A* (Liu *et al.*, 1998) in transgenic Arabidopsis significantly enhanced survival to cold and dehydration exposure. However, the constitutive high level of expression of transcription factors produced strong negative phenotypic effects, such as delayed flowering and short stature (Liu *et al.*, 1998). This negative pleiotropic effect was avoided by placing the *CBF/DREB* genes under control of the stress-regulated *rd29A* promoter, induced by cold and dehydration but not by ABA (Yamaguchi-Shinozaki and Shinozaki, 1994). Other stress-induced genes are expressed in the two ABA-independent pathways (Abe *et al.*, 1997). Indeed, engineered Arabidopsis plants expressing *CBF3* in a regulated way, showed a greatly improved appearance under control conditions, although they retained a slight growth retardation when compared to the wild type (Kasuga *et al.*, 1999; Seki *et al.*, 2001). In this way, tolerance genes are activated only when the stress event occurs, minimizing the negative pleiotropic side effect. In tomato, the overexpression of Arabidopsis *DREB1A* enhanced tolerance to cold, drought and oxidative stress, most likely through an overproduction of proline and catalase in transgenic plants (Hsieh *et al.*, 2002). Promising results have also been reported in bread wheat, where preliminary experiments have shown that plantlets engineered with the Arabidopsis *DREB1A* under the control of the *rd29* promoter survived a short and intensive water stress, while control plants were completely desiccated (Pellegrineschi *et al.*, 2001). Another promising gene encoding for a transcription factor capable of affecting tolerance to multiple stresses is the rice *Osmyb4*, whose expression is strongly enhanced by cold treatments, but not by ABA or other abiotic stresses (Pandolfi *et al.*, 1997). Preliminary evidence indicates that ectopic expression of *Osmyb4* in Arabidopsis greatly enhances survival to both biotic and abiotic stresses (Vannini *et al.*, 2003a,b; Immacolata Coraggio, personal communication).

The lack of a multidisciplinary approach and accurate field screening techniques coupled with an incomplete understanding of both the genetic basis of drought resistance and the complexity of the signalling pathways are major constraints to a more widespread and successful application of genetic engineering. Nonetheless, the examples herein reported clearly show the power of recombinant DNA technology to further our knowledge of the factors limiting WUE and yield. Although genetic engineering may entail more failures than successes and in many cases will not lead to applicable results, its role and contribution will be essential for facilitating future improvements in WUE and yield.

9.7 Conclusions

During the 1990s, the rate of population growth outpaced the rate of increase in food-grain production, thus indicating that unless this trend is offset by the release of cultivars with a higher yield potential and stability, food shortages will occur in the twenty-first century (Khush, 1999). The release of cultivars endowed with a higher yield potential and stability will largely depend on our ability to increase crops' water use and/or WUE. In the past decades, both empirical and analytical breeding have contributed to the improvement of seasonal WUE and yield of crops under both well-watered and water-limited conditions, mainly, but not exclusively, through better partitioning of photosynthates rather then an increased biomass production (Turner, 1997; Austin, 1999; Duvick and Cassman, 1999). When it has been possible to enhance biomass production, this has been mostly achieved through an increase in water extracted from the soil, rather than an increased intrinsic WUE. One notable exception is 'Drysdale' a cultivar recently released in Australia following a selection based on $\Delta^{13}C$ (CSIRO, 2002). In dry years, 'Drysdale' has a major advantage over comparable wheats, producing about 10% more grain despite receiving the same rainfall.

In a recent review, Passioura (2002) pointed out that the improvements in the management of health and nutrition achieved during the past decades have resulted in many crops now being more limited by water availability. Hence, a greater emphasis is being placed on the interactions between roots and the rhyzosphere, probably one of the most complex topics in plant biology and certainly one extremely relevant in the context of WUE, although still largely unexplored due to the difficulty in properly tackling this topic adopting conventional approaches. As compared to traditional breeding approaches, molecular genomics and genetic engineering allow us to identify and manipulate at will the single genes accounting for variation in traits affecting WUE and yield, ushering in new opportunities to further our understanding of the physiological cause-effect relationships governing such variation. Although the impact of molecular approaches on cultivar release has so far been negligible, QTL analysis and genetic engineering have both contributed to disentangle the genetic complexity governing plant adaptation to fluctuating environmental conditions and in some cases, almost paradoxically, to 'Mendelize' such complexity. The ectopic expression of single genes has already provided significant levels of tolerance to drought and other abiotic stresses in model species and crops grown under controlled conditions; these encouraging results await validation on a field scale, particularly in terms of yield. Mendelization of QTLs through the production of congenic strains should be regarded as a fundamental contribution toward a better understanding of the role of natural allelic variation in the adaptation of crops to environmental insults and provides an ideal ground for interdisciplinary research between different categories of plant scientists. QTL cloning is now a reality, although its applicability remains limited to only a handful of the myriad of QTLs affecting WUE and yield.

Even though certain aspects of molecular genomics may remain unfamiliar and intimidating, the genomic approach, when appropriately intersected with other relevant disciplines, will positively impact our understanding of plant metabolism and, eventually, our breeding activities (Morandini and Salamini, 2003). An interesting example is the successful attempt to conjugate physiological modelling and QTL information for predicting leaf elongation in maize irrespectively of fluctuating environmental factors whose effect can be accounted for by the model (Reymond *et al.*, 2003). Crop modelling provides a means to help resolving G × E interactions and to dissect yield into characters that might be under simpler genetic control (Asseng *et al.*, 2002, 2003). Further opportunities for collaboration between modellers and geneticists in ideotype breeding for high crop yield have been recently illustrated (Yin *et al.*, 2003).

As to the traits more amenable to future molecular approaches aimed at improving intrinsic and/or seasonal WUE, root architecture and photosynthetic efficiency will probably receive greater attention, in view of the difficulties of manipulating these traits through conventional approaches and also considering that biomass production can only be increased through a greater water use and/or a more efficient CO_2 fixation. Traits other than those herein surveyed may also provide meaningful contributions, should it be possible to devise appropriate screening techniques and identify sources of genetic variation suitable for QTL discovery and gene cloning. One example is offered by hydraulic lift (Caldwell *et al.*, 1998), a complex trait so far little explored and for which preliminary evidence has indicated the existence of genetic variability associated to drought tolerance in maize (Wan *et al.*, 2000). Mycorrhizal colonization is another complex trait whose manipulation may influence water use of crops; although QTLs for mycorrhizal responsiveness have been reported in maize (Kaeppler *et al.*, 2000), limited information is available on the genetic control of the interaction between crops' roots and mycorrhiza (Barker *et al.*, 2002).

On the molecular side, extensive EST databases and unigene sets derived from cDNA libraries of different drought-stressed tissues and organs are valuable sources of markers to construct functional maps that could more directly lead to the identification of candidates for QTLs (Cushman and Bohnert, 2000). High throughput genomic profiling based on the detection of SNPs (Single Nucleotide Polymorphisms; Kanazin *et al.*, 2002; Rafalski, 2002; Batley *et al.*, 2003) and insertion-deletion polymorphisms (Bhattramakki *et al.*, 2002) and other more advanced enabling platforms (McCallum *et al.*, 2000; Uetz, 2002; Hardenbol *et al.*, 2003) will greatly improve our capacity for allele mining and high-throughput profiling of the genome, thus facilitating QTL discovery and cloning procedures. Data mining, comparative genomics and bioinformatics already provide a powerful, albeit imperfect, operational framework to deploy the available sequence information (Kantety *et al.*, 2002; Sorrells *et al.*, 2003). From an applicative standpoint, it is important to underline that the new technological platforms and approaches should be utilized in the right biological context, preferably applicable also at the level of plant community grown in farmers' fields. In this 'DNAge', the real challenge faced by plant scientists is how to best and most effectively integrate into extant breeding programs the materials and the deluge of information generated through the innovative techniques and approaches herein presented. This integration will further our understanding of the physiology and genetics of both WUE and yield potential, and will enhance our ability to forge new improved varieties for a more sustainable agriculture.

References

Abe, H., Yamaguchi-Shinozaki, K., Urao, T., Iwasaki, T., Hosakawa, D. and Shinozaki, K. (1997) Role of Arabidopsis MYC and MYB homologs in drought- and abscisic acid-regulated gene expression. *Plant Cell*, **9**, 1859–1868.

Acevedo, E. (1993) Potential of carbon isotope discrimination as a selection criterion in barley breeding, in *Stable isotopes and plant carbon-water relations* (eds. J.R. Ehleringer, A.E. Hall and G.D. Farquhar), Academic Press, New York, pp. 399–416.

Akhtar, M.S., Goldschmidt, E.E., John, I., Rodoni, S., Matile, P. and Grierson, D. (1999) Altered patterns of senescence and ripening in gf, a stay green green mutant of tomato (*Lycopersicon esculentum* Mill.). *Journal of Experimental Botany*, **50**, 1115–1122.

Ali, M.L., Pathan, M.S., Zhang, J., Bai, G., Sarkarung, S. and Nguyen, H.T. (2000) Mapping QTLs for root traits in a recombinant inbred population from two indica ecotypes in rice. *Theoretical and Applied Genetics*, **101**, 756–766.

Araus, J.L., Brown, H.R., Febrero, A., Bort, J. and Serret, M.D. (1993) Ear photosynthesis, carbon isotope discrimination and the contribution of respiratory CO_2 to differences in grain mass in durum. *Plant, Cell and Environment*, **16**, 383–392.

Araus, J.L., Amaro, T., Zuhair, Y. and Nachit, M.M. (1997) Effect of leaf structure and water status on carbon isotope discrimination in field-grown durum wheat. *Plant, Cell and Environment*, **20**, 1484–1494.

Araus, J.L., Slafer, G.A., Reynolds, M.P. and Royo, C. (2002) Plant breeding and drought in C_3 cereals: what should we breed for? *Annals of Botany*, **89**, 925–940.

Asins, M.J. (2002) Present and future of quantitative trait locus analysis in plant breeding. *Plant Breeding*, **121**, 281–291.

Asseng, S., Turner, N., Ray, J. and Keating, B. (2002) A simulation analysis that predicts the influence of physiological traits on the potential yield of wheat. *European Journal of Agronomy*, **17**, 123–141.

Asseng, S., Turner, N.C., Botwright, Y. and Condon, A.G. (2003) Evaluating the impact of a trait for increased specific leaf area on wheat yields using a crop simulation model. *Agronomy Journal*, **95**, 10–19.

Atlin, G.N. and Lafitte, H.R. (2002) Marker-assisted breeding versus direct selection for drought tolerance in rice. *Field screening for drought tolerance in crop plants with emphasis on rice. Proceedings of an International Workshop on Field Screening for Drought Tolerance in Rice*, eds. N.P. Saxena and J.C. O'Toole, ICRISAT, Patancheru, India, Dec. 11–14, 2000, 71–81.

Austin, R. (1999) Yield of wheat in the United Kingdom: recent advances and prospects. *Crop Science*, **39**, 1604–1610.

Austin, R.B., Ford, M.A. and Morgan, C.L. (1989) Genetic improvement in the yield of winter wheat: a further evaluation. *Journal of Agricultural Science*, **112**, 295–301.

Azhiri-Sigari, T., Yamauchi, A., Kamoshita, A. and Wade, L.J. (2000) Genotypic variation in response of rainfed lowland rice to drought and rewatering. II. Root growth. *Plant Production Science*, **3**, 180–188.

Babu, R., Nguyen, B., Chamarerk, V., Shanmugasundaram, P., Chezhian, P., Jeyaprakash, P., Ganesh, S., Palchamy, A., Sadasivam, S., Sarkarung, S., Wade, L. and Nguyen, H. (2003) Genetic analysis of drought resistance in rice by molecular markers: Association between secondary traits and field performance. *Crop Science*, **43**, 1457–1469.

Bachem, C.W.B., van der Hoeven, R.S., de Bruijin, S.M., Vreugdenhil, D., Zabeau, M. and Visser, R.F. (1996) Visualization of differential gene expression using a novel method of RNA fingerprinting based on AFLP: analysis of gene expression during potato tuber development. *Plant Journal*, **9**, 745–753.

Bacon, M. (1999) Biochemical control of leaf expansion. *Plant Growth Regulation*, **29**, 101–112.

Bacon, M.A., Davies, W.J., Mingo, D. and Wilkinson, S. (2003) Root signals, in *Roots: the hidden half* (eds. Y. Waisel, A. Eshel and U. Kafkafi), Marcel Dekker, Inc., New York, USA, pp. 460–471.

Bajaj, S., Targolli, J., Liu, L.F., Ho, T.H.D. and Wu, R. (1999) Transgenic approaches to increase dehydration-stress tolerance in plants. *Molecular Breeding*, **5**, 493–503.

Barbour, M.M., Fischer, R.A., Sayre, K.D. and Farquhar, G.D. (2000) Oxygen isotope ratio of leaf and grain material correlates with stomatal conductance and grain yield in irrigated wheat. *Australian Journal of Plant Physiology*, **27**, 625–637.

Barker, S.J., Duplessis, S. and Tagu, D. (2002) The application of genetic approaches for investigations of mycorrhizal symbioses. *Plant and Soil*, **244**, 85–95.

Barrière, Y., Gibelin, C., Argillier, O. and Méchin, V. (2001) Genetics analysis in recombinant inbred lines of early dent forage maize. I – QTL mapping for yield, earliness, starch and crude protein contents from per se value and top cross experiments. *Maydica*, **46**, 253–266.

Bartels, D. (2001) Targeting detoxification pathways: An efficient approach to obtain plants with multiple stress tolerance? *Trends in Plant Science*, **6**, 284–286.

Bartels, D. and Salamini, F. (2001) Desiccation tolerance in the resurrection plant *Craterostigma plantagineum*. A contribution to the study of drought tolerance at the molecular level. *Plant Physiology*, **127**, 1346–1353.

Batley, J., Mogg, R., Edwards, D., O'Sullivan, H. and Edwards, K.J. (2003) A high-throughput SNuPE assay for genotyping SNPs in the flanking regions of *Zea mays* sequence tagged simple sequence repeats. *Molecular Breeding*, **11**, 111–120.

Beavis, W.D. (1998) QTL analysis: power, precision, and accuracy, in *Molecular dissection of complex traits* (ed. A.H. Paterson), CRC Press, Boca Raton, pp. 145–162.

Ben Haj Salah, H. and Tardieu, F. (1995) Temperature affects expansion rate of maize leaves without change in spatial distribution of cell length. Analysis of the coordination between cell division and cell expansion. *Plant Physiology*, **109**, 861–870.

Ben Haj Salah, H. and Tardieu, F. (1996) Quantitative analysis of the combined effects of temperature, evaporative demand and light on leaf elongation rate in well-watered field and laboratory-grown maize plants. *Journal of Experimental Botany*, **47**, 1689–1698.

Bernacchi, D., Beck-Bunn, T., Eshed, Y., Lopez, J., Petiard, V., Uhlig, J., Zamir, D. and Tanksley, S.D. (1998) Advanced backcross QTL analysis in tomato. Identification of QTLs for traits of agronomic importance from *Lycopersicon hirsutum*. *Theoretical and Applied Genetics*, **97**, 381–397.

Bhattramakki, D., Dolan, M., Hanafey, M., Wineland, R., Vaske, D., Register, J.C., Tingey, S.V. and Rafalski, A. (2002) Insertion-deletion polymorphisms in 3' regions of maize genes occur frequently and can be used as highly informative genetic markers. *Plant Molecular Biology*, **48**, 539–547.

Björkman, O., Nobs, M., Pearcy, R., Boynton, J. and Berry, J. (1971) Characteristics of hybrids between C_3 and C_4 species of *Atriplex*, in *Photosynthesis and photorespiration* (eds. M.D. Hatch, C.B. Osmond and R.O. Slayter), Wiley-Interscience, New York, pp. 105–119.

Blum, A. (1988) *Breeding for stress environments*, CRC Press, Boca Raton.

Blum, A. (1996) Crop responses to drought and the interpretation of adaptation. *Plant Growth Regulation*, **20**, 135–148.

Blum, A. (2002) Drought tolerance – is it a complex trait? *Field screening for drought tolerance in crop plants with emphasis on rice. Proceedings of an International Workshop on Field Screening for Drought Tolerance in Rice*, eds. N.P. Saxena and J.C. O'Toole, ICRISAT, Patancheru, India, Dec. 11–14, 2000, 17–22.

Blum, A., Munns, R., Passioura, J.B. and Turner, N. (1996) Genetically engineered plants resistant to soil drying and salt stress: How to interpret osmotic relations? *Plant Physiology*, **110**, 1051–1053.

Bohnert, H.J. and Bressan, R.A. (2001) Abiotic stresses, plant reactions, and approaches towards improving stress tolerance, in *Crop Science: Progress and Prospects* (ed. J. Nössberger), CABI International, Wallingford, pp. 81–100.

Bohnert, H.J. and Cushman, J.C. (2000) The ice plant cometh: Lessons in abiotic stress tolerance. *Journal of Plant Growth Regulation*, **19**, 334–346.

Bohnert, H. and Shen, B. (1999) Transformation and compatible solutes. *Scientia Horticulturae*, **78**, 237–260.

Bohnert, H.J. and Sheveleva, E. (1998) Plant stress adaptations, making metabolism move. *Current Opinion in Plant Biology*, **1**, 267–274.

Bohnert, H.J., Nelson, E.D. and Jensen, R.G. (1995) Adaptations to environmental stresses. *Plant Cell*, **7**, 1099–1111.

Bolaños, J. and Edmeades, G.O. (1993) Eight cycles of selection for drought tolerance in lowland tropical maize. I. Responses in grain yield, biomass, and radiation utilization. *Field Crops Research*, **31**, 233–252.

Bolaños, J. and Edmeades, G. (1996) The importance of anthesis-silking interval in breeding for drought tolerance in tropical maize. *Field Crops Research*, **48**, 65–80.

Bolaños, J., Edmeades, G. and Martinez, L. (1993) Eight cycles of selection for drought tolerance in lowland tropical maize. III. Responses in drought adaptive physiological and morphological traits. *Field Crops Research*, **31**, 269–286.

Borel, C., Simonneau, T., This, D. and Tardieu, F. (1997) Stomatal conductance and ABA concentration in the xylem sap of barley lines of contrasting genetic origins. *Australian Journal of Plant Physiology*, **24**, 607–615.

Borel, C., Audran, C., Frey, A., Marion Poll, A., Tardieu, F. and Simonneau, T. (2001a) N-plumbaginifolia zeaxanthin epoxidase transgenic lines have unaltered baseline ABA accumulations in roots and xylem sap, but contrasting sensitivities of ABA accumulation to water deficit. *Journal of Experimental Botany*, **52**, 427–434.

Borel, C., Frey, A., Marion, P.A., Tardieu, F. and Simonneau, T. (2001b) Does engineering abscisic acid biosynthesis in *Nicotiana plumbaginifolia* modify stomatal response to drought? *Plant Cell and Environment*, **24**, 477–489.

Bowers, J.E., Chapman, B.A., Rong, J.K. and Paterson, A.H. (2003) Unravelling angiosperm genome evolution by phylogenetic analysis of chromosomal duplication events. *Nature*, **422**, 433–438.

Boyer, J.S. (1982) Plant productivity and the environment. *Science*, **218**, 443–448.

Boyko, E., Kalendar, R., Korzun, V., Fellers, J., Korol, A., Schulman, A.H. and Gill, B.S. (2002) A high-density cytogenetic map of the *Aegilops tauschii* genome incorporating retrotransposons and defense-related genes: insights into cereal chromosome structure and function. *Plant Molecular Biology*, **48**, 767–790.

Bray, E.A. (1997) Plant responses to water deficit. *Trends in Plant Science*, **2**, 48–54.

Bray, E.A. (2002a) Abscisic acid regulation of gene expression during water-deficit stress in the era of the *Arabidopsis* genome. *Plant Cell Environment*, **25**, 153–161.

Bray, E.A. (2002b) Classification of genes differentially expressed during water-deficit stress in *Arabidopsis thaliana*: an analysis using microarray and differential expression data. *Annals of Botany*, **89**, 803–811.

Brenner, S., Williams, S.R., Vermaas, E.H., Storck, T., Moon, K., McCollum, C., Mao, J., Luo, S., Kirchner, J.J., Eletr, S., Robert, B., DuBridge, R.B., Burcham, T. and Albrecht, G. (2000) In vitro cloning of complex mixtures of mcrobeads: physical separation of differentially expressed cDNAs. *Proceedings of the National Academy of Science, USA*, **97**, 1665–1670.

Broman, K.W., Wu, H., Sen, S. and Churchill, G.A. (2003) QTL mapping in experimental crosses. *Bioinformatics*, **19**, 889–890.

Bruce, W., Desbons, P., Crast, O. and Folkerts, O. (2001) Gene expression profiling of two related maize inbred lines with contrasting root-lodging traits. *Journal of Experimental Botany*, **52**, 459–468.

Bruce, W.B., Edmeades, G.O. and Barker, T.C. (2002) Molecular and physiological approaches to maize improvement for drought tolerance. *Journal of Experimental Botany*, **53**, 13–25.

Buckler, E.S.I. and Thornsberry, J.M. (2002) Plant molecular diversity and applications to genomics. *Current Opinion in Plant Biology*, **5**, 107–111.

Caldwell, M.M., Dawson, T.E. and Richards, J.H. (1998) Hydraulic lift: consequences of water efflux from the roots of plants. *Oecologia*, **113**, 151–161.

Cattivelli, L., Baldi, P., Crosatti, C., Di Fonzo, N., Faccioli, P., Grossi, M., Mastrangelo, A.M., Pecchioni, N. and Stanca, A.M. (2002) Chromosome regions and stress-related sequences involved in resistance to abiotic stress in *Triticeae*. *Plant Molecular Biology*, **48**, 649–665.

Causse, M., Rocher, J.P. and Henry, A.M. (1995) Genetic detection of the relationship between carbon metabolism and early growth in maize with emphasis on key-enzyme loci. *Molecular Breeding*, **1**, 259–272.

Ceccarelli, S. (1984) Plant responses to water stress: a review. *Genetica Agraria*, **38**, 43–73.

Champoux, M.C., Wang, G., Sarkarung, S., Mackill, D.J., O'Toole, J.C., Huang, N. and McCouch, S.R. (1995) Locating genes associated with root morphology and drought avoidance in rice via linkage to molecular markers. *Theoretical and Applied Genetics*, **90**, 969–981.

Chen, T.H. and Murata, N. (2002) Enhancement of tolerance of abiotic stress by metabolic engineering of betaines and other compatible solutes. *Current Opinion in Plant Biology*, **5**, 250–257.

Chen, W., Provart, N.J., Glazebrook, J., Katagiri, F., Chang, H.S., Eulgem, T., Mauch, F., Luan, S., Zou, G., Whitham, S.A., Budworth, P.R., Tao, Y., Xie, Z., Chen, X., Lam, S., Kreps, J.A., Harper, J.F., Si-Ammour, A., Mauch-Mani, B., Heinlein, M., Kobayashi, K., Hohn, T., Dangl, J.L., Wang, X. and Zhu, T. (2002) Expression profile matrix of Arabidopsis transcription factor genes suggests their putative functions in response to environmental stresses. *Plant Cell*, **14**, 559–574.

Churchill, G.A. and Doerge, R.W. (1994) Empirical threshold values for quantitative trait mapping. *Genetics*, **138**, 963–971.

Close, T.J. (1997) Dehydrins: a commonality in the response of plants to dehydration and low temperature. *Physiologia Plantarum*, **100**, 291–296.

Condon, A.G., Farquar, G.D. and Richards, R.A. (1990) Genotypic variation in carbon isotope discrimination and transpiration efficiency in wheat. Leaf gas exchange and whole plant studies. *Australian Journal of Plant Physiology*, **17**, 9–22.

Condon, A.G., Richards, R.A. and Farquhar, G.D. (1993) Relationships between carbon isotope discrimination, water use efficiency and transpiration efficiency for dryland wheat. *Australian Journal of Agricultural Research*, **44**, 1693–1711.

Condon, A., Richards, R., Rebetzke, G. and Farquhar, G. (2002) Improving intrinsic water-use efficiency and crop yield. *Crop Science*, **42**, 122–131.

Consoli, L., Lefévre, C.L., Zivy, M., de Vienne, D. and Damerval, C. (2002) QTL analysis of proteome and transcriptome variations for dissecting the genetic architecture of complex traits in maize. *Plant Molecular Biology*, **48**, 575–581.

Conti, S., Frascaroli, E., Gherardi, F., Landi, P., Sanguineti, M.C., Stefanelli, S. and Tuberosa, R. (1994) Accumulation of and response to abscisic acid in maize. *Proceedings of the XVI Conference of the Eucarpia, Maize and Sorghum Section*, eds. A. Bianchi *et al.*, Bergamo, Italy, June 6–9, 1993, 212–223.

Courtois, B., McLaren, G., Sinha, P.K., Prasad, K., Yadav, R. and Shen, L. (2000) Mapping QTLs associated with drought avoidance in upland rice. *Molecular Breeding*, **6**, 55–66.

Crasta, O.R., Xu, W.W., Rosenow, D.T., Mullet, J. and Nguyen, H.T. (1999) Mapping of post flowering drought resistance traits in grain sorghum: association between QTLs influencing premature senescence and maturity. *Molecular General Genetics*, **262**, 579–588.

Crouch, J.H. and Serraj, R. (2002) DNA marker technology as a tool for genetic enhancement of drought tolerance at ICRISAT. *Field screening for drought tolerance in crop plants with emphasis on rice. Proceedings of an International Workshop on Field Screening for Drought Tolerance in Rice*, cds. N.P. Saxcna and J.C. O'Toolc, ICRISAT, Patancheru, India, Dec. 11–14, 2000, 155–170.

Cushman, J.C. and Bohnert, H.J. (2000) Genomic approaches to plant stress tolerance. *Current Opinions in Plant Biology*, **3**, 117–124.

Damerval, C., Maurice, A., Josse, J.M. and de Vienne, D. (1994) Quantitative trait loci underlying gene product variation: a novel perspective for analyzing regulation of genome expression. *Genetics*, **137**, 289–301.

Davis, G.L., McMullen, M.D., Baysdorfer, C., Musket, T., Grant, D., Staebell, M., Xu, G., Polacco, M., Koster, L., Melia, Hancock, S., Houchins, K., Chao, S. and Coe, E.H.J. (1999) A maize map standard with sequenced core markers, grass genome reference points and 932 expressed sequence tagged sites (ESTs) in a 1736 locus map. *Genetics*, **152**, 1137–1172.

de Vienne, D., Leonardi, A., Damerval, C. and Zivy, M. (1999) Genetics of proteome variation for QTL characterization: application to drought-stress responses in maize. *Journal of Experimental Botany*, **50**, 303–309.

Devos, K.M., Beales, J., Nagamura, Y. and Sasaki, T. (1999) Arabidopsis-rice: will colinearity allow gene prediction across the eudicot-monocot divide? *Genome Research*, **9**, 825–829.

Dodd, I.C. and Davies, W.J. (1996) The relationship between leaf growth and ABA accumulation in the grass leaf elongation zone. *Plant Cell and Environment*, **19**, 1047–1056.

Duvick, D.N. and Cassman, K.G. (1999) Post-green revolution trends in yield potential of temperate maize in the north-central United States. *Crop Science*, **39**, 1622–1630.

Ehdaie, B. and Waines, J.G. (1994) Genetic analysis of carbon isotope discrimination and agronomic characters in a bread wheat cross. *Theoretical and Applied Genetics*, **88**, 1023–1028.

El Hafid, R., Smith, D.H., Karrou, M. and Samir, K. (1998) Physiological attributes associated with early-season drought resistance in spring durum wheat cultivars. *Canadian Journal of Plant Science*, **78**, 227–237.

Ellis, R.P., Forster, B.P., Robinson, D., Handley, L.L., Gordon, D.C., Russell, J. and Powell, W. (2000) Wild barley: a source of genes for crop improvement in the 21st century? *Journal of Experimental Botany*, **51**, 9–17.

Ellis, R.P., Forster, B.P., Gordon, D.C., Handley, L.L., Keith, R.P., Lawrence, P., Meyer, R., Powell, W., Robinson, D., Scrimgeour, C.M., Young, G.R. and Thomas, W.T.B. (2002) Phenotype/genotype associations for yield and salt tolerance in a mapping population segregating for two dwarfing genes. *Journal of Experimental Botany*, **53**, 1–14.

Falconer, D.S. (1981) *Introduction to quantitative genetics*, 2nd edn., Longman Inc., London.

Farquhar, G.D., Ehleringer, J.R. and Hubick, K.T. (1989) Carbon isotope discrimination and photosynthesis. *Annual Review of Plant Physiology and Plant Molecular Biology*, **40**, 503–537.

Fell, D.A. (2001) Beyond genomics. *Trends in Genetics*, **17**, 680–682.

Fiehn, O. (2002) Metabolomics – the link between genotypes and phenotypes. *Plant Molecular Biology*, **48**, 155–171.

Fischer, R.A., Rees, D., Sayre, K.D., Lu, Z.M., Condon, A.G. and Larque Saavedra, A. (1998) Wheat yield progress associated with higher stomatal conductance and photosynthetic rate, and cooler canopies. *Crop Science*, **38**, 1467–1475.

Forster, B.P., Ellis, R.P., Thomas, W.T.B., Newton, A.C., Tuberosa, R., This, D., El Enein, R.A., Bahri, M.H. and Ben Salem, M. (2000) The development and application of molecular markers for abiotic stress tolerance in barley. *Journal of Experimental Botany*, **51**, 19–27.

Foyer, C.H. and Noctor, G. (2000) Oxygen processing in photosynthesis: regulation and signalling. *New Phytologist*, **146**, 359–388.

Frascaroli, E. and Tuberosa, R. (1993) Effect of abscisic acid on pollen germination and tube growth of maize genotypes. *Plant Breeding*, **110**, 250–254.

Gale, M.D. and Devos, K.M. (1998) Plant comparative genetics after 10 years. *Science*, **282**, 656–659.

Gilmour, S.J., Sebolt, A.M., Salazar, M.P., Everard, J.D. and Thomashow, M.F. (2000) Overexpression of the Arabidopsis CBF3 transcriptional activator mimics multiple biochemical changes associated with cold acclimation. *Plant Physiology*, **124**, 1854–1865.

Gimeno, V., Fernandez-Martinez, J.M. and Fereres, E. (1989) Winter planting as a means of drought escape in sunflower. *Field Crops Research*, **22**, 307–316.

Goff, S.A.*et al.* (2002) A draft sequence of the rice genome (*Oryza sativa* L. ssp. *japonica*). *Science*, **296**, 92–100.

Gowda, M., Venu, R.C., Roopalakshmi, K., Sreerekha, M.V. and Kulkarni, R.S. (2003) Advances in rice breeding, genetics and genomics. *Molecular Breeding*, **11**, 337–352.

Greene, E.A., C.C., Taylor, N.E., Henikoff, J.G., Till, B.J., Reynolds, S.H., Enns, L.C., Burtner, C., Johnson, J.E., Odden, A.R., Comai, L. and Henikoff, S. (2003) Spectrum of chemically induced mutations from a large-scale reverse-genetic screen in Arabidopsis. *Genetics*, **164**, 731–740.

Hackett, C. (2002) Statistical methods for QTL mapping in cereals. *Plant Molecular Biology*, **48**, 585–599.

Hallauer, A.R. and Miranda Fo, J.B. (1988) *Quantitative genetics in maize breeding*, 2nd edn., Iowa State University Press, Ames.

Handley, L.L., Nevo, E., Raven, J.A., Martinez-Carrasco, R., Scrimgeour, C.M., Pakniyat, H. and Forster, B.P. (1994) Chromosome 4 controls potential water use efficiency ($\delta^{13}C$) in barley. *Journal of Experimental Botany*, **45**, 1661–1663.

Hardenbol, P., Baner, J., Jain, M., Nilsson, M., Namsaraev, E.A., Karlin-Neumann, G.A., Fakhrai-Rad, H., Ronaghi, M., Willis, T.D., Landegren, U. and Davis, R.W. (2003) Multiplexed genotyping with sequence-tagged molecular inversion probes. *Nature Biotechnology*, **21**, 673–678.

Hartung, W., Sauter, A. and Hose, E. (2002) Abscisic acid in the xylem: where does it come from, where does it go to? *Journal of Experimental Botany*, **53**, 27–32.

Hash, C.T., Folkertsma, R.T., Ramu, P., Reddy, B.V.S., Mahalakshmi, V., Sharma, H.C., Rattunde, H.F.W., Weltzien, E.R., Haussmann, B.I.G., Ferguson, M.E. and Crouch, J.H. (2003) Marker-assisted breeding across ICRISAT for terminal drought tolerance and resistance to shoot fly and striga in sorghum. *Abstracts of the Congress 'In the wake of the double helix from green revolution to the gene revolution'*, Bologna, Italy, May 27–31, 2003, 81.

Hausler, R.E., Hirsch, H.J., Kreuzaler, F. and Peterhansel, C. (2002) Overexpression of C_4-cycle enzymes in transgenic C_3 plants: a biotechnological approach to improve C_3-photosynthesis. *Journal of Experimental Botany*, **53**, 591–607.

Hayes, P.M., Blake, T., Chen, T.H., Tragoonrung, S., Chen, F., Pan, A. and Liu, B. (1993) Quantitative trait loci on barley (*Hordeum vulgare* L.) chromosome 7 associated with components of winterhardiness. *Genome*, **36**, 66–71.

Hazen, S.P. and Kay, S.A. (2003) Gene arrays are not just for measuring gene expression. *Trends in Plant Science*, in press.

Hazen, S.P., Wu, Y.J. and Kreps, J.A. (2003) Gene expression profiling of plant responses to abiotic stress. *Funct. Integr. Genomics*, **3**, 105–111.

Herve, D., Fabre, F., Berrios, E.F., Leroux, N., Al Chaarani, G., Planchon, C., Sarrafi, A. and Gentzbittel, L. (2001) QTL analysis of photosynthesis and water status traits in sunflower (*Helianthus annuus* L.) under greenhouse conditions. *Journal of Experimental Botany*, **52**, 1857–1864.

Hirel, B., Bertin, P., Quilleré, I., Bourdoncle, W., Attagnant, C., Dellay, C., Gouy, A., Cadiou, S., Retailliau, C., Falque, M. and Gallais, A. (2001) Towards a better understanding of the genetic and physiological basis for nitrogen use efficiency in maize. *Plant Physiology*, **125**, 1258–1270.

Holmberg, N. and Bülow, L. (1998) Improving stress tolerance in plants by gene transfer. *Trends in Plant Science*, **3**, 61–66.

Holmstrom, K.O., Mantyla, E., Welin, B., Mandal, A. and Palva, E.T. (1996) Drought tolerance in tobacco. *Nature*, **379**, 683.

Hong, Z., Lakkineni, K., Zhang, Z. and Verma, D.P.S. (2000) Removal of feedback inhibition of DELTA-pyrroline-5–carboxylate synthetase results in increased proline accumulation and protection from osmotic stress. *Plant Physiology*, **122**, 1129–1136.

Hose, E., Clarkson, D.T., Steudle, E., Schreiber, L. and Hartung, W. (2001) The exodermis: a variable apoplastic barrier. *Journal of Experimental Botany*, **52**, 2245–2264.

Hose, E., Steudle, E. and Hartung, W. (2000) Abscisic acid and hydraulic conductivity of maize roots: a study using cell and root pressure probes. *Planta*, **211**, 874–882.

Hospital, R. and Charcosset, A. (1997) Marker-assisted introgression of quantitative trait loci. *Genetics*, **147**, 1469–1485.

Hoth, S., Morgante, M., Sanchez, J.-P., Hanafey, M.K., Tingey, S.V. and Chua, N.H. (2002) Genome-wide gene expression profiling in *Arabidopsis thaliana* reveals new targets of abscisic acid and largely impaired gene regulation in the abi1–1 mutant. *Journal of Cell Science*, **115**, 1–10.

Hsieh, T.H., Lee, J.T., Yang, P.T., Chiu, L.H., Charng, Y.Y., Wang, Y.C. and Chan, M.T. (2002) Heterology expression of the Arabidopsis C-repeat/Dehydration Response Element Binding factors 1 gene confers elevated tolerance to chilling and oxidative stresses in transgenic tomato. *Plant Physiology*, **129**, 1086–1094.

Hwang, J.U. and Lee, Y. (2001) Abscisic acid-induced actin reorganization in guard cells of dayflower is mediated by cytosolic calcium levels and by protein kinase and protein phosphatase activities. *Plant Physiology*, **125**, 2120–2128.

Innes, P., Blackwell, R.D. and Quarrie, S.A. (1984) Some effects of genetic variation in drought-induced abscisic acid accumulation on the yield and water use of spring wheat. *Journal of Agricultural Science*, **102**, 341–351.

WATER USE EFFICIENCY IN PLANT BIOLOGY

Ishimaru, K., Shirota, K., Iliga, M. and Kawamitsu, Y. (2001) Identification of quantitative trait loci for adaxial and abaxial stomatal frequencies in *Oryza sativa*. *Plant Physiology and Biochemistry*, **39**, 173–177.

Ito, O., O' Toole, J. and Hardy, B. (1999) Genetic improvement of rice for water-limited environments. *Workshop on Genetic Improvement of Rice for Water-Limited Environment*, Los Baños, Philippines, International Rice Research Institute, Dec. 1–3, 1998.

Ivandic, V., Hackett, C.A., Zhang, Z.J., Staub, J.E., Nevo, E., Thomas, W.T.B. and Forster, B.P. (2000) Phenotypic responses of wild barley to experimentally imposed water stress. *Journal of Experimental Botany*, **51**, 2021–2029.

Jaglo Ottosen, K.R., Gilmour, S.J., Zarka, D.G., Schabenberger, O. and Thomashow, M.F. (1998) Arabidopsis CBF1 overexpression induces COR genes and enhances freezing tolerance. *Science*, **280**, 104–106.

Jansen, R.C. and Stam, P. (1994) High resolution of quantitative traits into multiple loci via interval mapping. *Genetics*, **136**, 1447–1455.

Jeanneau, M., Gerentesb, D., Foueillassarc, X., Zivyd, M., Vidala, J., Toppand, A. and Perez, P. (2002a) Improvement of drought tolerance in maize: towards the functional validation of the *Zm-Asr1* gene and increase of water use efficiency by over-expressing C4PEPC. *Biochimie*, **84**, 1127–1135.

Jeanneau, M., Vidal, J., Gousset-Dupont, A., Lebouteiller, B., Hodges, M., Gerentes, D. and Perez, P. (2002b) Manipulating PEPC levels in plants. *Journal of Experimental Botany*, **53**, 1837–1845.

Jensen, A.B., Busk, P.K., Figueras, M., Albà, M.M., Peracchia, G., Messeguer, R., Goday, A. and Pagès, M. (1996) Drought signal transduction in plants. *Plant Growth Regulation*, **20**, 105–110.

Johnson, W.C., Jackson, L.E., Ochoa, O., Peleman, J., St Clair, D.A., Michelmore, R.W. and van Wijk, R. (2000) Lettuce, a shallow-rooted crop, and *Lactuca serriola*, its wild progenitor, differ at QTL determining root architecture and deep soil water exploitation. *Theoretical and Applied Genetics*, **101**, 1066–1073.

Jones, H.G., Archer, N., Rotenberg, E. and Casa, R. (2003) Radiation measurement for plant ecophysiology. *Journal of Experimental Botany*, **54**, 879–889.

Jongdee, B., Fukai, S. and Cooper, M. (2002) Leaf water potential and osmotic adjustment as physiological traits to improve drought tolerance in rice. *Field Crops Research*, **76**, 153–163.

Kaeppler, S.M., Parke, J.L., Mueller, S.M., Senior, L., Stuber, C. and Tracy, W.F. (2000) Variation among maize inbred lines and detection of quantitative trait loci for growth at low phosphorus and responsiveness to arbuscular mycorrhizal fungi. *Crop Science*, **40**, 358–364.

Kamoshita, A., Wade, L.J., Ali, M.L., Pathan, M.S., Zhang, J., Sarkarung, S. and Nguyen, H.T. (2002) Mapping QTLs for root morphology of a rice population adapted to rainfed lowland conditions. *Theoretical and Applied Genetics*, **104**, 880–893.

Kanazin, V., Talbert, H., See, D., DeCamp, P., Nevo, E. and Blake, T. (2002) Discovery and assay of single-nucleotide polymorphisms in barley (*Hordeum vulgare*). *Plant Molecular Biology*, **48**, 529–537.

Kantety, R.V., La Rota, M., Matthews, D.E. and Sorrells, M.E. (2002) Data mining for simple sequence repeats in expressed sequence tags from barley, maize, rice, sorghum and wheat. *Plant Molecular Biology*, **48**, 501–510.

Kasuga, M., Liu, Q., Miura, S., Yamaguchi-Shinozaki, K. and Shinozaki, K. (1999) Improving plant drought, salt, and freezing tolerance by gene transfer of a single stress-inducible transcription factor. *Nature Biotechnology*, **17**, 287–291.

Khush, G.S. (1999) Green revolution: preparing for the 21st century. *Genome*, **42**, 646–655.

Klein, M., Perfus-Barbeoch, L., Frelet, A., Gaedeke, N., Reinhardt, D., Mueller-Roeber, B., Martinoia, E. and Forestier, C. (2003) The plant multidrug resistance ABC transporter AtMRP5 is involved in guard cell hormonal signalling and water use. *Plant Journal*, **33**, 119–129.

Kojima, S., Takahashi, Y., Kobayashi, Y., Monna, L., Sasaki, T., Araki, T. and Yano, M. (2002) *Hd3a*, a rice orthologue of the Arabidopsis FT gene, promotes transition to flowering downstream of *Hd1* under short-day conditions. *Plant and Cell Physiology*, **43**, 1096–1105.

Kondo, M., Pablico, P., Aragones, D., Agsibit, R., Abe, J., Morita, S. and Courtois, B. (2003)

Genotypic and environmental variations in root morphology in rice genotypes under upland field conditions. *Plant and Soil*, in press.

Koumproglou, R., Wilkes, T.M., Townson, P., Wang, X.Y., Beynon, J., Pooni, H.S., Newbury, H.J. and Kearsey, M.J. (2002) STAIRS: a new genetic resource for functional genomic studies of Arabidopsis. *Plant Journal*, **31**, 355–364.

Kreps, J.A., Wu, Y.J., Chang, H.S., Zhu, T., Wang, X. and Harper, J.F. (2002) Transcriptome changes for Arabidopsis in response to salt, osmotic, and cold stress. *Plant Physiology*, **130**, 2129–2141.

Ku, M., Kano-Murakami, Y. and Matsuoka, M. (1996) Evolution and expression of C_4 photosynthesis genes. *Plant Physiology*, **111**, 949–957.

Ku, M.S.B., Agarie, S., Nomura, M., Fukayama, H., Tsuchida, H., Ono, K., Hirose, S., Toki, S., Miyao, M. and Matsuoka, M. (1999) High-level expression of maize phosphoenolpyruvate carboxylase in transgenic rice plants. *Nature Biotechnology*, **17**, 76–80.

Kulwal, P.L., Roy, J.K., Balyan, H.S. and Gupta, P.K. (2003) QTL mapping for growth and leaf characters in bread wheat. *Plant Science*, **164**, 267–277.

Kumar, A. and Singh, D.P. (1998) Use of physiological indices as a screening technique for drought tolerance in oilseed *Brassica* species. *Annals of Botany*, **81**, 413–420.

Kurata, N., Umehara, Y., Tanoue, H. and Sasaki, T. (1997) Physical mapping of the rice genome with YAC clones. *Plant Molecular Biology*, **35**, 101–113.

Lafitte, H.R., Courtois, B. and Arraudeau, M. (2002) Genetic improvement of rice in aerobic systems: progress from yield to genes. *Field Crop Research*, **75**, 171–190.

Landi, P., Sanguineti, M.C., Conti, S. and Tuberosa, R. (2001) Direct and correlated responses to divergent selection for leaf abscisic acid concentration in two maize populations. *Crop Science*, **41**, 335–344.

Landi, P., Salvi, S., Sanguineti, M.C., Stefanelli, S. and Tuberosa, R. (2002a) Development and preliminary evaluation of near-isogenic lines differing for a QTL which affects leaf ABA concentration. *Maize Genetics Cooperation Newsletter*, **76**, 7–8.

Landi, P., Sanguineti, M.C., Darrah, L.L., Giuliani, M.M., Salvi, S. and Tuberosa, R. (2002b) Detection of QTLs for vertical root pulling resistance in maize and overlaps with QTLs for root traits in hydroponics and for grain yield at different water regimes. *Maydica*, **47**, 233–243.

Laporte, M.M., Shen, B. and Tarczynski, M.C. (2002) Engineering for drought avoidance: expression of maize NADP-malic enzyme in tobacco results in altered stomatal function. *Journal of Experimental Botany*, **53**, 699–705.

Laurie, D.A. (1997) Comparative genetics of flowering time. *Plant Molecular Biology*, **35**, 167–177.

Laurie, D.A., Pratchett, N., Bezant, J.H. and Snape, J.W. (1995) RFLP mapping of five major genes and eight QTL controlling flowering time in a winter × spring barley (*Hordeum vulgare* L.) cross. *Genome*, **38**, 575–585.

Lebreton, C., Lazic-Jancic, V., Steed, A., Pekic, S. and Quarrie, S.A. (1995) Identification of QTL for drought responses in maize and their use in testing causal relationships between traits. *Journal of Experimental Botany*, **46**, 853–865.

Lee, M. (1995) DNA markers and plant breeding programs. *Advances in Agronomy*, **55**, 265–344.

Lee, H.J., Hübel, A. and Scöffl, F. (1995) Derepression of the activity of genetically engineered heat shock factor causes constitutive synthesis of heat shock proteins and increased thermotolerance in transgenic Arabidopsis. *Plant Journal*, **8**, 603–612.

Lee, J.H., Van Montagu, M. and Verbruggen, N. (1999) A highly conserved kinase is an essential component for stress tolerance in yeast and plant cells. *Proceedings of the National Academy of Science USA*, **96**, 5873–5877.

Lee, S.B., Kwon, H.B., Kwon, S.J., Park, S.C., Jeong, M.J., Han, S.E., Byun, M.O. and Daniell, H. (2003) Accumulation of trehalose within transgenic chloroplasts confers drought tolerance. *Molecular Breeding*, **11**, 1–13.

Leegood, R.C. (2002) C_4 photosynthesis: principles of CO_2 concentration and prospects for its introduction into C_3 plants. *Journal of Experimental Botany*, **53**, 581–590.

Li, Z.K., Khush, G., Brar, D. and Hardy, B. (2001) QTL mapping in rice: a few critical considerations. *Fourth International Rice Genetics Symposium* (eds. G. Khush and D. Brar), Los Banos, Philippines, pp. 153–171.

Li, Z.K., Ito, O., O'Toole, J. and Hardy, B. (1999) Genetic improvement of rice for water limited environments, in *Genetic improvement of rice for water limited environments* (eds. O. Ito and J. O'Toole), pp. 157–172.

Liang, P. and Pardee, A.B. (1992) Differential display of eukaryotic messenger RNA by means of the polymerase chain reaction. *Science*, **257**, 967–970.

Lilley, J.M., Ludlow, M.M., McCouch, S.R. and O'Toole, J.C. (1996) Locating QTL for osmotic adjustment and dehydration tolerance in rice. *Journal of Experimental Botany*, **47**, 1427–1436.

Liu, B.H. (1998) *Statistical genomics: linkage, mapping and QTL analysis*, CRC Press, Boca Raton.

Liu, Q., Kasuga, M., Sakuma, Y., Abe, H., Miura, S., Yamaguchi-Shinozaki, K. and Shinozaki, K. (1998) Two transcriptional factors, DREB1 and DREB2, with an EREBP/AP2 DNA binding domain separate two cellular signal transduction pathways in drought- and low-temperature-responsive gene expression, respectively in Arabidopsis. *Plant Cell*, **10**, 1391–1406.

López-Castañeda, C., Richards, R.A., Farquhar, G.D. and Williamson, R.E. (1996) Seed and seedling characteristics contributing to variation in early vigour between wheat and barley. *Crop Science*, **36**, 1257–1266.

Lu, Z., Percy, R.G., Qualset, C.O. and Zeiger, E. (1998) Stomatal conductance predicts yields in irrigated Pima cotton and bread wheat grown at high temperatures. *Journal of Experimental Botany*, **49**, 543–560.

Luan, S. (1998) Protein phosphatases and signaling cascades in higher plants. *Trends in Plant Science*, **3**, 271–275.

Ludlow, M.M. and Muchow, R.C. (1990) A critical evaluation of traits for improving crop yields in water-limited environments. *Advances in Agronomy*, **43**, 107–153.

Mackill, D.J., Nguyen, H.T. and Zhang, J.X. (1999) Use of molecular markers in plant improvement programs for rainfed lowland rice. *Field Crops Research*, **64**, 177–185.

Maggio, A., Hasegawa, P.M., Bressan, R.A., Consiglio, M.F. and Joly, R.J. (2001) Unravelling the functional relationship between root anatomy and stress tolerance. *Australian Journal of Plant Physiology*, **28**, 999–1004.

Martin, B., Nienhuis, J., King, G. and Schaefer, A. (1989) Restriction fragment length polymorphisms associated with water use efficiency in tomato. *Science*, **243**, 1725–1728.

Matsumura, H., Nirasawa, S. and Terauchi, R. (1999) Transcript profiling in rice (*Oryza sativa* L.) seedling using serial analysis of gene expression (SAGE). *The Plant Journal*, **20**, 719–726.

Matsuoka, M., Fukayama, H., Tsuchida, H., Nomura, M., Agarie, S., Ku, M.S.B. and Miyao, M. (2000) How to express some C_4 photosynthesis genes at high levels in rice, in *Redesigning rice photosynthesis to increase yield*, eds. J. Sheehy, P. Mitchell and B. Hardy, Elsevier Science Publishers, Amsterdam, pp. 167–175.

Matsuoka, M., Furbank, R.T., Fukayama, H. and Miyao, M. (2001) Molecular engineering of C_4 photosynthesis. *Annual Review of Plant Physiology and Plant Molecular Biology*, **52**, 297–314.

Matus, A., Slinkard, A.E. and Van Kassel, C. (1996) Carbon isotope discrimination and indirect selection for transpiration efficiency at flowering in lentil (*Lens culinaris* Medikus), spring wheat (*Triticum aestivum* L.), durum wheat (*T. turgidum* L.) and canola (*Brassica napus* L.). *Euphytica*, **87**, 141–151.

Maurel, C. and Chrispeels, M.J. (2001) Aquaporins: a molecular entry into plant water relations. *Plant Physiology*, **125**, 135–138.

McCallum, C.M., Comai, L., Greene, E.A. and Henikoff, S. (2000) Targeted screening for induced mutations. *Nature Biotechnology*, **18**, 455–457.

McDonald, A.J.S. and Davies, W.J. (1996) Keeping in touch: responses of the whole plant to deficits in water and nitrogen supply. *Advances in Botanical Research*, **22**, 229–300.

McKay, J.K., Richards, J.H. and Mitchell-Olds, T. (2003) Genetics of drought adaptation in *Arabidopsis thaliana*: I. Pleiotropy contributes to genetic correlations among ecological traits. *Molecular Ecology*, **12**, 1137–1151.

McKersie, B.D., Bowley, S.R. and Jones, K.S. (1996) Winter survival of transgenic alfalfa overexpressing superoxide dismustase. *Plant Physiology*, **119**, 839–848.

Mei, H.W., Luo, L.J., Ying, C.S., Wang, Y.P., Yu, X.Q., Guo, L.B., Paterson, A.H. and Li, Z.K. (2003) Gene actions of QTLs affecting several agronomic traits resolved in a recombinant inbred rice population and two testcross populations. *Theoretical and Applied Genetics*, **107**, 89–101.

Menz, M.A., Klein, R.R., Mullet, J.E., Obert, J.A., Unruh, N.C. and Klein, P.E. (2002) A high density genetic map of *Sorghum bicolor* (L.) Moench based on 2926 AFLP, RFLP and SSR markers. *Plant Molecular Biology*, **48**, 483–499.

Merah, O., Deléens, E. and Monneveux, P. (2001) Relationships between carbon isotope discrimination, dry matter production, and harvest index in durum wheat. *Journal of Plant Physiology*, **158**, 723–729.

Meyre, D., Leonardi, A., Brisson, G. and Vartanian, N. (2001) Drought-adaptive mechanisms involved in the escape/tolerance strategies of *Arabidopsis* Landsberg *erecta* and Columbia ecotypes and their F1 reciprocal progeny. *Journal of Plant Physiology*, **158**, 1145–1152.

Mian, M.A.R., Bailey, M.A., Ashley, D.A., Wells, R., Carter, T.J., Parrott, W.A. and Boerma, H.R. (1996) Molecular markers associated with water use efficiency and leaf ash in soybean. *Crop Science*, **36**, 1252–1257.

Mian, M.A.R., Ashley, D.A. and Boerma, H.R. (1998) An additional QTL for water use efficiency in soybean. *Crop Science*, **38**, 390–393.

Michelmore, R.W., Paran, I. and Kesseli, R.V. (1991) Identification of markers linked to disease resistance genes by bulked segregant analysis: a rapid method to detect markers in specific genomic regions by using segregating population. *Proceedings of the National Academy of Sciences USA*, **88**, 9828–9832.

Miflin, B. (2000) Crop improvement in the 21st century. *Journal of Experimental Botany*, **51**, 1–8.

Milborrow, B.W. (2001) The pathway of biosynthesis of abscisic acid in vascular plants: a review of the present state of knowledge of ABA biosynthesis. *Journal of Experimental Botany*, **52**, 1145–1164.

Mitra, J. (2001) Genetics and genetic improvement of drought resistance in crop plants. *Current Science*, **80**, 758–763.

Miyao, M. (2003) Molecular evolution and genetic engineering of C_4 photosynthetic enzymes. *Journal of Experimental Botany*, **54**, 179–189.

Mizoguchi, T., Ichimura, K. and Shinozaki, K. (1997) Environmental stress response in plants: The role of mitogen-activated protein kinases. *Trends in Biotechnology*, **15**, 15–19.

Mizoguchi, T., Ichimura, K., Yoshida, R. and Shinozaki, K. (2000) MAP Kinase cascade in Arabidopsis: their role in stress and hormone response. *Results Probl. Cell Differ.*, **27**, 29–38.

Mohamed, M.F., Keutgen, N., Tawfik, A.A. and Noga, G. (2002) Dehydration-avoidance responses of tepary bean lines differing in drought resistance. *Journal of Plant Physiology*, **159**, 31–38.

Moncada, P., Martinez, C., Borrero, J.M.C., Gauch, H.J., Guimaraes, E., Tohme, J. and McCouch, S. (2001) Quantitative trait loci for yield and yield components in an *Oryza sativa Oryza rufipogon* BC_2F_2 population evaluated in an upland environment. *Theoretical and Applied Genetics*, **102**, 41–52.

Morandini, P. and Salamini, F. (2003) Plant biotechnology and breeding: allied for years to come. *Trends in Plant Science*, **8**, 70–75.

Morgan, J.W. (1984) Osmoregulation and water stress in higher plants. *Annual Review of Plant Physiology*, **35**, 299–319.

Morgante, M. and Salamini, F. (2003) From plant genomics to breeding practice. *Current Opinion in Biotechnology*, **14**, 214–219.

Morris, M., Dreher, K., Ribaut, J. and Khairallah, M. (2003) Money matters (II): costs of maize inbred line conversion schemes at CIMMYT using conventional and marker-assisted selection. *Molecular Breeding*, **11**, 235–247.

Munns, R. (1988) Why measure osmotic adjustment? *Australian Journal of Plant Physiology*, **15**, 717–726.

Munns, R., Passioura, J.B., Guo, J., Chazen, O. and Cramer, G.R. (2000) Water elations and leaf expansion: importance of timing. *Journal of Experimental Botany*, **51**, 1495–1504.

Nagy, Z., Tuba, Z., Zsoldos, F. and Erdei, L. (1995) CO_2-exchange and water relation responses of sorghum and maize during water and salt stress. *Journal of Plant Physiology*, **145**, 539–544.

Nal, B., Mohr, E. and Ferrier, P. (2001) Location analysis of DNA-bound proteins at the whole genome level: untangling transcriptional regulatory networks. *Bioessays*, **23**, 473–476.

Nguyen, H.T., Babu, R.C. and Blum, A. (1997) Breeding for drought resistance in rice: physiology and molecular genetics considerations. *Crop Science*, **37**, 1426–1434.

Nienhuis, J., Sills, G.R., Martin, B. and King, G. (1994) Variance for water-use efficiency among ecotypes and recombinant inbred lines of *Arabidopsis thaliana* (*Brassicaceae*). *American Journal of Botany*, **81**, 943–947.

Nuccio, M.L., Rhodes, D., McNeil, S.D. and Hanson, A.D. (1999) Metabolic engineering of plants for osmotic stress resistance. *Current Opinion in Plant Biology*, **2**, 128–134.

O'Toole, J.C. and Bland, W.L. (1987) Genotypic variation in crop plant root systems. *Advances in Agronomy*, **41**, 91–145.

Oberschall, A., Deak, M., Torok, K., Sass, L., Vass, I., Kovacs, I., Feher, A., Audits, D. and Horvath, G.V. (2000) A novel aldose/aldehyde reductase protects transgenic plants against lipid peroxidation under chemical and drought stresses. *Plant Journal*, **24**, 437–446.

Oono, Y., Seki, M., Nanjo, T., Narusaka, M., Fujita, M., Satoh, R., Satou, M., Sakurai, T., Ishida, J., Akiyama, K., Lida, K., Maruyama, K., Satoh, S., Yamaguchi, Shinozaki, K. and Shinozaki, K. (2003) Monitoring expression profiles of Arabidopsis gene expression during rehydration process after dehydration using ca. 7000 full-length cDNA microarray. *Plant Journal*, **34**, 868–887.

Ozturk, Z.N., Talamé, V., Deyholos, M., Michalowski, C.B., Galbraith, D.W., Gozukirmizi, N., Tuberosa, R. and Bohnert, H.J. (2002) Monitoring large-scale changes in transcript abundance in drought- and salt-stressed barley. *Plant Molecular Biology*, **48**, 551–573.

Pandolfi.D., Solinas, G., Valle, G. and Coraggio, I. (1997) Cloning of a cDNA encoding a novel *myb* gene (Accession No. Y11414) highly expressed in cold stressed rice coleoptiles. *Plant Physiology*, **114**, 747.

Passioura, J.B. (1977) Grain yield, harvest index and water use of wheat. *Journal of the Australian Institute for Agricultural Science*, **43**, 117–120.

Passioura, J.B. (1996) Drought and drought tolerance. *Plant Growth Regulation*, **20**, 79–83.

Passioura, J.B. (2002) Environmental biology and crop improvement. *Functional Plant Biology*, **29**, 537–546.

Pastori, G.M. and Foyer, C.H. (2002) Common components, networks, and pathways of cross-tolerance to stress. The central role of 'redox' and abscisic acid-mediated controls. *Plant Physiology*, **129**, 460–468.

Patel, J., McLeod, L.E., Vries, R.G.J., Flynn, A., Wang, X.M. and Proud, C.G. (2002) Cellular stresses profoundly inhibit protein synthesis and modulate the states of phosphorylation of multiple translation factors. *European Journal of Biochemistry*, **269**, 3076–3085.

Paterson, A.H., Lan, T.-H., Reischmann, K.P., Chang, C., Lin, Y.-R., Liu, S.-C., Burow, M.D., Kowalski, S.P., Katsar, C.S. and Del Monte, T.A. (1996) Toward a unified genetic map of higher plants, transcending the monocot-dicot divergence. *Nature Genetics*, **14**, 380–382.

Paterson, A.H., Saranga, Y., Menz, M., Jiang, C.X. and Wright, R.J. (2003) QTL analysis of genotype × environment interactions affecting cotton fiber quality. *Theoretical and Applied Genetics*, **106**, 384–396.

Pellegrineschi, A., Ribaut, J.M., Trethowan, R., Yamaguchi-Shinozaki, K. and Hoisington, D. (2001) Progress in the genetic engineering of wheat for water-limited conditions. *Genetic engineering of crop plants for abiotic stress JIRCAS-Working-Report. Proceedings of an APEC-JIRCAS joint symposium and workshop*, Bangkok, Thailand, 23, 55–60.

Pelleschi, S., Guy, S., Kim, J.Y., Pointe, C., Mahe, A., Barthes, L., Leonardi, A. and Prioul, J.L. (1999) *Ivr2*, a candidate gene for a QTL of vacuolar invertase activity in maize leaves. Gene-specific expression under water stress. *Plant Molecular Biology*, **39**, 373–380.

Perry, M.W. and D'Antuono, L.F. (1989) Yield improvement and associated characteristics of some Australian spring wheat cultivars introduced between 1860 and 1982. *Australian Journal of Agricultural Research*, **40**, 457–472.

Pfeiffer, W.H., Sayre, K.D. and Reynolds, M.P. (2000) Enhancing genetic grain yield potential and yield stability in durum wheat, in *Durum Wheat Improvement in the Mediterranean Region: New Challenges* (eds. C. Royo, M.M. Nachit, N. Di Fonzo and J.L. Araus), Options Méditerranéennes, pp. 83–93.

Pflieger, S., Lefebvre, V. and Causse, M. (2001) The candidate gene approach in plant genetics: a review. *Molecular Breeding*, **7**, 275–291.

Pilon-Smits, E., Ebskamp, M., Paul, M., Jeuken, M., Weisbeek, P. and Smeekens, S. (1995) Improved performance of transgenic fructan-accumulating tobacco under drought stress. *Plant Physiology*, **107**, 125–130.

Pilon-Smits, E.H., Terry, N., Sears, T., Kim, H., Zayed, A., Hwang, S.B., van Dun, K., Voogd, E., Verwoerd, T.C., Krutawagen, R.W.H.H. and Goddijn, O.J.M. (1998) Trehalose-producing transgenic tobacco plants show improved growth performance under drought stress. *Journal of Plant Physiology*, **152**, 525–532.

Pilon-Smits, E., Terry, N., Sears Tobin, K. and Van Dun, K. (1999) Enhanced drought resistance in fructan-producing sugar beet. *Plant Physiology and Biochemistry*, **37**, 313–317.

Popova, L.P., Tsonev, T.D., Lazova, G.N. and Stoinova, Z.G. (1996) Drought- and ABA-induced changes in photosynthesis of barley plants. *Physiologia Plantarum*, **4**, 623–629.

Price, A.H. and Tomos, A.D. (1997) Genetic dissection of root growth in rice (*Oryza sativa* L.). II: Mapping quantitative trait loci using molecular markers. *Theoretical and Applied Genetics*, **95**, 143–152.

Price, A., Young, E. and Tomos, A. (1997) Quantitative trait loci associated with stomatal conductance, leaf rolling and heading date mapped in upland rice (*Oryza sativa*). *New Phytologist*, **137**, 83–91.

Price, A., Steele, K., Moore, B., Barraclough, P. and Clark, L. (2000) A combined RFLP and AFLP linkage map of upland rice (*Oryza sativa* L.) used to identify QTLs for root-penetration ability. *Theoretical and Applied Genetics*, **100**, 49–56.

Price, A., Cairns, J., Horton, P., Jones, H. and Griffiths, H. (2002a) Linking drought-resistance mechanisms to drought avoidance in upland rice using a QTL approach: progress and new opportunities to integrate stomatal and mesophyll responses. *Journal of Experimental Botany*, **53**, 989–1004.

Price, A.H., Steele, K.A., Gorham, J., Bridges, J.M., Moore, B.J., Evans, J.L., Richardson, P. and Jones, R.G.W. (2002b) Upland rice grown in soil-filled chambers and exposed to contrasting water-deficit regimes I. Root distribution, water use and plant water status. *Field Crop Research*, **76**, 11–24.

Price, A.H., Townend, J., Jones, M.P., Audebert, A. and Courtois, B. (2002c) Mapping QTLs associated with drought avoidance in upland rice approach grown in the Philippines and West Africa. *Plant Molecular Biology*, **48**, 683–695.

Prioul, J.L., Quarrie, S., Causse, M. and de Vienne, D. (1997) Dissecting complex physiological functions through the use of molecular quantitative genetics. *Journal of Experimental Botany*, **48**, 1151–1163.

Quarrie, S.A. (1991) Implications of genetic differences in ABA accumulation for crop production, in *Abscisic acid: physiology and biochemistry* (eds. W.J. Davies and H.G. Jones), Bios Scientific Publishers, Oxford, UK, pp. 227–243.

Quarrie, S.A. (1996) New molecular tools to improve the efficiency of breeding for increased drought resistance. *Plant Growth Regulation*, **20**, 167–178.

Quarrie, S. and Jones, H. (1977) Effects of abscisic acid and water stress on development and morphology of wheat. *Journal of Experimental Botany*, **28**, 192–203.

Quarrie, S., Steed, A., Semikhdoski, A., Lebreton, C., Calestani, C., Clarkson, D., Tuberosa, R., Sanguineti, M., Melchiorre, R. and Prioul, J. (1995) Identification of quantitative trait loci regulating water and nitrogen-use efficiency in wheat. *2nd STRESSNET Conference*, Salsomaggiore, Italy, 175–180.

Quarrie, S., Lazic Jancic, V., Kovacevic, D., Steed, A. and Pekic, S. (1999a) Bulk segregant analysis with molecular markers and its use for improving drought resistance in maize. *Journal of Experimental Botany*, **50**, 1299–1306.

Quarrie, S.A., Stojanovic, J. and Pekic, S. (1999b) Improving drought resistance in small grained cereals: A case study, progress and prospects. *Plant Growth Regulation*, **29**, 1–21.

Quarrie, S.A., Steed, A., Calestani, C., Semikhodskii, A., Lebreton, C., Chinoy, C., Steele, N., Pljevljakusic, D., Waterman, E., Weyen, J., Schondelmaier, J., Farmer, P., Saker, L., Clarkson, D.T., Abugalieva, A., Yessimbekova, M., Turuspekov, Y., Abugalieva, S., Hollington, P., Aragues, R., Royo, A., Habash, D., Tuberosa, R. and Dodig, D. (2004) A

294   WATER USE EFFICIENCY IN PLANT BIOLOGY

genetic map of hexaploid wheat (*Triticum aestivum* L.) from the cross Chinese Spring × SQ1 and its use to compare QTLs for grain yield across a range of environments. *Theoretical and Applied Genetics* (submitted).

Radin, J.W., Lu, Z., Percy, R.G. and Zeiger, E. (1994) Genetic variability for stomatal conductance in Pima cotton and its relation to improvements of heat adaptation. *Proceedings of the National Academy of Sciences USA*, **91**, 7217–7221.

Rafalski, A. (2002) Applications of single nucleotide polymorphisms in crop genetics. *Current Opinion in Plant Biology*, **5**, 94–100.

Ramanjulu, A. and Bartels, D. (2002) Drought- and desiccation-induced modulation of gene expression in plants. *Plant Cell and Environment*, **25**, 141–151.

Rebetzke, G., Read, J., Barbour, M., Condon, A. and Rawson, H. (2000) A hand held porometer for rapid assessment of leaf conductance in wheat. *Crop Science*, **40**, 277–280.

Rebetzke, G.J., Condon, A.G., Richards, R.A. and Farquhar, G.D. (2002) Selection for reduced carbon isotope discrimination increases aerial biomass and grain yield of rainfed bread wheat. *Crop Science*, **42**, 739–745.

Rebetzke, G., Condon, A., Richards, R. and Farquhar, G. (2003) Gene action for leaf conductance in three wheat crosses. *Australian Journal of Agricultural Research*, **54**, 381–387.

Reymond, M., Muller, B., Leonardi, A., Charcosset, A. and Tardieu, F. (2003) Combining quantitative trait loci analysis and an ecophysiological model to analyze the genetic variability of the responses of maize leaf growth to temperature and water deficit. *Plant Physiology*, **131**, 664–675.

Reynolds, M.P. and Pfeiffer, W.H. (2000) Applying physiological strategies to improve yield potential, in *Durum wheat improvement in the Mediterranean region: New challenges* (eds. C. Royo, M.M. Nachit, N. Di Fonzo and J.L. Araus), Options Méditerranéennes, pp. 95–103.

Reynolds, M.P., Rajaram, S. and Sayre, K.D. (1999) Physiological and genetic changes of irrigated wheat in the post-green revolution period and approaches for meeting projected global demand. *Crop Science*, **39**, 1611–1621.

Ribaut, J.M. and Betran, J. (2000) Single large-scale marker-assisted selection (SLS-MAS). *Molecular Breeding*, **5**, 531–541.

Ribaut, J.M. and Hoisington, D. (1998) Marker-assisted selection: new tools and strategies. *Trends in Plant Science*, **3**, 236–239.

Ribaut, J.M., Hu, X., Hoisington, D. and Gonzales-de-Leon, D. (1997a) Use of STSs and SSRs as rapid and reliable preselection tools in a marker-assisted selection backcross scheme. *Plant Molecular Biology Report*, **15**, 156–164.

Ribaut, J.M., Jiang, C., Gonzalez-de-Leon, D., Edmeades, G. and Hoisington, D.A. (1997b) Identification of quantitative trait loci under drought conditions in tropical maize. 2. Yield components and marker-assisted selection strategies. *Theoretical of Applied Genetics*, **94**, 887–896.

Ribaut, J.M., Banziger, M., Betran, J., Jiang, C., Edmeades, G.O., Dreher, K. and Hoisington, D. (2002) Use of molecular markers in plant breeding: drought tolerance improvement in tropical maize, in *Quantitative Genetics, Genomics, and Plant Breeding* (ed. M.S. Kang), CABI Publishing, Wallingford, pp. 85–99.

Riccardi, F., Gazeau, P., de Vienne, D. and Zivy, M. (1998) Protein changes in response to progressive water deficit in maize: quantitative variation and polypeptide identification. *Plant Physiology*, **117**, 1253–1263.

Richards, R.A. (1991) Crop improvement for temperate Australia: future opportunities. *Field Crops Research*, **96**, 141–169.

Richards, R.A. (1996) Defining selection criteria to improve yield under drought. *Plant Growth Regulation*, **20**, 157–166.

Richards, R.A. (2000) Selectable traits to increase crop photosynthesis and yield of grain crops. *Journal of Experimental Botany*, **51**, 447–458.

Richards, R.A. and Condon, A.G. (1993) Challenges ahead in using carbon isotope discrimination in plant-breeding programs, in *Stable isotopes and plant carbon-water relations* (eds. J.R. Ehleringer, A.E. Hall and G.D. Farquhar), Academic Press, San Diego, CA, USA, pp. 451–462.

Richards, R.A. and Passioura, J.B. (1981a) Seminal root morphology and water-use in wheat. I. Environmental effects. *Crop Science*, **21**, 249–252.

Richards, R.A. and Passioura, J.B. (1981b) Seminal root morphology and water-use in wheat. II. Genetic variation. *Crop Science*, **21**, 253–255.

Richards, R.A. and Passioura, J.B. (1989) A breeding program to reduce the diameter of the major xylem vessel in the seminal roots of wheat and its effect on grain yield in rain-fed environments. *Australian Journal of Agricultural Research*, **40**, 943–950.

Richards, R.A., Rebetzke, G., Condon, A. and van Herwaarden, A. (2002) Breeding opportunities for increasing the efficiency of water use and crop yield in temperate cereals. *Crop Science*, **42**, 111–121.

Roberts, J.K.M. (2002) Proteomics and a future generation of plant molecular biologists. *Plant Molecular Biology*, **48**, 143–154.

Robertson, D.S. (1985) A possible technique for isolating genic DNA for quantitative traits in plant. *Journal of Theoretical Biology*, **117**, 1–10.

Robinson, D., Handley, L.L., Scrimgeour, C.M., Gordon, D.C., Forster, B.P. and Ellis, R.P. (2000) Using stable isotope natural abundances (δ15N and δ13C) to integrate the stress responses of wild barley (*Hordeum spontaneum* C. Koch) genotypes. *Journal of Experimental Botany*, **51**, 41–50.

Rontein, D., Basset, G. and Hanson, A.D. (2002) Metabolic engineering of osmoprotectant accumulation in plants metabolic engineering. *Metabolic Engineering*, **4**, 49–56.

Roxas, V.P., Smith, R.K., Allen, E.R. and Allen, R.D. (1997) Overexpression of glutathione S-transferase/glutathione peroxidase enhances the growth of transgenic tobacco seedlings during stress. *Nature Biotechnology*, **15**, 988–991.

Royo, R., Villegas, D., García del Moral, L.F., El Hani, S., Aparicio, N., Rharrabti, Y. and Araus, J.L. (2001) Comparative performance of carbon isotope discrimination and canopy temperature depression as predictors of genotype differences in durum wheat yield in Spain. *Australian Journal of Agricultural Research*, **53**, 1–9.

Saijo, Y., Hata, S., Kyozuka, J., Shimamoto, K. and Izui, K. (2000) Over-expression of a single Ca^{2+}-dependent protein kinase confers both cold and salt/drought tolerance on rice plants. *Plant Journal*, **23**, 319–327.

Saijo, Y., Kinoshita, N., Ishiyama, K., Hata, S., Kyozuka, J., Hayakawa, T., Nakamura, T., Shimamoto, K., Yamaya, T. and Izui, K. (2001) A Ca^{2+}-dependent protein kinase that endows rice plants with cold- and salt-stress tolerance functions in vascular bundles. *Plant and Cell Physiology*, **42**, 1228–1233.

Salekdeh, G.H., Siopongco, J., J, W.L., B, G. and J, B. (2002) A proteomic approach to analyzing drought- and salt-responsiveness in rice. *Field Crops Research*, **76**, 199–219.

Salvi, S., Morgante, M., Fengler, K., Meeley, B., Ananiev, E., Svitashev, S., Bruggemann, E., Niu, X., Li, B., Tingey, S.C., Tomes, D., Guo, Hua Miao, G.H., Phillips, R.L. and Tuberosa, R. (2003) Progress in the positional cloning of *Vgt1*, a QTL controlling flowering time in maize. *57th Annual Corn and Sorghum Research Conference*, American Seed Trade Association, 1–18.

Salvi, S., Tuberosa, R., Sanguineti, M.C., Landi, P. and Conti, S. (1997) Molecular market analysis of maize populations divergently selected for abscisic acid concentration in the leaf. *Maize Genetics Cooperation Newsletter*, **71**, 15–16.

Salvi, S., Tuberosa, R. and Phillips, R.L. (2001) Development of PCR-based assays for allelic discrimination in maize by using the 5-nuclease procedure. *Molecular Breeding*, **8**, 169–176.

Salvi, S., Tuberosa, R., Chiapparino, E., Maccarerri, M., Veillet, S., Van Beuningen, L., Isaac, P., Edwards, K. and Phillips, R.L. (2002) Toward positional cloning of *Vgt1*, a QTL controlling the transition from the vegetative to the reproductive phase in maize. *Plant Molecular Biology*, **48**, 601–613.

Sanchez, A.C., Subudhi, P.K., Rosenow, D.T. and Nguyen, H.T. (2002) Mapping QTLs associated with drought resistance in sorghum (*Sorghum bicolor* L. Moench). *Plant Molecular Biology*, **48**, 713–726.

Sanguineti, M.C., Conti, S., Landi, P. and Tuberosa, R. (1996) Abscisic acid concentration in maize leaves: genetic control and response to divergent selection in two populations. *Maydica*, **41**, 193–203.

Sanguineti, M., Tuberosa, R., Landi, P., Salvi, S., Maccaferri, M., Casarini, E. and Conti, S. (1999) QTL analysis of drought-related traits and grain yield in relation to genetic variation for leaf abscisic acid concentration in field-grown maize. *Journal of Experimental Botany*, **50**, 1289–1297.

Saranga, Y., Menz, M., Jiang, C.X., Wright, R.J., Yakir, D. and Paterson, A.H. (2001) Genomic dissection of genotype × environment interactions conferring adaptation of cotton to arid conditions. *Genome Research*, **11**, 1988–1995.

Sarkarung, S., Pantuwan, G. and Hardy, B. (1999) Improving rice for drought-prone rainfed lowland environments, in *Genetic improvement of rice for water-limited environments* (eds. O. Ito and J. O' Toole), pp. 57–70.

Sasahara, H., Fukuta, Y. and Fukuyama, T. (1999) Mapping of QTLs for vascular bundle system and spike morphology in rice, *Oryza sativa* L. *Breeding Science*, **49**, 75–81.

Saxena, N.P. and O'Toole, J.C. (2002) *Field screening for drought tolerance in crop plants with emphasis on rice. Proceedings of an International Workshop on Field Screening for Drought Tolerance in Rice*, eds. N.P. Saxena and J.C. O'Toole, ICRISAT, Patancheru, India, Dec. 11–14, 2000.

Schadt, E.E., Monks, S.A., Drake, T.A., Lusis, A.J., Che, N., Colinayo, V., Ruff, T.G., Milligan, S.B., Lamb, J.R., Cavet, G., Linsley, P.S., Mao, M., Stoughton, R.B. and Friend, S.H. (2003) Genetics of gene expression surveyed in maize, mouse and man. *Nature*, **422**, 297–302.

Schmid, K.J., Sorensen, T.R., Stracke, R., Torjek, O., Altmann, T., Mitchell-Olds, T. and Weisshaar, B. (2003) Large-scale identification and analysis of genome-wide single-nucleotide polymorphisms for mapping in *Arabidopsis thaliana*. *Genome Research*, **13**, 1250–1257.

Schwartz, S.H., Tan, B.C., Gage, D.A., Zeevaart, J.A.D. and McCarty, D.R. (1997) Specific oxidative cleavage of carotenoids by VP14 of maize. *Science*, **276**, 1872–1874.

Seki, M., Kamei, A., Yamaguchi-Shinozaki, K. and Shinozaki, K. (2003) Molecular responses to drought, salinity and frost: common and different paths for plant protection. *Current Opinion in Biotechnology*, **14**, 194–199.

Seki, M., Narusaka, M., Abe, H., Kasuga, M., Yamaguchi-Shinozaki, K., Carninci, P., Hayashizaki, Y. and Shinozaki, K. (2001) Monitoring the expression pattern of Arabidopsis genes underdrought and cold stresses by using a full-length cDNA microarray. *Plant Cell*, **13**, 61–72.

Seki, M., Narusaka, M., Ishida, J., Nanjo, T., Fujita, M., Oono, Y., Kamiya, A., Nakajima, M., Enju, A., Sakurai, T., Satou, M., Akiyama, K., Taji, T., Yamaguchi-Shinozaki, K., Carninci, P., Kawai, J., Hayashizaki, Y. and Shinozaki, K. (2002) Monitoring the expression profiles of 7000 Arabidopsis genes under drought, cold and high-salinity stresses using a full-length cDNA microarray. *Plant Journal*, **31**, 279–292.

Serageldin, I. (1999) Biotechnology and food security in the 21st century. *Science*, **285**, 387–389.

Setter, T.L. (1997) Role of the phytohormone ABA in drought tolerance: potential utility as a selection tool. *Developing drought- and low N-tolerant maize. Proceedings of a Symposium*, eds. G.O. Edmeades, M. Bänziger, H.R. Mickelson and C.B. Peña-Valdivia, CIMMYT, El Batán, Mexico, March 25–29, 1996, 142–150.

Sharp, R.E. (1996) Regulation of plant growth responses to low soil water potentials. *Horticulture Science*, **31**, 36–39.

Sharp, R.E. (2002) Interaction with ethylene: changing views on the role of abscisic acid in root and shoot growth responses to water stress. *Plant Cell and Environment*, **25**, 211–222.

Sharp, R. and Davies, W. (1985) Root growth and water uptake by maize plants in drying soil. *Journal of Experimental Botany*, **36**, 1441–1456.

Sharp, R., Boyer, J., Nguyen, H. and Hsio, T. (1996) Abscisic acid accumulation maintains maize primary root elongation at low water potentials by restricting ethylene production. *Plant Physiology*, **110**, 1051–1053.

Sharp, R., Wu, Y., Voetberg, G., Saab, I., LeNoble, M., Wang, T., Davies, W. and Pollock, C. (1994) Confirmation that abscisic acid accumulation is required for maize primary root elongation at low water potentials. *Journal of Experimental Botany*, **45**, 1743–1751.

Shen, B., Jensen, R.G. and Bohnert, H.J. (1997) Increased resistance to oxidative stress in transgenic plants by targeting mannitol biosynthesis to chloroplasts. *Plant Physiology*, **113**, 1177–1183.

Shen, L., Courtois, B., McNally, K.L., Robin, S. and Li, Z. (2001) Evaluation of near-isogenic lines of rice introgressed with QTLs for root depth through marker-aided selection. *Theoretical and Applied Genetics*, **103**, 75–83.

Shinozaki, K. and Yamaguchi-Shinozaki, K. (1996) Molecular responses to drought and cold stress. *Current Opinion in Biotechnology*, **7**, 161–167.

Shinozaki, K. and Yamaguchi-Shinozaki, K. (2000) Molecular responses to dehydration and low temperature: differences and cross-talk between two stress signaling pathways. *Current Opinion in Plant Biology*, **3**, 217–223.

Siddique, K.H., Kirby, E.J.M. and Perry, M.W. (1989) Ear:stem ratio in old and modern wheat varieties; relationship with improvement in number of grains per ear and yield. *Field Crops Research*, **21**, 59–78.

Siddique, K.H.M., Tennant, D., Perry, M.W. and Belford, R.K. (1990) Water use and water use efficiency of old and modern wheat cultivars in a mediterranean-type environment. *Australian Journal of Agricultural Research*, **41**, 431–447.

Simko, I., McMurry, S., Yang, H.M., Manschot, A., Davies, P.J. and Ewing, E.E. (1997) Evidence from polygene mapping for a causal relationship between potato tuber dormancy and abscisic acid content. *Plant Physiology*, **115**, 1453–1459.

Sivamani, E., Bahieldin, A., Wraith, J.M., Al-Niemi, T., Dyer, W.E., Ho, T.H.D. and Qu, R.D. (2000) Improved biomass productivity and water use efficiency under water deficit conditions in transgenic wheat constitutively expressing the barley *HVA1* gene. *Plant Science*, **155**, 1–9.

Smirnoff, N. and Pallanca, J.E. (1996) Ascorbate metabolism in relation to oxidative stress. *Biochemical Society Transactions*, **24**, 472–478.

Smith, R.D. and Walker, J.C. (1996) Plant protein phosphatases. *Annual Review of Plant Physiology and Plant Molecular Biology*, **47**, 101–125.

Snape, J.W., Sarma, R., Quarrie, S.A., Fish, L., Galiba, G. and Sutka, J. (2001) Mapping genes for flowering time and frost tolerance in cereals using precise genetic stocks. *Euphytica*, **120**, 309–315.

Sorrells, M.E., La Rota, M., Bermudez-Kandianis, C.E., Greene, R.A., Kantety, R., Munkvold, J.D., Miftahudin, Mahmoud, A., Ma, X., Gustafson, P.J., Qi, L.L., Echalier, B., Gill, B.S., Matthews, D.E., Lazo, G.R., Chao, S., Anderson, O.D., Edwards, H., Linkiewicz, A.M., Dubcovsky, J., Akhunov, E.D., Dvorak, J., Zhang, D., Nguyen, H.T., Peng, J., Lapitan, N.L.V., Gonzalez-Hernandez, J.L., Anderson, J.A., Hossain, K., Kalavacharla, V., Kianian, S.F., Choi, D.-W., Close, T.J., Dilbirligi, M., Gill, K.S., Steber, C., Walker-Simmons, M.K., McGuire, P.E. and Qualset, C.O. (2003) Comparative DNA sequence analysis of wheat and rice genomes. *Genome Research*, **13**, 1818–1827.

Specht, J.E., Chase, K., Macrander, M., Graef, G.L., Chung, J., Markwell, J.P., Germann, M., Orf, J.H. and Lark, K.G. (2001) Soybean response to water: a QTL analysis of drought tolerance. *Crop Science*, **41**, 493–509.

Spollen, W.G., LeNoble, M.E., Samuels, T.D., Bernstein, N. and Sharp, R.E. (2000) Abscisic acid accumulation maintains maize primary root elongation at low water potentials by restricting ethylene production. *Plant Physiology*, **122**, 967–976.

Steuer, R., Kurths, J., Fiehn, O. and Weckwerth, W. (2003) Observing and interpreting correlations in metabolomic networks. *Bioinformatics*, **19**, 1019–1026.

Stockinger, E.J., Gilmour, S.J. and Thomashow, M.F. (1997) *Arabidopsis thaliana* CBF1 encodes an AP2 domain containing transcriptional activator that binds to the C-repeat/DRE, a cis-acting DNA regulatory element that stimulates transcription in response to low temperature and water deficit. *Proceedings of the National Academy of Sciences, USA*, **94**, 1035–1040.

Stuber, C.W., Edwards, M.D. and Wendel, J.F. (1987) Molecular marker facilitated investigations of quantitative trait loci in maize. II. Factors influencing yield and its component traits. *Crop Science*, **27**, 639–648.

Stuber, C.W., Polacco, M. and Senior, M.L. (1999) Synergy of empirical breeding, marker-assisted selection, and genomics to increase crop yield potential. *Crop Science*, **39**, 1571–1583.

Subudhi, P.K., Rosenow, D.T. and Nguyen, H.T. (2000) Quantitative trait loci for the stay green trait in sorghum (*Sorghum bicolor* L. Moench): consistency across genetic backgrounds and environments. *Theoretical and Applied Genetics*, **101**, 733–741.

Takahashi, Y., Shomura, A., Sasaki, T. and Yano, M. (2001) Hd6, a rice quantitative trait locus involved in photoperiod sensitivity, encodes the a subunit of protein kinase CK2. *Proceedings of the National Academy of Sciences USA*, **98**, 7922–7927.

Talamè, V., Sanguineti, M.C., Chiapparino, E., Bahri, H., Ben Salem, M., Ellis, R., Forster, B.P., Rhouma, S., Zoumarou, W. and Tuberosa, R. (2004) Identification of agronomically valuable QTL alleles in wild barley (*Hordeum spontaneum*). *Annals of Botany*, submitted.

Tan, B.C., Schwartz, S.H., Zeevaart, J.A.D. and McCarty, D.R. (1997) Genetic control of abscisic acid biosynthesis in maize. *Proceedings of the National Academy of Sciences USA*, 94, 12235–12240.

Tangpremsri, T., Fukai, S. and Fischer, K.S. (1995) Growth and yield of sorghum lines extracted from a population for differences in osmotic adjustment. *Australian Journal of Agricultural Research*, 46, 61–74.

Tanksley, S.D. (1993) Mapping polygenes. *Annual Review of Genetics*, 27, 205–233.

Tanksley, S. and Nelson, J. (1996) Advanced backcross QTL analysis: a method for the simultaneous discovery and transfer of valuable QTLs from unadapted germplasm into elite breeding lines. *Theoretical and Applied Genetics*, 92, 191–203.

Tanksley, S.D., Ganal, M.W. and Martin, G.B. (1995) Chromosome landing: a paradigm for map-based gene cloning in plants with large genomes. *Trends in Genetics*, 11, 63–68.

Tanksley, S.D., Grandillo, S., Fulton, T., Zamir, D., Eshed, T., Petiard, V., Lopez, J. and Beck, B.T. (1996) Advanced backcross QTL analysis in a cross between an elite processing line of tomato and its wild relative L. *pimpinellifolium. Theoretical and Applied Genetics*, 92, 213–224.

Tarczynski, M.C., Jensenn, R. and Bohnert, H.J. (1993) Stress protecyion and transgenic tobacco by production of osmolyte of the osmolyte mannitol. *Science*, 259, 508–510.

Tardieu, F., Zhang, J., Katerji, N., Bethenod, O., Palmer, S. and Davies, W.J. (1992) Xylem ABA controls the stomatal conductance of field-grown maize subjected to soil compaction or soil drying. *Plant Cell Environment*, 15, 193–197.

Tardieu, F., Gowing, D. and Davies, W.J. (1993) A model of stomatal control by both ABA concentration in the xylem sap and leaf water status: test of the model and of alternative mechanisms for droughted and ABA-fed field-grown maize. *Plant Cell Environment*, 16, 413–420.

Teulat, B., Monneveux, P., Wery, J., Borries, C., Souyris, I., Charrier, A. and This, D. (1997) Relationships between relative water content and growth parameters under water stress in barley: a QTL study. *New Phytologist*, 137, 99–107.

Teulat, B., This, D., Khairallah, M., Borries, C., Ragot, C., Sourdille, P., Leroy, P., Monneveux, P. and Charrier, A. (1998) Several QTLs involved in osmotic-adjustment trait variation in barley (*Hordeum vulgare* L.). *Theoretical and Applied Genetics*, 96, 688–698.

Teulat, B., Borries, C. and This, D. (2001a) New QTLs identified for plant water status, water-soluble carbohydrate and osmotic adjustment in a barley population grown in a growth-chamber under two water regimes. *Theoretical and Applied Genetics*, 103, 161–170.

Teulat, B., Merah, O., Souyris, I. and This, D. (2001b) QTLs for agronomic traits from a Mediterranean barley progeny grown in several environments. *Theoretical and Applied Genetics*, 103, 774–787.

Thiellement, H., Bahrman, N., Damerval, C., Plomion, C., Rossignol, M., Santoni, V., de Vienne, D. and Zivy, M. (1999) Proteomics for genetic and physiological studies in plants. *Electrophoresis*, 20, 2013–2026.

Thiellement, H., Zivy, M. and Plomion, C. (2002) Combining proteomic and genetic studies in plants. *Journal of Chromatography Analytical Technologies in the Biomedical and Life Sciences*, 782, 137–149.

Thomas, H. and Howarth, C.J. (2000) Five ways to stay green. *Journal of Experimental Botany*, 51, 329–337.

Thomas, J.C., Sepahi, M., Arendall, B. and Bohnert, H.J. (1995) Enhancement of seed germination in high salinity by engineering mannitol expression in *Arabidopsis thaliana*. *Plant Cell Environment*, 18, 801–806.

Thomashow, M.F. (1999) Plant cold acclimation: freezing tolerance genes and regulatory mechanisms. *Annual Review Plant Physiology Plant Molecular Biology*, 50, 571–599.

Thomashow, M.F. (2001) So what's new in the field of plant cold acclimation? Lots! *Plant Physiology*, 125, 89–93.

Thumma, B.R., Naidu, B.P., Chandra, A., Cameron, D., Bahnisch, L. and Liu, C.J. (2001) Identification of causal relationships among traits related to drought resistance in *Stylosanthes scabra* using QTL analysis. *Journal of Experimental Botany*, 52, 203–214.

Tollenaar, M. and Wu, J. (1999) Yield improvement in temperate maize is attributable to greater stress tolerance. *Crop Science*, **39**, 1597–1604.

Touzet, P., Winkler, R.G. and Helentjaris, T. (1995) Combined genetic and physiological analysis of a locus contributing to quantitative variation. *Theoretical and Applied Genetics*, **91**, 200–205.

Tripathy, J.N., Zhang, J., Robin, S., Nguyen, T.T. and Nguyen, H.T. (2000) QTLs for cell-membrane stability mapped in rice (*Oryza sativa* L.) under drought stress. *Theoretical and Applied Genetics*, **100**, 1197–1202.

Tuberosa, R., Sanguineti, M.C. and Conti, S. (1986) Divergent selection for heading date in barley. *Plant Breeding*, **97**, 345–351.

Tuberosa, R., Sanguineti, M., Stefanelli, S. and Quarrie, S.A. (1992) Number of endosperm cells and weight of barley kernels in relation to endosperm abscisic acid content. *European Journal of Agronomy*, **1**, 125–132.

Tuberosa, R., Sanguineti, M. and Landi, P. (1994) Abscisic acid concentration in the leaf and xylem sap, leaf water potential, and stomatal conductance in drought-stressed maize. *Crop Science*, **34**, 1557–1563.

Tuberosa, R., Parentoni, S., Kim, T.S., Sanguineti, M.C. and Phillips, R.L. (1998a) Mapping QTLs for ABA concentration in leaves of a maize cross segregating for anthesis date. *Maize Genetics Cooperation Newsletter*, **72**, 72–73.

Tuberosa, R., Sanguineti, M., Landi, P., Salvi, S. and Conti, S. (1998b) RFLP mapping of quantitative trait loci controlling abscisic acid concentration in leaves of drought-stressed maize (*Zea mays* L.). *Theoretical and Applied Genetics*, **97**, 744–755.

Tuberosa, R., Gill, B. and Quarrie, S. (2002a) Cereal genomics: ushering in a brave new world. *Plant Molecular Biology*, **48**, 445–449.

Tuberosa, R., Sanguineti, M., Landi, P., Giuliani, M., Salvi, S. and Conti, S. (2002b) Identification of QTLs for root characteristics in maize grown in hydroponics and analysis of their overlap with QTLs for grain yield in the field at two water regimes. *Plant Molecular Biology*, **48**, 697–712.

Tuberosa, R., Salvi, S., Sanguineti, M.C., Landi, P., Maccaferri, M. and Conti, S. (2002c) Mapping QTLs regulating morpho-physiological traits and yield in drought-stressed maize: case studies, shortcomings and perspectives. *Annals of Botany*, **89**, 941–963.

Tuinstra, M., Grote, E., Goldsbrough, P. and Ejeta, G. (1997) Genetic analysis of post-flowering drought tolerance and components of grain development in *Sorghum bicolor* (L.) Moench. *Molecular Breeding*, **3**, 439–448.

Tuinstra, M.R., Ejeta, G. and Goldbrough, P. (1998) Evaluation of near-isogenic sorghum lines contrasting for QTL markers associated with drought tolerance. *Crop Science*, **38**, 835–842.

Turner, N.C. (1990) Plant water relations and irrigation management. *Agricultural Water Management*, **17**, 59–73.

Turner, N.C. (1997) Further progress in crop water relations, in *Advances in Agronomy*, ed. D.L. Sparks, Academic Press, San Diego, 58, pp. 293–338.

Turner, N.C., Nicolas, M.E., Hubick, K.T. and Farquhar, G.D. (1989) Evaluation of traits for the improvement of water use efficiency and harvest index, in *Drought Resistance in Cereals* (ed. F.W.G. Baker), ICSU Press, Paris, pp. 177–189.

Turpeinen, T., Tenhola, T., Manninen, O., Nevo, E. and Nissila, E. (2001) Microsatellite diversity associated with ecological factors in *Hordeum spontaneum* populations in Israel. *Molecular Ecology*, **10**, 1577–1591.

Tyerman, S., Niemietz, C. and Bramley, H. (2002) Plant aquaporins: multifunctional water and solute channels with expanding roles. *Plant, Cell and Environment*, **25**, 173–194.

Tyerman, S.D., Bohnert, H.J., Maurel, C., Steudle, E. and Smith, J.A.C. (1999) Plant aquaporins: their molecular biology biophysics and significance for plant water relations. *Journal of Experimental Botany*, **50**, 1055–1071.

Uesono, Y. and Toh, E.A. (2002) Transient inhibition of translation initiation by osmotic stress. *Journal of Biology and Chemistry*, **277**, 13848–13855.

Uetz, P. (2002) Two-hybrid arrays. *Current Opinion in Chemistry and Biology*, **6**, 57–62.

Ulloa, M., Cantrell, R.G., Percy, R.G., Zeiger, E. and Lu, Z.M. (2000) QTL analysis of stomatal conductance and relationship to lint yield in an interspecific cotton. *Journal of Cotton Science*, **4**, 10–18.

Vagujfalvi, A., Galiba, G., Cattivelli, L. and Dubcovsky, J. (2003) The cold-regulated transcriptional activator Cbf3 is linked to the frost-tolerance locus *Fr-A2* on wheat chromosome 5A. *Molecular General Genetics*, **269**, 60–67.

Van Buuren, M., Salvi, S., Morgante, M., Serhani, B. and Tuberosa, R. (2002) Comparative genomic mapping between a 754 kb region flanking DREB1A in *Arabidopsis thaliana* and maize. *Plant Molecular Biology*, **48**, 741–750.

Van Camp, W., Capian, K., Van Montagu, M., Inzé, D. and Slooten, L. (1996) Enhancement of oxidative stress tolerance in transgenic tobacco plants overproducing Fe-superoxide dismutase in chloroplasts. *Plant Physiology*, **112**, 1703–1714.

Vannini, C., Locatelli, F., Bracale, M., Magnani, E., Marsoni, M., Mattana, M., Faoro, F. and Coraggio, I. (2003a) The rice *osmyb4* gene increases the tolerance to biotic and abiotic stresses in transgenic *Arabidopsis thaliana* plants. *7th International Congress of Plant Molecular Biology*, Barcelona, June 23–28, 2003, 212.

Vannini, C., Locatelli, F., Bracale, M., Magnani, M., Marsoni, M., Osnato, M., Mattana, M. and Coraggio, I. (2003b) Overexpression of the rice Osmyb4 gene increases the tolerance to chilling and freezing in transgenic Arabidopsis thaliana plants. *Plant Journal*, submitted.

Velculescu, V.E., Zhang, L., Vogelstein, B. and Kinzler, K.W. (1995) Serial analysis of gene expression. *Science*, **270**, 484–487.

Villegas, D., Aparicio, N., Nachit, M.M., Araus, J.L. and Royo, C. (2000) Photosynthetic and developmental traits associated with genotypic differences in durum wheat yield across the Mediterranean Basin. *Australian Journal of Agricultural Research*, **51**, 891–901.

Wade, L.J., Kamoshita, A., Yamauchi, A. and Azhiri-Sigari, T. (2000) Genotypic variation in response of rainfed lowland rice to drought and rewatering. I. Growth and water use. *Plant Production Science*, **3**, 173–179.

Walker-Simmons, M.K., Reaney, M.J.T., Quarrie, S.A., Perata, P., Vernieri, P. and Abrams, S.R. (1991) Monoclonal antibody recognition of abscisic acid analogs. *Plant Physiology*, **95**, 46–51.

Wan, C.G., Xu, W.W., Sosebee, R.E., Machado, S., Archer, T., Wan, C.G. and Xu, W.W. (2000) Hydraulic lift in drought-tolerant and -susceptible maize hybrids. *Plant and Soil*, **219**, 117–126.

Ware, D., Jaiswal, P., Ni, J.J., Pan, X.K., Chang, K., Clark, K., Teytelman, L., Schmidt, S., Zhao, W., Cartinhour, S., McCouch, S. and Stein, L. (2002) Gramene: a resource for comparative grass genomics. *Nucleic Acids Research*, **30**, 103–105.

Wayne, M.L. and McIntyre, L. (2002) Combining mapping and arraying: an approach to candidate gene identification. *Proceedings of the National Academy of Science USA*, **99**, 14903–14906.

Wechwerth, W. and Fiehn, O. (2002) Can we discover novel pathways using metabolomic analysis? *Current Opinion in Biotechnology*, **13**, 156–160.

Weinmann, A.S., Yan, P.S., Oberley, M.J., Huang, T.H.M. and Farnham, P.F. (2002) Isolating human transcription factor targets by coupling chromatin immunoprecipitation and CpG island microarray analysis. *Genes Development*, **16**, 234–244.

Westgate, M.E. and Boyer, J.S. (1985) Osmotic adjustment and the inhibition of leaf, root, stem and silk growth at low water potentials in maize. *Planta*, **164**, 540–549.

Widawsky, D.A. and O'Toole, J.C. (1990) *Prioritizing the rice biotechnology research agenda for Eastern India*, edn, The Rockefeller Foundation, New York, USA.

Winkler, R. and Helentjaris, T. (1995) The maize *dwarf3* gene encodes a cytochrome P450 mediated early step in gibberellin biosynthesis. *Plant, Cell and Environment*, **7**, 1307–1317.

Xiao, J.H., Li, J.M., Yuan, L.P. and Tanksley, S.D. (1996) Identification of QTLs affecting traits of agronomic importance in a recombinant inbred population derived from a subspecific rice cross. *Theoretical and Applied Genetics*, **92**, 230–244.

Xiao, J.H., Li, J., Grandillo, S., Ahn, S.N., Yuan, L., Tanksley, S.D. and McCouch, S.R. (1998) Identification of trait-improving quantitative trait loci alleles from a wild rice relative, *Oryza rufipogon. Genetics*, **150**, 899–909.

Xu, W.W., Subudhi, P.K., Crasta, O.R., Rosenow, D.T., Mullet, J.E. and Nguyen, H.T. (2000) Molecular mapping of QTLs conferring stay-green in grain sorghum (*Sorghum bicolor* L. Moench). *Genome*, **43**, 461–469.

Yadav, R., Courtois, B., Huang, N. and McLaren, G. (1997) Mapping genes controlling root morphology and root distribution in a doubled-haploid population of rice. *Theoretical and Applied Genetics*, **94**, 619–632.

Yamaguchi-Shinozaki, K. and Shinozaki, K. (1994) A novel cis-acting element in an Arabidopsis gene is involved in responsiveness to drought, low-temperature, or high-salt stress. *Plant Cell*, **6**, 251–264.

Yano, M., Katayose, Y., Ashikari, M., Yamanouchi, U., Monna, L., Fuse, T., Baba, T., Yamamoto, K., Umehara, Y., Nagamura, Y. and Sasaki, T. (2000) *Hd1*, a major photoperiod sensitivity quantitative trait locus in rice, is closely related to the Arabidopsis flowering time gene *CONSTANS*. *Plant Cell*, **12**, 2473–2484.

Yim, Y., Davis, G., Duru, N., Musket, T., Linton, E., Messing, J., McMullen, M., Soderlund, C., Polacco, M., Gardiner, J. and Coe, E. (2002) Characterization of three maize bacterial artificial chromosome libraries toward anchoring of the physical map to the genetic map using high-density bacterial artificial chromosome filter hybridization. *Plant Physiology*, **130**, 1686–1696.

Yin, X.Y., Stam, P., Kropff, M.J. and Schapendonk, A.H.C.M. (2003) Crop modeling, QTL mapping, and their complementary role in plant breeding. *Agronomy Journal*, **95**, 90–98.

Young, N.D. (1999) A cautiously optimistic vision for marker-assisted breeding. *Molecular Breeding*, **5**, 505–510.

Yu, L.X. and Setter, T.L. (2003) Comparative transcriptional profiling of placenta and endosperm in developing maize kernels in response to water deficit. *Plant Physiology*, **131**, 568–582.

Zeevaart, J.A.D. and Creelman, R.A. (1988) Metabolism and physiology of abscisic acid. *Annual Review of Plant Physiology and Plant Molecular Biology*, **39**, 439–473.

Zeng, Z.B. (1994) Precision mapping of quantitative trait loci. *Genetics*, **136**, 1457–1468.

Zhang, J. and Davies, W.J. (1990) Does ABA in the xylem control the rate of leaf growth in soil dried maize and sunflower plants? *Journal of Experimental Botany*, **41**, 1125–1132.

Zhang, W.P., Shen, X.Y., Wu, P., Hu, B. and Liao, C.Y. (2001a) QTLs and epistasis for seminal root length under a different water supply in rice (*Oryza sativa* L.). *Theoretical and Applied Genetics*, **103**, 118–123.

Zhang, J., Zheng, H., Aarti, A., Pantuwan, G., Nguyen, T., Tripathy, J., Sarial, A., Robin, S., Babu, R., Nguyen, B., Sarkarung, S., Blum, A. and Nguyen, H. (2001b) Locating genomic regions associated with components of drought resistance in rice: comparative mapping within and across species. *Theoretical and Applied Genetics*, **103**, 19–29.

Zhang, Z.B. and Xu, P. (2002) Reviews of wheat genome. *Hereditas-Beijing*, **24**, 389–394.

Zhu, J.K. (2002) Salt and drought stress signal transduction in plants. *Annual Review of Plant Biology*, **53**, 247–273.

Zhu, J.K., Hasegawa, P. and Bressan, R. (1997) Molecular aspects of osmotic stress in plants. *Critical Review of Plant Science*, **16**, 253–277.

Zinselmeier, C., Westgate, M.E., Schussler, J.R. and Jones, R.J. (1995) Low water potential disrupts carbohydrate metabolism in maize (*Zea mays* L.) ovaries. *Plant Physiology*, **107**, 385–391.

Zinselmeier, C., Jeong, B.R. and Boyer, J.S. (1999) Starch and the control of kernel number in maize at low water potentials. *Plant Physiology*, **121**, 25–36.

Zinselmeier, C., Sun, Y., Helentjaris, T., Beatty, M., Yang, S., Smith, H. and Habben, J. (2002) The use of gene expression profiling to dissect the stress sensitivity of reproductive development in maize. *Field Crop Research*, **75**, 111–121.

Zivy, M. and de Vienne, D. (2000) Proteomics: a link between genomics, genetics and physiology. *Plant Molecular Biology*, **44**, 575–580.

10 Water use efficiency in the farmers' fields

John Passioura

10.1 Introduction

The notion of water-use efficiency (WUE) had its origins in the field and much of its theoretical development depended on ideas at the scale of a field or at least of an extensive canopy (Tanner and Sinclair, 1983). Physiological and biochemical analysis of major genetic variation in WUE led to the distinction between C3, C4, and CAM metabolism and thence to the idea of internal CO_2 concentration and its dependence on the interplay between stomatal conductance and photosynthetic rate. There has subsequently been much superb work on the use of stable isotopes of carbon and oxygen that have elaborated the processes controlling the exchange of carbon and water by leaves, that is, WUE defined at the level of a leaf. That work led to a targeted breeding program, based on selection involving carbon-isotope discrimination, that has produced a cultivar of wheat ('Drysdale') now in commercial use in Australia, one of the very few examples of physiological analysis leading directly to a new cultivar.

To most physiologists and biochemists, WUE is an idea that concerns the gas exchange of a leaf, extended by crop physiologists and micro-meteorologists to gas exchange of a canopy. But to farmers and agronomists, who are interested in what can be harvested, WUE refers to the amount of grain or fruit that can be produced using a given water supply. To translate between these two notions requires knowledge of the processes that affect harvest index (HI, the ratio of harvestable yield to total dry matter), and processes that affect the acquisition of water by plant roots in competition with other processes that may lead to the loss of water by direct evaporation from the soil or by drainage beyond the reach of roots.

In effect, the agronomist sees the study and manipulation of WUE as a problem in resource economics rather than instantaneous gas exchange, and asks the question: given a fixed and limiting supply of water, what is the maximum harvestable yield that can be obtained? This question can be dissected into three subsidiary questions (Passioura, 1977; Richards *et al.*, 2002), namely, how can one manage or improve the genetic makeup of a crop to:

- capture more of the water supply
- exchange transpired water for CO_2 more effectively in producing biomass, and,
- convert more of the biomass into grain or other harvestable product.

Furthermore, the part of the water supply that is not captured by the crop because of runoff or drainage can have environmental consequences, including eutrophication of waterways, pollution of groundwater, and hydrologic changes that may lead to waterlogging or salinity in lower parts of the landscape. The genetics, biochemistry and physiology of crops impacts on these economic and environmental problems in sometimes subtle, but often profound, ways.

My aim is to explore the interactions and issues involved in extrapolating from WUE at the level of the leaf to what may be better called the 'effectiveness of water use' in dryland agriculture, especially in relation to crops in mediterranean environments in which the cropping season lasts from autumn to late spring or early summer with winter-dominant rainfall. Table 10.1 provides an outline of the issues involved in influencing the agronomist's WUE, based on yield of grain, and the time and spatial scales that they concern. It deals with crops that are free of disease, but may be subject to other stresses that influence the effective use of water. The percentages in the columns denoting 'influence' are very approximate, but serve to highlight the likely maximal relative importance of various processes in determining the effective use of water.

10.2   WUE on farm

It is notable that the idea of WUE strongly influences many farming communities in semi-arid agriculture, especially in the Australian wheat belt. It sets a benchmark against which farmers assess the performance of their crops. Figure 10.1, adapted from data of French and Schultz (1984), Cornish and Murray (1989), and Angus and van Herwaarden (2001) illustrates this influence. It shows the effective upper bound of farmers' wheat yields, a line of slope 20 kg ha^{-1} of grain per millimetre of water supply, offset on the x-axis by an approximate loss of water by direct evaporation from the soil. The illustrative points on or below the line cover the range of yields experienced in practice. If yields are substantially below the line, then farmers take that as prima facie evidence of poor management that should be corrected. A striking example of that attitude is that a competition amongst farmers in Western Australia for high yield, which used to be called the '5 tonne club' (because only those who had achieved 5 tonne ha^{-1} yields of wheat could be members), was renamed to be the '20kg per hectare-millimetre club' in

Table 10.1 The effects of processes at various scales on the effective use of water by crop plants in producing grain.

Spatial Scale	Temporal scale	Maximal influence on water-limited yield	Issues, processes
chloroplast	seconds	5%	Carbon fixation rate at constant stomatal conductance
stomata	minutes to hours	5%	Instantaneous exchange rate of carbon and water
leaf	seconds to hours	5%	Boundary layer effects, orientation
floral organs	hours to days or weeks	100%	Harvest index, matching phenology to water supply, impact of water deficits on fertility and on supply of assimilate to the grain
canopy, root system	weeks to months	50%	Trajectory of leaf area through time; ratio of water use of plant to other evaporative losses (soil, weeds) and to drainage; effective depth of roots
field	one to several growing seasons	30%	Lateral movement of water: run-on and run-off; spatial variability in soil properties and plant growth; carry-over effects of different crops between seasons

recognition of that deeper understanding of the interaction between weather and yield.

Figure 10.1 summarises myriad influences on yield, including poor nutrition, infestations with weeds, and catastrophes such as sterility arising from severe weather at flowering time – frost, very high temperatures, or severe water deficits. It is notable that the farmers have made much progress, especially in the last 15 years, in dealing with these various maladies and moving their average yields towards that of the water-limited line. Figure 10.2 shows how average wheat yield has changed in Australia since the 1860s. After a sustained fall in the late nineteenth century during which nutrients in the soil became exhausted, every decade in the twentieth century produced an increase in average yield, despite substantial extension of cropping into more marginal land at various times. Given that rainfall altered little during that

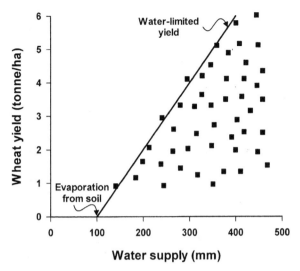

Figure 10.1   Wheat yield in southern Australia in relation to water supply (rainfall during the growing season plus available water in the soil at the time of sowing). The solid line, which has a slope of about 20 kg ha^{-1}mm^{-1}, depicts water-limited yield. Its intercept on the x-axis is the putative loss of water by direct evaporation from the soil. The points illustrate the range of farmers' experience and are typically well below the solid line because of difficulties with weeds, disease, poor nutrition, frost, and other problems.

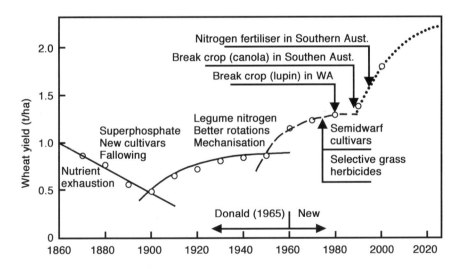

Figure 10.2   Average decadal wheat yields in Australia since 1860, an extension by Angus (2001) of an earlier analysis by Donald (1965). Copyright CSIRO 2002. Reproduced from the *Australian Journal of Experimental Agriculture* **41**, 277–288 (Angus J.F., 2001) by permission of CSIRO Publishing.

time, the improvements in yield last century also represent improvements in water-use efficiency of grain yield. The pattern of these increases was strikingly uneven, marked by sharp rises followed by periods of little change. The first of the rises was due mainly to better nutrition, and the second to the introduction of highly productive pasture legumes which provided a rich source of nitrogen to following crops. The third and most spectacular was due to a more complex set of changes that resulted in better control of endemic root diseases which gave farmers the confidence to apply more nitrogen, with consequent large responses in yield (Passioura, 2002b). Despite these advances, though, the average yield in the 1990s, of 1.8 tonne/ha, is still well below the estimated water-limited yield of 3.2 tonne/ha (John Angus, personal communication).

Improvements in cultivars over this period have been of roughly equal importance to those in agronomy, though typically they have been steady through time. Figure 10.3 shows relative yields of a range of Australian wheat cultivars, spanning more than a century in their times of release, grown simultaneously in well-managed experimental plots in the field (Richards 1991) over a range of seasons in the 1980s. The rate of improvement with time of release averaged roughly 1% per year, with variation around the trend line mostly due to wheat breeders' having to focus more on disease resistance than on yield. Again, given that rainfall has altered little during this time, these improvements also reflect improvements in water-use efficiency of grain yield.

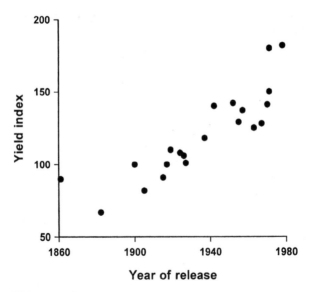

Figure 10.3 Yield of the best wheat cultivars grown in New South Wales versus year of release. The data are derived from yields of these cultivars grown together over several seasons in southern NSW in the 1980s. Adapted from Richards (1991).

In terms of water economy, an effective way of analysing these past improvements and prospects for future ones is by means of the three dot points in the Introduction, as follows.

10.3   Accessing more of the water supply

10.3.1   Canopy development to reduce evaporative losses from the soil

A dryland crop's water supply is essentially the sum of available water in the soil at the time of sowing and rainfall during the growing season. The main losses from this supply are by direct evaporation from the soil surface and vertical drainage of water beyond the reach of the roots. Run-off from the soil surface may be substantial during heavy rain, but much of that run-off becomes run-on in lower parts of a field (Eastham *et al.*, 2000; Batchelor *et al.*, 2002), so that there may be little net loss from the field as a whole, except perhaps through greater vertical drainage from low-lying parts.

In Mediterranean climates, where much of the water supply comes from rain during the autumn to spring growing season and the soil surface is frequently wet, much water can be lost by direct evaporation from the soil (Leuning *et al.*, 1994). The intercept of 100 mm shown in Figure 10.1 is a rough average. The actual loss varies with the seasonal weather as well as with the rate of development of the leaf canopy and the presence of mulches such as stubble from the previous crop. Figure 10.4 illustrates the impact of canopy size on evaporative loss from the soil in a mediterranean environment in which the size of the canopy was manipulated by varying the crop's nitrogen supply. The seasonal evaporation from the soil surface was determined using the technique of Cooper *et al.* (1983), and ranged from about 60 to 160 mm. The corresponding overall WUE of grain yield ranged, with increasing nitrogen supply, from 11 to 20 kg ha^{-1}mm^{-1}, though WUE is not necessarily higher at luxurious nitrogen supply because of the possibility that the crop may use too much water during its vegetative phase and run out of water during grain filling, a point that I cover in some detail later.

Another way of interpreting the interaction between canopy development and water loss by direct evaporation is to look at the marginal rate of extra evapotranspiration induced by additional leaf area. This marginal rate can approach zero if the distribution of rain is such that the soil surface remains wet during much of the crop's vegetative phase, for the evaporative demand by the environment may be met no matter what the ratio of transpiration to direct evaporation.

Agronomy and breeding can both strongly affect development of a crop's canopy. Poor establishment of seedlings results in slow development, and is an especial problem with the semi-dwarf wheats that are used widely throughout the world. The difficulty is that these wheats, because their dwarfism arises from insensitivity to gibberellic acid (GA) conferred by the genes Rht1 and Rht2, have short coleoptiles that do not extend to the soil surface if the seed is sown more deeply than about 60 mm. The leaf emerging from the still-buried coleoptile is then trapped within the soil and the seedling dies. In well-managed well-structured soils with reliable weather this is no problem because depth of sowing can be precisely controlled. But in semi-arid environments, especially where there are soil structural problems, and especially where direct sowing of seed into untilled soil is practised, the timing of sowing is often dictated by the weather, and the seed bed is often rough. The imperative for getting the crop sown quickly in an unreliable environment means that the sowing machines are large (10 to 15 m wide) and fast (travelling at speeds of up to 10 km h^{-1}), with the sowing boots, which place the seed in the soil, bouncing around a lot. A large proportion of seeds can then be sown too deeply. To solve this problem, wheat breeders are searching for dwarfing genes other than Rht1 or Rht2 that provide the benefits of short stems but do not overly restrict the maximum length of the coleoptile. A good candidate is Rht8 which in experimental breeding lines enables emergence from sowing depths as great as 120 mm but yet provides adequate dwarfing of the canopy (Rebetzke et al., 1999; Botwright et al., 2001).

Even if emergence of seedlings is good, the rate of development of leaf area may be suboptimal. Leaf growth is strongly affected by temperature of both air and soil, so sowing winter-growing crops early, when soil and air are still warm, leads to good canopy cover during late autumn and winter with consequently less evaporative losses from the soil surface. Changes in agronomic technology during the last twenty years have enabled farmers to sow their crops earlier, where appropriate. These changes include: directly sowing ('drilling') seed into the soil without the need for prior ploughing; using large and fast machinery that can sow large areas quickly; using powerful general herbicides such as glyphosate for killing emerged weeds just before or during sowing; and using specific herbicides for controlling weeds once the crop has established. Repeated cultivation of the soil to control weeds and to make a fine seed bed is no longer needed. These techniques have greatly improved the timeliness of sowing and have enabled sowing to take place many days, sometimes weeks, earlier than the traditional ones, whenever conditions are appropriate to do so. They thereby substantially improve yields in water-limited environments where the weather is often fickle at the start of the growing season and opportunities for sowing are best taken when they arise. Breeders have been stimulated by these agronomic changes to produce cultivars that can be sown early yet still flower at the optimal time. Farmers are now able to tune cultivar to season, cultivars that are specifically suitable for early, mid, or late starts to the season (Shakley and Anderson, 1995).

Other environmental influences on how fast the leaf canopy develops

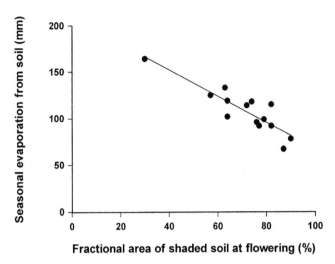

Figure 10.4 Evaporative loss of water from soil under wheat canopies of different size. The size of the canopy was varied by varying nitrogen fertiliser. (Adapted from van Herwaarden and Passioura, 2001).

include nutrition, especially nitrogen (Figure 10.4), and the physical state of the soil, especially during establishment. There is intriguing evidence that there is an optimal hardness of soil for rapid leaf growth – that is, soil may not only be too hard, which is well known, but also too soft (Passioura, 2002a). Further the availability of substantial macroporosity, which one might expect to enable easy root growth, may also inhibit leaf growth if the general soil matrix is too hard for roots to enter easily (Stirzaker et al., 1996). It is likely that root signals may be involved, for these effects occur even if the supplies of nutrients and water to the leaves are unaffected (Masle and Passioura, 1986; Donald et al., 1987).

While excessive ploughing of the soil may lead to problems of soil structure and thence poor establishment, lack of ploughing may allow soil particles to coalesce, with consequent problems for root growth even when there is seemingly adequate macroporosity (Cockroft and Olsson, 2000). Direct sowing of crops into unploughed soil often results in slow early leaf growth (Kirkegaard et al., 1994a), an effect that disappears if the soil is sterilised (Simpfendorfer et al., 2001). Evidently the physical and microbiological properties of the soil interact. Inhibitory *Pseudomonas spp.* that build up in the rhizosphere in untilled soil may be the main cause of this effect (Simpfendorfer et al., 2002). Roots grow more slowly in untilled than in cultivated soil and that may be a predisposing factor to this microbiological interaction: Watt et al. (2003) have shown that slow-growing roots in the field carry 20 times more pseudomonad bacteria per unit root length than do fast-growing ones.

These major agronomic effects are matched by equally strong genetic ones. Richards *et al.* (2002) have comprehensively reviewed the prospects for breeding for faster leaf area development in wheat. Selecting for large early leaves has produced breeding lines that develop leaf area twice as fast as standard cultivars in common use. Physiologically, the main trait is the size of the first leaf in the embryo – that is, the starting capital is of most importance, there being little variation in relative growth rate. The second most important trait is specific leaf area, the more vigorous lines having a larger leaf area per unit biomass. Because of the danger that these new genotypes may lead to excessive vegetative growth (see later), it is important that this rapid development of the main stem that provides good early ground cover is not accompanied by an excessive proliferation of tillers. The incorporation of a gene for inhibiting tiller development may prevent mid-season canopy development getting out of hand (Richards *et al.*, 2002), and ensure that resources are not used in producing unproductive tillers.

The upshot of this discussion is that the trajectory of leaf area development through time is an important determinant of how effectively a crop can use water. Rapid early development of leaf area can greatly reduce evaporative losses from the soil surface, effectively enhancing the productive flow of water through the plants. In mediterranean environments it can also improve the terms of trade of water for carbon dioxide by leaves during the cool winter months. There are, however, trade-offs: too much water use during vegetative growth may mean too little during grain filling.

10.3.2 *Reducing losses of water by deep drainage*

Water lost by drainage beyond the reach of crop roots is usually much less than that lost by direct evaporation from the soil. Nevertheless it is of equal or even greater importance in foregone yield (Angus and van Herwaarden, 2001). It is hard to measure, but is likely, in a semi-arid environment, to vary from zero to 100 mm per year depending on soil, management, and season, with an average of about 30 mm (Dunin *et al.*, 2001).

Capturing water that may otherwise drain may have two great benefits. First, if roots do access it, they do so late in the season, after anthesis, when the products of the photosynthesis it enables can go almost entirely towards filling the grain with little respiratory or other losses. An extra 30 mm could be translated into an increased yield of 600 kg ha^{-1} if converted at the rate of 20 kg ha^{-1}mm^{-1} as in the line of water-limited yield in Figure 10.1, or even 1000 kg ha^{-1} if the rate of 33 kg ha^{-1}mm^{-1} estimated by Angus and van Herwaarden (2001) for post-anthesis growth pertains. This would be a very substantial increase in water-limited environments in which average yields may be less than 3 tonnes per hectare. Further, this water is often rich in mineral nitrogen, leached from the topsoil earlier in the season when the plants were too young with root systems too small to use it (Angus, 2001). This nitrogen can boost the quality and possibly the amount of the developing grain if it keeps the nitrogen content of the leaves high.

The second benefit of capturing this water is environmental. Water lost to deep drainage may contribute to offsite environmental damage such as dryland salinity, as the escaped water mobilises salt and brings it to the surface lower in the landscape, or eutrophication, if the water carries nutrients to discharge areas where they may generate algal blooms.

Active deep roots help reduce drainage losses, but many soils in semi-arid areas are beset by subsoils that are inhospitable to roots for one or more of the following reasons: saline, sodic, too hard, too alkaline, too acid, too high in boron or too low in zinc and other nutrients that roots need locally for their adequate growth. Many crops may fail to send roots deeper than about 50 cm in such soils despite water penetrating to a metre or more in average to wet seasons. Plant breeding has made little impact on such problems, with the notable exception of tolerance to high boron (Paull et al., 1991).

Roots typically penetrate inhospitable subsoils through biopores, large extended pores made and repeatedly recolonised by successive generations of roots. These pores differ chemically and microbiologically as well as physically from the surrounding soil matrix (Pierret et al., 1999), and act as highways from which lateral roots can explore the adjacent soil matrix. They are plentiful under perennial vegetation, and are important as conduits for water as well as roots (Nulsen et al., 1986), but they tend to get blocked in tilled agricultural soils. Given the environmental complexity of the subsoil, breeding crop plants whose roots can better exploit it will be extraordinarily difficult. Creating and maintaining a network of accessible biopores by agronomic means is more feasible (McCallum et al., 2004).

Crops that are vigorous when young tend to extract more water from the subsoil, presumably because their roots grow deeper, though this is not proven (Angus et al., 2001). There are substantial effects of cropping history – the sequence of earlier crop species – on the abilities of following crops to extract water from the subsoil, which may be through effects on early vigour but may also be due to other still unknown mechanisms (Kirkegaard et al., 1995; Angus et al., 2001). Premature senescence, premature in the sense that the plants shut down their leaves despite some available water being accessible in the soil, may be common. The 'stay green' character in sorghum (Borrell et al., 2001) suggests this may be so; it seems to arise from positive feedback in nitrogen acquisition: plants that maintain nitrogen in their leaves during grain filling (and hence stay green) fix more carbon, which in turn enables roots to continue extracting soil nitrogen, so that the system is self-reinforcing.

Over a run of seasons, water that has escaped the roots of a succession of annual crops accumulates in deep subsoil, and may be accessible there to the roots of deep-rooted perennial agricultural plants such as lucerne grown for two to three years (Latta *et al.*, 2001; Ridley *et al.*, 2001). Reclaiming water in this way during droughts is especially valuable, though it relies on having a substantial proportion of a farm sown to lucerne or similar plants before the start of the drought.

In summary, there is substantial variation in the ability of crop roots to capture water that may otherwise drain beyond reach. This variation arises mostly from agronomic effects, including little understood results of cropping history and season, possibly involving interactions with soil organisms. There may well be genetic variation in the ability of crop roots to exploit subsoils, but there are as yet no obvious traits that breeders could realistically select for. Punctuating sequences of annual crops with phases of deep-rooted perennial pasture plants such as lucerne can reclaim much of the water that has escaped the roots of annual crops and put it to productive use as well as reducing the risks of environmental damage such as dryland salinity and eutrophication of discharge areas.

10.4   Improving transpiration efficiency, the exchange of water for CO_2

Several other chapters in this volume cover this topic thoroughly. In brief, the transpiration efficiency of leaves, i.e. the amount of carbon fixed per unit of water transpired, depends on both evaporative demand by the environment and the CO_2 concentration within the leaves (Condon *et al.*, 2002). In mediterranean environments, transpiration efficiency is higher during the winter, when daily pan evaporation rates are in the range of 2 to 3 mm than in the late spring and summer when these rates are 6 to 8 mm. It is therefore doubly important for crops to have established a large leaf area by the onset of winter, firstly, to reduce evaporative losses from the soil, as outlined earlier; and secondly to make good use of the favourable terms of trade of CO_2 for water during this evaporatively undemanding time (Richards *et al.*, 2002).

At a given evaporative demand and stomatal conductance, the lower is the concentration of CO_2 within a leaf the larger is the transpiration efficiency – the lower concentration drives faster diffusion of CO_2 into the leaf without much change in the diffusion rate of water vapour out of the leaf. Because ^{13}C, the heavy stable isotope of carbon, is discriminated against during photosynthesis in favour of the more abundant ^{12}C, the lower is the CO_2 concentration within a leaf, the more demanding are the carboxylating enzymes for CO_2 and therefore the less discrimination there is against ^{13}C during photosynthesis. These two relationships together have provided an effective tool for estimating the internal CO_2 concentration within leaves, and thence the intrinsic transpiration efficiency (Farquhar and Richards, 1984); isotopic analysis of plant tissue enables selection for intrinsically transpiration-efficient plants. Crop breeding programs have been making effective use of

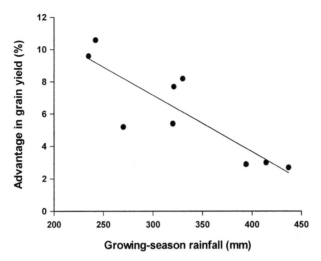

Figure 10.5   The advantage in grain yield of lines selected for low discrimination against ^{13}C during photosynthesis (i.e. high intrinsic transpiration efficiency) over those selected for high discrimination, as a function of growing-season rainfall. Adapted from Rebetzke *et al.* (2002).

this technique (Rebetzke *et al.*, 2002). Figure 10.5 shows the yield advantage of breeding lines selected for intrinsically higher transpiration efficiency. As expected, this trait has greater impact the lower is the rainfall. One of these breeding lines has culminated in the release of a new Australian wheat cultivar, 'Drysdale', which promises to increase water-limited yields by about 10% above those of the widely sown cultivar ('Hartog') from which it was derived, at least in years of moderate to low rainfall. Thus, in terms of Figure 10.1, it promises to increase the slope of the lower end of the water-limited line by up to 10%.

10.5   Converting biomass into grain

10.5.1   Adapting phenology to environment

The timing of flowering is the most important trait that plant breeders select for when targeting water-limited environments. In Mediterranean environments, crops that flower too early may not have built sufficient vegetative capacity to set and fill a large number of seeds, and are also at greater risk of being damaged by frost. Those that flower too late, while they may have set a large number of grains per unit area and thereby have attained a large yield potential, may fail to fill their grain adequately because they have too little water left in the soil and are exposed to the heat and aridity of late spring and early summer (Richards, 1991).

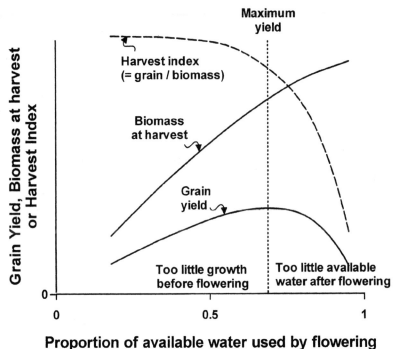

Proportion of available water used by flowering

Figure 10.6 Schematic graph of grain yield of wheat, biomass at harvest, and harvest index, in relation to proportion of the available water supply used by flowering. The scale of the y-axis is arbitrary, though the maximal harvest index is typically 0.5. Copyright CSIRO 2002. Reproduced from *Functional Plant Biology* **29**, 537–546 (Passioura, 2002b) with permission of CSIRO Publishing.

Figure 10.6 schematically illustrates these points for wheat. There is an optimal flowering time at which there is an appropriate balance between water used during canopy development and water used during grain filling. Crops that flower before the optimal time may achieve large harvest indices unless damaged by frost, but do not produce enough biomass to set a large enough number of seeds to generate a good yield potential (Fischer, 1979). Those that flower too late are at risk of severe water deficits that can lead to sterility, and may have too little water left to allow for adequate post-flowering photosynthesis or time to mobilise stores of carbohydrate accumulated before flowering and transfer them to the grain.

Plant breeders have generally produced a range of cultivars that flower close to the optimal time in a given environment. This optimum is necessarily an average, rather than a sharp point, for depending on the pattern of rainfall during the growing season earlier flowering crops may do better in one season, and later flowering ones may do better in another. As an average there is little room for further genetic improvement, though breeders have been producing slower maturing cultivars that can be sown earlier in the season while still flowering at the optimal time (Anderson *et al.*, 1996). Such cultivars allow farmers to capitalise on the flexibility in sowing time that their modern machinery and agronomic techniques enable, as mentioned earlier. Global warming may, over the next few decades, alter the optimal time of flowering, but breeders are so attuned to getting the phenology right that they will probably make the necessary adjustments without being consciously aware that they are doing so.

Nutrient management also strongly affects the pattern of water use by a crop. Too much nitrogen, whether from fertiliser or from excessive mineralisation of soil organic matter, can result in crops that are too vigorous and that use too much water before flowering; they set a large number of seeds but are unable to produce enough carbohydrate to fill these adequately, neither from photosynthesis after flowering nor from carbohydrate stored before flowering and available for retranslocation (van Herwaarden *et al.*, 1998; Angus and van Herwaarden, 2001). The result is a low yield of grain of poor quality. The syndrome is known to farmers as 'haying off' because the crop senesces prematurely. Farmers can get around this potential problem by applying nitrogen fertiliser tactically, in mid-season, once they have a better idea of how much water their crops are likely to get, rather than applying large amounts at sowing.

10.5.2   *Effects of drought on fertility*

Water deficits during specific stages of floral development can severely damage seed set, through pollen sterility or abortion of embryos, or can prematurely end grain filling (Saini and Westgate, 2000). Low water potentials during pollen mother cell meiosis (about 3 weeks before anthesis in wheat) can induce severe pollen sterility and thence low yields in the cereals even though subsequent conditions might be benign. Because the water status of the floral tissue is maintained despite the low water potential of the leaves (Westgate *et al.*, 1996), it is likely that a sporicide, transported from vegetative tissue to the reproductive tissue, causes the damage (Morgan and King, 1984) possibly disrupting invertase activity and thereby preventing proper utilisation of sucrose by the developing anthers (Dorion *et al.*, 1996). ABA may be such a sporicide (Morgan, 1980), though that remains equivocal (Dembinska *et al.*, 1992).

Low water potentials around the time of anthesis can induce severe embryo abortion, especially in maize, where abortion can be almost complete. Remarkably, abortion can be largely prevented by infusing stem internodes with sucrose solutions that essentially replace the assimilate that would have been produced by photosynthesis had the plants not been water-stressed (Boyle *et al.*, 1991; Zinselmeier *et al.* 1995a). However, it seems that it is not the paucity of assimilate alone that is the problem, for the requirement of the ear for assimilate at this stage is small (Schussler and Westgate, 1995). There is also a metabolic disruption of carbohydrate metabolism in the ovary, especially of acid invertase, that leads to the failure of the embryos to develop (Zinselmeier et al., 1995b). The similarity of metabolic disruption between this and what happens in developing wheat anthers is intriguing (Saini and Westgate, 2000).

Water deficits in maize can also bring about a mismatch in the timing of anthesis and silking, such that silking is delayed until after the pollen has been shed, leading to lack of fertilisation (Bolaños and Edmeades, 1993). The genetics of this effect are simple enough to have enabled the development of hybrids with markedly better yields during drought.

These various effects of water deficits on fertility can lead to severe, sometimes complete, loss of yield in droughted grain crops, which is why Table 10.1 shows 100% in this category, much higher than any other maximal influences on water-limited grain yield. While total loss is rare, it is likely that drought-induced infertility can unnecessarily reduce yields in seasons in which there is a reasonable water supply but in which water deficits occur at these especially sensitive times. Further unravelling of the processes involved is a promising way of laying a foundation for genetically improving grain yields in such droughts.

10.5.3   Mobilising pre-anthesis reserves during grain filling

Crops that suffer drought during grain filling may produce a large biomass but be unable to match that with a good harvest index. Excessive vegetative growth, especially that induced by an oversupply of mineral nitrogen can worsen the effects of such droughts by using too much water before flowering (Figure 10.6). The result is 'haying off' (section 10.5.1), in which the crop senesces prematurely and its yield responds negatively to nitrogen supply. Van Herwaarden et al. (1998) have argued that an oversupply of nitrogen worsens this imbalance in water use by reducing the amount of storage carbohydrates available for retranslocation to the grain. There is a negative correlation between nitrogen level and storage carbohydrate, not necessarily related to water deficits (Batten et al., 1993), with the implication that excess nitrogen results in the investment of photosynthate into structural rather than storage carbohydrate when stimulating excessive vegetative growth. Breeders may be able to help solve this problem, for there looks to be useful genetic variation in the ability to store mobilisable carbohydrate in the stems before flowering (Richards et al., 2002).

10.6 The impact of spatial variability

Farmers are well aware that their fields vary spatially in yield, but until recently they did not have the tools for quantifying that variability and possibly managing it to improve the performance of their crops. The new tools are yield monitors and global positioning (GPS) meters on harvesters which can generate yield maps, coupled with geographic information system (GIS) databases that enable such information to be stored electronically and compared across different seasons and with other spatially referenced information that may be available, such as satellite imagery of the paddock during the growing season (e.g. Batchelor *et al.*, 2002).

Although the location of the lowest yielding parts of a field may vary across seasons, for example, areas of low elevation may do best in dry years and worst in wet years, there are many cases in which particular areas may yield poorly in most years, with consequently low average WUE. Further, economic analysis of such areas often shows that farmers may be losing money by farming them – the returns may be less than the costs of the inputs (Passioura, 2002b). When that happens it is likely that the soil in these areas is especially poor and that it would be worthwhile excluding them from production. Doing so would not only increase the average WUE of the crop, but could also have substantial environmental benefits if the excluded land were sown to perennial plants that would reduce the risks of erosion and of offsite effects due to the excessive leakage of water and nutrients that is likely to occur under poorly growing crops.

10.7 Concluding comments

The notion of WUE defined in terms of grain yield has had a big impact on farmers, at least in Australia. It has provided a benchmark against which they can assess their performance independently of how wet the growing season might be. If their yields are well below the line of upper limit in Figure 10.1, with which they are familiar, they seek out reasons for the shortfall and try to improve their management accordingly. The spectacular rise in Australian average wheat yields of about 35% during the 1990s (Angus, 2001) is also a rise of about 35% in average WUE, given that there was no noticeable change in rainfall. It came about largely through better management predicated on a deeper understanding of what was limiting yield (Passioura, 2002b). In the context of water economy, it was associated largely with more vigorous crops with healthier root systems that allowed less evaporation from the soil and less drainage beyond the reach of their roots. The crops were better fertilised, especially with nitrogen, and hence there would have also been improvements in transpiration efficiency, i.e. WUE defined in terms of gas exchange, though that is likely to have been minor by comparison.

What are the prospects for further improvement? In the context of Table 10.1, there remains scope for improving transpiration efficiency along the lines of development of new cultivars such as 'Drysdale', adapted to a greater range of environments. Such cultivars have the potential to increase the slope of the line of upper limit in Figure 10.1 by about 10%, at least in seasons with moderate to low rainfall and in which evaporative losses from the soil are likely to be low, as in crops growing during autumn to spring largely on water stored in the soil from summer rainfall.

There still remains much scope for optimising the trajectories of leaf area and root development through time, for reducing evaporative losses from the soil and capturing water that may otherwise drain beyond reach. Improvements will require better matching of current genotypes and management, and new genotypes and management techniques that are even better attuned to each other. One development that will affect this interaction is improvement in seasonal weather forecasting, which while it is unlikely to become very accurate, can nevertheless take some of the risk out of the decisions that farmers must make at the beginning of each season, thereby enabling them to make better use, on average, of the season's water supply.

An area of great opportunity is that of the control of sterility in water-stressed plants. In many instances crop plants seem to be overly sensitive to water-deficits during pollen mother cell meiosis or anthesis – overly, in the sense that yields would probably have been better had the plants been more fertile. Despite the importance of this area, there is remarkably little research being carried out in it. The thought-provoking review by Saini and Westgate (2000) on reproductive development in grain crops during drought thoroughly addresses these issues. The processes involved are amenable to modern techniques of plant physiology, biochemistry and molecular biology in ways that they have not been before. Unravelling them will enable breeders to make targeted progress in this area with consequent substantial improvements in WUE of grain yield on farm.

Acknowledgments

I thank Tony Condon, John Kirkegaard, Rana Munns and Richard Richards, for pertinent criticisms of the manuscript.

References

Anderson, W.K., Heinrich, A., Abbotts, R. (1996) Long-season wheats extend sowing opportunities in the central wheat belt of Western Australia. *Australian Journal of Experimental Agriculture*, **36**, 203–208.

Angus, J. F. (2001) Nitrogen supply and demand in Australian agriculture. *Australian Journal of Experimental Agriculture*, **41**, 277–288.

Angus, J. F., Gault, R. R., Peoples, M.B., Stapper, M., and van Herwaarden, A.F. (2001) Soil water extraction by dryland crops, annual pastures, and lucerne in south-eastern Australia. *Australian Journal of Agricultural Research*, **52**, 183–192.

Angus, J.F. and van Herwaarden, A.F. (2001) Increasing water use and water use efficiency in dryland wheat. *Agronomy Journal*, **93**, 290–298.

Batchelor, W.D., Basso B. and Paz, J.O. (2002) Examples of strategies to analyze spatial and temporal yield variability using crop models. *European Journal of Agronomy*, **18**, 141–158.

Batten, G.D., Blakeney, A.B., McGrath, V.B., Ciavarella, S. and Barrow, N.J. (1993) Non-structural carbohydrate: analysis by near infrared reflectance spectroscopy and its importance as an indicator of plant growth. *Plant and Soil*, **155/156**, 243–246.

Bolaños, J. and Edmeades, G.O. (1993) Eight cycles of selection for drought tolerance in lowland tropical maize. II. Responses in reproductive behavior. *Field Crops Research*, **31**, 253–268.

Borrell, A., Hammer, G., and van Oosterom, E. (2001) Stay-green: A consequence of the balance between supply and demand for nitrogen during grain filling? *Annals of Applied Biology*, **138**, 91–95.

Botwright, T., Rebetzke, G., Condon, T., and Richards, R. (2001) The effect of rht genotype and temperature on coleoptile growth and dry matter partitioning in young wheat seedlings. *Australian Journal of Plant Physiology*, **28**, 417–423.

Boyle, M.G., Boyer, J.S. and Morgan, P.W. (1991) Stem infusion of liquid culture medium prevents reproductive failure of maize at low water potential. *Crop Science*, **31**, 1246–1252.

Cockroft, B. and Olsson, K.A. (2000) Degradation of soil structure due to coalescence of aggregates in no-till, no-traffic beds in irrigated crops. *Australian Journal of Soil Research*, **38**, 61–70.

Condon, A.G., Richards, R.A., Rebetzke, G.J. and Farquhar, G.D. (2002) Improving intrinsic water-use efficiency. *Crop Science*, **42**, 122–131.

Cooper, P.J.M., Keatinge, J.D.H. and Hughes, G. (1983) Crop evapotranspiration – a technique for calculation of its components by field measurements. *Field Crops Research*, **7**, 299–312.

Cornish, P.S. and Murray, G.M. (1989) Low rainfall rarely limits wheat yields in southern New South Wales. *Australian Journal of Experimental Agriculture*, **29**, 77–83.

Dembinska, O., Lalonde, S. and Saini, H.S. (1992) Evidence against the regulation of grain set by spikelet abscisic acid levels in water-stressed wheat. *Plant Physiology*, **100**, 1599–1602.

Donald, C.M. (1965) The progress of Australian agriculture and the role of pastures in environmental change. *Australian Journal of Science*, **27**, 187–198.

Donald, R.G., Kay, B.D. and Miller, M.H. (1987) The effect of soil aggregate size on early shoot and root growth of maize (*Zea mays* L.). *Plant and Soil*, **103**, 251–259.

Dorion, S., Lalonde, S. and Saini, H.S.(1996) Induction of male sterility in wheat by meiotic-stage water deficit is preceded by a decline in invertase activity and changes in carbohydrate metabolism in anthers. *Plant Physiology*, **111**, 137–145.

Dunin, F.X., Smith, C.J., Zegelin, S.J., Leuning, R., Denmead, O.T. and Poss, R. (2001) Water balance changes in a crop sequence with lucerne. *Australian Journal of Agricultural Research*, **52**, 247–261.

Eastham, J., Gregory, P.J. and Williamson, D.R. (2000) A spatial analysis of lateral and vertical fluxes of water associated with a perched watertable in a duplex soil. *Australian Journal of Soil Research*, **38**, 879–890.

Farquhar, G.D., Richards, R.A. (1984) Isotopic composition of plant carbon correlates with water-use efficiency of wheat genotypes. *Australian Journal of Plant Physiology*, **11**, 539–552.

Fischer, R.A. (1979) Growth and water limitation to dryland wheat yield in Australia: a physiological framework. *Journal of the Australian Institute of Agricultural Science*, **45**, 83–94.

French, R.J. and Schultz, J.E. (1984) Water use efficiency of wheat in a mediterranean-type environment. I. The relation between yield, water use and climate. *Australian Journal of Agricultural Research*, **35**, 743–764.

Kirkegaard, J.A., Angus, J.F., Gardner, P.A. and Muller, W. (1994) Reduced growth and yield of wheat with conservation cropping. I. Field studies in the first year of the cropping phase. *Australian Journal of Agricultural Research*, **45**, 511–528.

Kirkegaard, J.A., Munns, R., James, R.A., Gardner, P.A. and Angus, J.F. (1995) Reduced growth and yield of wheat with conservation cropping. II. Soil biological factors limit growth under direct drilling. *Australian Journal of Agricultural Research*, **46**, 75–88.

Latta, R.A., Blacklow, L.J. and Cocks, P.S. (2001) Comparative soil water, pasture production, and crop yields in phase farming systems with lucerne and annual pasture in Western Australia. *Australian Journal of Agricultural Research*, **52**, 295–304.

Leuning, R., Condon, A.G., Dunin, F.X., Zegelin, S. and Denmead, O.T. (1994) Rainfall interception and evaporation from soil below a wheat canopy. *Agricultural and Forest Meteorology*, **67**, 221–238.

McCallum, M.H., Kirkegaard, J.A., Green, T., Cresswell, H.P., Davies, S.L., Angus, J.F. and Peoples, M.B. (2004) Evidence of improved subsoil macro-porosity following perennial pastures on a duplex soil. *Australian Journal of Experimental Agriculture* (in press).

Masle, J., Passioura, J.B. (1987) The effect of soil strength on the growth of young wheat plants. *Australian Journal of Plant Physiology*, **14**, 643–656.

Morgan, J.M. (1980) Possible role of abscisic acid in reducing seed set in water-stressed wheat plants. *Nature* (London), **289**, 655–657.

Morgan, J.M. and King, R.W. (1984) Association between loss of leaf turgor, abscisic acid levels and seed set in two wheat cultivars. *Australian Journal of Plant Physiology*, **11**, 143–150.

Nulsen, R.A., Bligh, K.J., Baxter, I.N., Solin, E.J. and Imrie, D.H. (1986) The fate of rainfall in a mallee and heath vegetated catchment in southern Western Australia. *Australian Journal of Ecology*, **11**, 361–371.

Passioura, J.B. (1977) Grain yield, harvest index, and water use of wheat. *Journal of the Australian Institute of Agricultural Science*, **43**, 117–121.

Passioura, J.B. (2002a) Soil conditions and plant growth. *Plant, Cell and Environment*, **25**, 311–318.

Passioura, J.B. (2002b) Environmental biology and crop improvement. *Functional Plant Biology*, **29**, 537–546.

Paull, J.G., Rathjen, A.J. and Cartwright, B. (1991) Major gene control of tolerance of bread wheat (*Triticum aestivum* L.) to high concentrations of soil boron. *Euphytica*, **55**, 217–228.

Pierret, A., Moran, C.J. and Pankhurst, C.E. (1999) Differentiation of soil properties related to the spatial association of wheat roots and soil macropores. *Plant and Soil*, **211**, 51–58.

Rebetzke, G J., Condon, A.G., Richards, R A., and Farquhar, G.D. (2002) Selection for reduced carbon isotope discrimination increases aerial biomass and grain yield of rainfed bread wheat. *Crop Science*, **42**, 739–745.

Rebetzke, G.J., Richards, R.A., Fischer, V.M., and Mickelson, B.J. (1999) Breeding long coleoptile, reduced height wheats. *Euphytica*, **106**, 159–168.

Richards, R.A. (1991) Crop improvement for temperate Australia – future opportunities. *Field Crops Research*, **26**, 141–169.

Richards, R.A., Rebetzke, G.J., Condon, A.G. and van Herwaarden, A.F. (2002) Breeding opportunities for increasing the efficiency of water use and crop yield in temperate cereals. *Crop Science*, **42**, 111–121.

Ridley, A.M., Christy, B., Dunin, F.X., Haines, P.J., Wilson, K.F., and Ellington, A. (2001) Lucerne in crop rotations on the Riverine Plains 1. The soil water balance. *Australian Journal of Agricultural Research*, **52**, 263–277.

Saini, H.S. and Westgate, M.E. (2000) Reproductive development in grain crops during drought. *Advances in Agronomy*, **68**, 59–96.

Schussler, J.R. and Westgate, M.E. (1995) Assimilate flux determines kernel set at low water potential in maize. *Crop Science*, **35**, 1074–1080.

Shackley, B.J. and Anderson, W.K. (1995) Responses of wheat cultivars to time of sowing in the southern wheatbelt of Western Australia. *Australian Journal of Experimental Agriculture*, **35**, 579–587.

Simpfendorfer, S., Kirkegaard, J.A, Heenan, D.P. and Wong, P.T.W. (2001) Involvement of root inhibitory Pseudomonas spp. in the poor early growth of direct drilled wheat: studies in intact cores. *Australian Journal of Agricultural Research*, **52**, 845–853.

Simpfendorfer, S., Kirkegaard, J.A., Heenan, D.P. and Wong, P.T.W. (2002) Reduced early growth of direct drilled wheat in southern New South Wales – role of root inhibitory pseudomonads. *Australian Journal of Agricultural Research*, **53**, 323–331.

Stirzaker, R.J., Passioura, J.B. and Wilms, Y. (1996) Soil structure and plant growth: impact of bulk density and biopores. *Plant and Soil*, **185**, 151–162.

Tanner, C.B., Sinclair, T.R. (1983) Efficient water use in crop production: research or re-search? In *Limitations to Efficient Water Use in Crop Production* (eds. H.M. Taylor, W.R. Jordan and T.R. Sinclair) American Society of Agronomy, Wisconsin, USA, pp. 1–27.

van Herwaarden, A.F., Angus, J.F., Richards, R.A. and Farquhar, G.D. (1998) Haying-off, the negative grain yields response of dryland wheat to nitrogen fertiliser. II. Carbohydrate and protein dynamics. *Australian Journal of Agricultural Research*, **49**, 1083–1093.

van Herwaarden, A.F. and Passioura, J.B. (2001) Improving estimates of water-use efficiency in wheat. *Australian Grain*, **11**(4), 3–5.

Watt, M., McCully, M.E. and Kirkegaard, J.A. (2003) Soil strength and rate of root elongation alter the distribution of total bacteria and *Pseudomonas* spp. in the rhizosphere of wheat *Functional Plant Biology* (in press)

Westgate, M.E., Passioura, J.B. and Munns, R. (1996) Water status and ABA content of floral organs in drought-stressed wheat. *Australian Journal of Plant Physiology*, **23**, 763–772.

Zinselmeier, C., Lauer, M.J. and Boyer, J.S. (1995a) Reversing drought-induced losses in grain yield: sucrose maintains embryo growth in maize. *Crop Science*, **35**, 1390–1400.

Zinselmeier, C., Westgate, M.E., Schussler, J.R., and Jones, R.J. (1995b) Low water potential disrupts carbohydrate metabolism in maize (*Zea mays* L.) ovaries. *Plant Physiology*, **107**, 385–391.

Index

pulses 3, 32
pyruvate 275

quantitative trait loci (QTL) 15–17, 231,
235–56, 261–74, 280–81
Quercus 44, 47, 48–50, 62, 177–8

rapeseed 267
Regulated Deficit Irrigation (RDI)
122–5, 135
relative growth rate 55, 171, 176, 184,
310
relative water content 236, 246–7, 250
restriction fragment length polymorphism
(RFLP) 241–5, 250, 252, 256
Rhamnus 63
rhizobia 184
rhizosphere 88, 91, 117, 127, 176, 181,
186, 309
Ribulose 1,5 bisphosphate carboxylase/
oxygenase 17, 49, 54, 56–7, 145,
179, 181, 273
ribulose bisphosphate 43
rice 6, 16, 114, 146, 159, 198–208,
211–18, 221–2, 228, 236–40, 245,
247–50, 257, 261–2, 265–7, 270–71,
274, 278–9
Ricinus 92–3, 178, 185
root(s) 117, 120
root architecture 237, 248, 266, 270,
281
root development 125, 318
root diameter 253
root elongation 235
root growth 12, 40, 76–7, 80, 99, 102,
149, 158, 238, 249, 309
root length 158, 229, 249, 253, 309
root morphology 186, 248–50
root pulling force 253
root signals 21, 309
Rubisco *see* Ribuslose 1,5 bisphopshate
carboxylase/oxygenase
Rumex 50
rye 31

salinity 153, 178, 258, 260, 303, 311–12
salt 240, 276, 278, 311
salt tolerance 276
Schoenefeldia 45, 48
seed 28, 64, 80, 88, 153, 218, 235, 252,
308, 315

senescence 81, 98, 202, 204–8, 222, 251,
258, 311
signal transduction 79, 101, 258–60, 277
silking 316
soil 3, 10–17, 20–22, 27, 29, 35, 39–40,
43, 48–9, 62, 66, 75–94, 97–104,
117–22, 124, 129, 131–5, 137, 142,
144, 146, 148–66, 171–6, 179–83, 186,
188–9, 198, 200, 202–3, 206–8,
211–15, 218–22, 234–9, 247, 250, 260,
269, 272, 280, 302–5, 307–13, 315,
317–18
soil properties 142, 304
soil structure 12–13, 237, 309
sorghum 3, 46, 145, 154, 157, 186, 236,
239, 251, 270, 274, 311
soybean 86, 114, 128, 245–6
sporicide 315
stable isotopes 37, 60–61, 312
starch 61, 213, 218, 221–2, 259, 262–3
starch phosphorylase 221
stay green 206–9, 213, 223, 251–2, 311
sterility 304, 309, 314–15, 318
Stipa 177
stomatal behaviour 37, 90, 124, 187, 275
stomatal closure 15, 28, 39, 45–7, 62,
64, 66, 75–85, 89–90, 96, 100, 102,
105, 107, 129, 134, 235, 262
stomatal conductance 4, 6, 9, 11, 13, 15,
20, 30–31, 35–6, 39, 42–8, 50–1, 53,
56, 61–3, 85–7, 94–5, 104, 113,
115–21, 123–37, 176, 181–3, 230–36,
239, 246–7, 250, 252, 273–5, 302, 304,
312
stomatal frequency 248
stomatal sensitivity 66, 96, 104–5, 126,
235
Stylosanthes scabra 246
sucrose 52, 218, 221–2, 276, 315
sugar beet 32, 276
sunflower 56, 92, 124, 147, 149, 239, 247
symbioses 173–4, 176, 184

Thelephora 185
thermal imaging 10, 234
tillers 271, 310
tobacco 5, 17, 50–51, 274–8
tomato 15–16, 79–80, 85, 86–92,
98–100, 118–19, 128, 244, 255, 279
transcription 257–8, 260, 263, 265, 274,
277–9